Relativistic Wave Mechanics

Lectures by

E. CORINALDESI

Edited by

F. STROCCHI

English text prepared with assistance from

R. HILTON

Dover Publications, Inc.

Mineola, New York

Bibliographical Note

This Dover edition, first published in 2015, is an unabridged republication of the work originally published by North-Holland Publishing Company, Amsterdam, and John Wiley & Sons, Inc., New York, in 1963. The English text was prepared with assistance from R. Hilton.

Library of Congress Cataloging-in-Publication Data

Corinaldesi, E., author.
Relativistic wave mechanics / E. Corinaldesi; edited by F. Strocchi; English text prepared with assistance from R. Hilton.
p. cm.
"This Dover edition, first published in 2015, is an unabridged republication of the work originally published by North-Holland Publishing Company, Amsterdam, and John Wiley & Sons, Inc., New York, in 1963. The English text was prepared with assistance from R. Hilton"—Title page verso.
Includes bibliographical references and index.
ISBN-13: 978-0-486-79377-1
ISBN-10: 0-486-79377-X
1. Wave mechanics. 2. Relativity (Physics) I. Strocchi, F., editor. II. Title.

QC174.2.C645 2015
530.11—dc23

2014048064

Manufactured in the United States by Courier Corporation
79377X01 2015
www.doverpublications.com

PREFACE

During the academic years 1959–60 and 1960–61, Prof. L. A. Radicati was on leave from the University of Pisa and asked me to lecture in his place to fourth-year students on Advanced Quantum Mechanics. This book originates from those lectures.

Dr. Franco Strocchi, then a student in my class, volunteered to go through and amplify the notes I handed to him at the end of each lecture and edited them in mimeographed form (in Italian) for the use of his fellow students.

In the Spring of 1961, while visiting University College London, I made a rough translation into English of the Introduction and Part I and asked Roger Hilton, a research student in Physics, to improve the style. Later on, Mr. Hilton read my translation of Parts II and III. The manuscript was ready by the end of the Summer.

The helpful and constructive criticism of the referee, Prof. S. A. Wouthuysen, led to improvements and to my writing a number of additional sections during the year 1961–62 while I was at the Institute for Advanced Study at Princeton.

Finally, I used the whole contents of this book as the basis for a lecture course at the State University of Iowa during the Winter of 1962. Each member of the class generously helped me and Strocchi, while we were correcting the galley proofs, with innumerable suggestions leading to many amendments. One of them, Louis A. Frank, deserves special thanks for his enthusiastic interest.

This book is thus the result of close collaboration and discussions with students. Its purpose is to take the reader, who already has some knowledge of non-relativistic wave mechanics, as far as the borderline between relativistic wave mechanics and field theory, so that he may more confidently attack the study of the latter. No problems are set for the reader to work out on his own. Such problems as were assigned to the students who followed the courses have been incorporated in the text as Notes. Those on pp. 106–110 are due to Dr. A. Di Giacomo of Pisa.

The inclusion of a list of notations has seemed unnecessary. Those used are self-explanatory: for example, four-vectors are boldface sans-serif, three-vectors are boldface italics.

E. C.

Department of Physics, University of Toronto
September, 1963

CONTENTS

INTRODUCTION

PARTICLES AND WAVE FUNCTIONS

PART I

PARTICLES OF SPIN ZERO

PART III

COLLISION AND RADIATION PROCESSES

INTRODUCTION

PARTICLES AND WAVE FUNCTIONS

1. De Broglie's relation

At the end of the last century it seemed possible to interpret all known physical phenomena in terms of electromagnetic waves and charged particles, but the discoveries of the twentieth century have brought about a radical revision of classical concepts. The theory of *blackbody radiation* (Planck 1900)* and the explanation of the *photoelectric effect* (Einstein 1905) were based on the hypothesis that exchanges of energy between matter and radiation took place in quanta

$$E = \hbar\omega \,, \tag{1}$$

$$\hbar = h/2\pi = 1.054 \times 10^{-27}\ \text{erg} \times \text{sec} = 0.658 \times 10^{-15}\ \text{eV} \times \text{sec}\,.$$

The subsequent discovery of the *Compton effect* (Compton 1923) showed that a quantum associated with a plane wave has a *momentum* $\boldsymbol{p}$ given by

$$\boldsymbol{p} = \hbar\boldsymbol{k} \,, \tag{2}$$

where $\boldsymbol{k}$ is the wave *propagation vector*, (briefly, *wavevector*), $|\boldsymbol{k}| = 2\pi/\lambda = \omega/c$. The relation between the wave properties of radiation and the particle properties of the associated quanta is relativistically invariant. In fact k_1, k_2, k_3 and $i\omega/c$ are the components of a four-vector, as may be shown from the invariance of the phase of a monochromatic plane wave. On the other hand it is known that the components of the momentum and the energy of a particle can be regarded as the space and time components of the four-momentum $p_\mu \equiv (\boldsymbol{p}, iE/c)$. Therefore equations (1) and (2) may be condensed in the four-vector equation

$$\mathfrak{p} = \hbar\mathfrak{k} \,,$$

equivalent to the four equations $p_\mu = \hbar k_\mu$, $(\mu = 1, 2, 3, 4)$.

---

* The theory of blackbody radiation leads to a somewhat indirect proof of the quantum nature of energy exchanges, since it involves the laws of statistical mechanics.

One notices now that the relativistic relation between energy and momentum of a particle of rest mass m *

$$E^2 = c^2(\boldsymbol{p}^2 + m^2c^2)$$

can be written in the form **

$$p^2 = p_\mu p_\mu = - m^2c^2$$

and, since

$$k^2 = k_\mu k_\mu = 0 ,$$

we have

$$m^2c^2 = - p^2 = - \hbar^2 k^2 = 0$$

for the quanta of the electromagnetic radiation. This shows that the rest mass of a photon is zero.

De Broglie's postulate of matter waves, which was based on purely theoretical arguments, can be condensed in the statement that the relation $p = \hbar k$, where

$$k \equiv \left(\boldsymbol{k} = \frac{2\pi}{\lambda} \frac{\boldsymbol{k}}{|\boldsymbol{k}|}, \mathrm{i}\frac{\omega}{c} \right),$$

between the wave four-vector and the four-momentum holds not only for photons, but also for particles of mass not equal to zero. In the case of electromagnetic radiation the wave aspect was discovered much earlier than the corpuscular, for material particles the opposite has been the case.***

Note 1. Using the conservation of four-momentum, the expression

$$\Delta\lambda = 4\pi \frac{\hbar}{mc} \sin^2 \tfrac{1}{2}\theta$$

* Note that, according to this formula, for a particular momentum there are two values of the energy, $E = \pm c(\boldsymbol{p}^2 + m^2c^2)^{\frac{1}{2}}$; for the moment we will not consider the negative root.

** The definition of the inner product of two four-vectors $a \equiv (\boldsymbol{a}, a_4 = \mathrm{i}a_0)$, $b \equiv (\boldsymbol{b}, b_4 = \mathrm{i}b_0)$, which we denote by (a, b), is

$$(a, b) = \sum_{\mu=1}^{4} a_\mu b_\mu = \boldsymbol{a} \cdot \boldsymbol{b} - a_0 b_0 = a_\mu b_\mu .$$

We shall continually make use of the convention of dummy indices: if any index occurs twice in a tensor product, this is to be understood as the sum over all its possible values. The square of the modulus of a four-vector a will be denoted by $a^2 = (a, a) = a_\mu a_\mu$, while the square of the modulus of a spatial vector will be

$$\boldsymbol{a}^2 = \sum_{i=1}^{3} a_i^2 = a_i a_i ,$$

Greek indices running from *1* to *4*, Roman indices from *1* to *3*.

*** L. de Broglie, Nature **112** (1923), 540; Ann. d. Phys. (10), **3** (1925), 22; Thesis (Paris, 1924).

for the change in wavelength in the Compton effect, may be cast into a Lorentz invariant form.

This amounts to finding an invariant equation for the four-momenta p and p' of the electron, and $\hbar k$ and $\hbar k'$ of the photon, before and after collision, which reduces to the given formula in the reference frame in which the electron is initially at rest ($\boldsymbol{p} = 0$).

From conservation of energy and momentum we have

$$p' = p + \hbar k - \hbar k' ,$$

which, on squaring, becomes

$$p'^2 = p^2 + \hbar^2(k^2 + k'^2) - 2\hbar^2(k, k') + 2\hbar(k - k', p) .$$

Since

$$p^2 = p'^2 = - m^2c^2 , \quad k^2 = k'^2 = 0 ,$$

we obtain the invariant equation

$$\hbar(k, k') + (k' - k, p) = 0 . \tag{α}$$

In the reference frame in which $\boldsymbol{p} = 0$, one has

$$(k' - k, p) = mc(|\boldsymbol{k}| - |\boldsymbol{k}'|) ,$$

$$(k, k') = - |\boldsymbol{k}| \cdot |\boldsymbol{k}'| (1 - \cos\theta) = - 2\, |\boldsymbol{k}| \cdot |\boldsymbol{k}'| \sin^2 \tfrac{1}{2}\theta ,$$

where θ is the angle formed by $\boldsymbol{k}$ and $\boldsymbol{k}'$.

Thus we have

$$\Delta\lambda = \lambda' - \lambda = 4\pi \frac{\hbar}{mc} \sin^2 \tfrac{1}{2}\theta .$$

As a further application of eq. (α), we consider the cases where the electron has:

a) initial non-vanishing momentum in the direction of propagation of the incident photon;

b) initial momentum $- \frac{1}{2}\boldsymbol{\Delta}$ and final momentum $\frac{1}{2}\boldsymbol{\Delta}$ (*brick wall system*). One can show that $|\boldsymbol{\Delta}|$ has the meaning of an *invariant momentum transfer*.

a) From (α) it follows that

$$4\pi\hbar \sin^2 \tfrac{1}{2}\theta + (\lambda - \lambda')E/c + (\lambda' - \lambda)|\boldsymbol{p}| + \lambda(1 - \cos\theta)|\boldsymbol{p}| = 0 ,$$

from which

$$\Delta\lambda = 2 \sin^2 \tfrac{1}{2}\theta \frac{c(h + \lambda|\boldsymbol{p}|)}{E - c|\boldsymbol{p}|} ;$$

b) Since the energy of the electron does not change in the collision,

$$p_0 = E/c = \tfrac{1}{2}(\Delta^2 + 4m^2c^2)^{\frac{1}{2}} = p'_0 ,$$

also the energy of the photon remains unchanged; therefore

$$k_0 = k'_0 \quad \text{i.e. } \lambda = \lambda' .$$

It is convenient to put $\hbar \boldsymbol{k} = \boldsymbol{\pi} + \frac{1}{2}\boldsymbol{\Delta}$ and $\hbar \boldsymbol{k}' = \boldsymbol{\pi} - \frac{1}{2}\boldsymbol{\Delta}$. The condition $|\boldsymbol{k}| = |\boldsymbol{k}'|$ implies that $\boldsymbol{\pi} \cdot \boldsymbol{\Delta} = 0$, so that $\boldsymbol{\pi}$ and $\boldsymbol{\Delta}$ are orthogonal (Fig. 1).

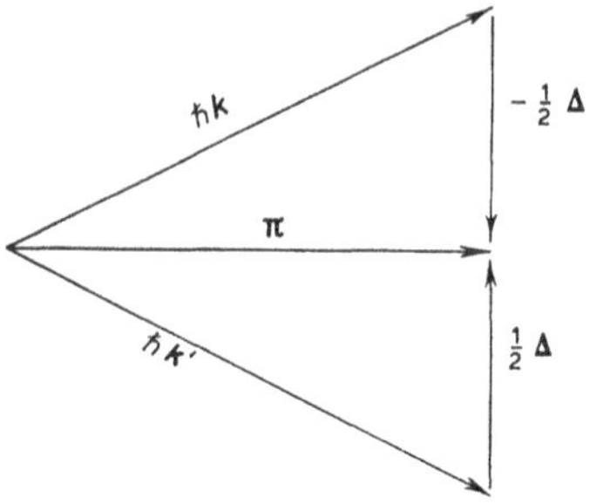

Fig. 1

The momentum transfer is

$$\boldsymbol{p}' - \boldsymbol{p} = \hbar(\boldsymbol{k} - \boldsymbol{k}') .$$

On the other hand the invariant

$$(p' - p)^2 = (\boldsymbol{p}' - \boldsymbol{p})^2 - (p'_0 - p_0)^2$$

may be calculated in any reference frame, in particular in the brick wall system, in which it has the value $|\boldsymbol{\Delta}|^2$.

2. Phase and group velocity

Waves associated with particles of non-zero mass have properties different from those of electromagnetic waves. For instance the *dispersive law* $\lambda = 2\pi c/\omega$ cannot be valid for matter waves. In fact it has been shown above that this is only consistent with de Broglie's relation if the photon rest mass is zero.

Since $E = \hbar\omega = c(\boldsymbol{p}^2 + m^2c^2)^{\frac{1}{2}}$, it follows that for a particle of mass m

$$\omega^2 = c^2(\boldsymbol{k}^2 + \kappa^2) , \quad (\kappa = mc/\hbar) , \tag{3}$$

and also

$$\lambda = \frac{2\pi c}{(\omega^2 - c^2\kappa^2)^{\frac{1}{2}}} = \frac{h}{mc}\frac{\sqrt{1-\beta^2}}{\beta}, \quad (\beta = v/c) .$$

From this it is seen that the *phase velocity* v_p of a monochromatic matter wave is a function of $|\boldsymbol{k}|$ since

$$v_\mathrm{p} = \frac{\omega}{|\boldsymbol{k}|} = c\left(1 + \frac{\kappa^2}{\boldsymbol{k}^2}\right)^{\frac{1}{2}} > c , \tag{4}$$

whereas for electromagnetic waves in vacuum the phase velocity is always c ($m = 0$ in eq. (4)). Therefore, for matter waves associated with a force-free particle, the vacuum is equivalent to a homogeneous isotropic medium of refractive index n given by

$$n = \frac{c}{v_\mathrm{p}} = \left(1 - \frac{c^2\kappa^2}{\omega^2}\right)^{\frac{1}{2}} < 1\,.$$

In contrast to this, for electromagnetic waves n is $\geqslant 1$ for most dispersive materials. The corresponding phase velocity is therefore less than c.

The fact, that the phase velocity of matter waves is greater than the speed of light, is at first sight somewhat strange. Although the formalism used so far is invariant under Lorentz transformations, it might seem that matter waves infringe one of the basic postulates of relativity i.e. that a signal cannot propagate with a speed greater than c. On closer inspection, though, it is seen that the phase velocity characterizes a plane wave of infinite extension, which therefore cannot be regarded as a signal. In order to speak of a signal one must have a *wave function* $\psi(\boldsymbol{x}, t)$ which is zero outside a certain region of space, the location (and generally shape and size) of which changes with time.

A *wave packet* may be constructed as a superposition of plane waves and expressed in the form of a triple Fourier integral *

$$\psi(\boldsymbol{x}, t) = (2\pi)^{-\frac{3}{2}} \int A(\boldsymbol{k})\, \mathrm{e}^{\mathrm{i}(\boldsymbol{k}\cdot\boldsymbol{x} - \omega t)}\, \mathrm{d}^3\mathrm{k}\,. \tag{5}$$

It must be stressed that (5) is not a four-dimensional Fourier integral as might be used to give an integral representation of any function of the variables x_μ, ($\mu = 1, 2, 3, 4$), (see *Note 4* below).

In writing (5) it has tacitly been assumed that the principle of superposition holds also for matter waves and that the general ideas on interference and diffraction of electromagnetic waves may be extended to matter waves.

In order to invert equation (5), one must multiply both sides by $(2\pi)^{-\frac{3}{2}}$ $\mathrm{e}^{-\mathrm{i}\boldsymbol{k}\cdot\boldsymbol{x}}$ and integrate with respect to $\boldsymbol{x}$

$$(2\pi)^{-\frac{3}{2}} \int \psi(\boldsymbol{x}, t)\, \mathrm{e}^{-\mathrm{i}\boldsymbol{k}\cdot\boldsymbol{x}}\, \mathrm{d}^3\mathrm{x} = \int A(\boldsymbol{k}')\, \mathrm{e}^{-\mathrm{i}\omega(k')t} \left((2\pi)^{-3} \int \mathrm{e}^{\mathrm{i}(\boldsymbol{k}'-\boldsymbol{k})\cdot\boldsymbol{x}}\, \mathrm{d}^3\mathrm{x}\right) \mathrm{d}^3\mathrm{k}'\,.$$

Remembering the properties of the three-dimensional δ function (see *Note 2*), one has

$$A(\boldsymbol{k})\, \mathrm{e}^{-\mathrm{i}\omega(k)t} = (2\pi)^{-\frac{3}{2}} \int \psi(\boldsymbol{x}, t)\, \mathrm{e}^{-\mathrm{i}\boldsymbol{k}\cdot\boldsymbol{x}}\, \mathrm{d}^3\mathrm{x}\,.$$

---

* In this expression and throughout the sequel, we denote by $\mathrm{d}^3\mathrm{k} = \mathrm{d}k_1\mathrm{d}k_2\mathrm{d}k_3$ the volume element in the three-dimensional wavevector space (which we may call momentum space), and simply by $\int$ the integral extended over all space.
In this section it is assumed that particles are free, and so $A(\boldsymbol{k})$ does not depend on time.

The problem of finding the Fourier transform of a given wave function $\psi(\boldsymbol{x}, t)$ is thus reduced to a simple integration.

Defining

$$\varphi(\boldsymbol{k}, t) = A(\boldsymbol{k})\, \mathrm{e}^{-\mathrm{i}\omega(k)t},$$

which may be interpreted as the wave function in $\boldsymbol{k}$ space, one can write more symmetrically

$$\begin{aligned} \psi(\boldsymbol{x}, t) &= (2\pi)^{-\frac{3}{2}} \int \varphi(\boldsymbol{k}, t)\, \mathrm{e}^{\mathrm{i}\boldsymbol{k}\cdot\boldsymbol{x}}\, \mathrm{d}^3\mathrm{k}\,, \\ \varphi(\boldsymbol{k}, t) &= (2\pi)^{-\frac{3}{2}} \int \psi(\boldsymbol{x}, t)\, \mathrm{e}^{-\mathrm{i}\boldsymbol{k}\cdot\boldsymbol{x}}\, \mathrm{d}^3\mathrm{x}\,. \end{aligned} \tag{6}$$

NOTE 2. We summarize some of the basic properties of the Dirac δ function. For a more complete treatment see P. A. M. Dirac, *The Principles of Quantum Mechanics* (Oxford, 1958), p. 58, and L. Schwartz, *Théorie des Distributions* (Hermann, Paris, 1950).

The Dirac function may be defined by the properties

$$\delta(x - x_0) = 0 \quad \text{for } x \neq x_0, \quad \int_{-\infty}^{\infty} \delta(x - x_0)\, \mathrm{d}x = 1\,,$$

or by the equivalent property

$$\int_{-\infty}^{\infty} f(x)\, \delta(x - x_0)\, \mathrm{d}x = f(x_0)$$

($f(x)$ is a continuous function). No ordinary function with these properties exists. $\delta(x - x_0)$ may be conceived as the limit of an ordinary function, e.g.

$$\lim_{h\to 0} \exp\left[-(x - x_0)^2/h^2\right]/\sqrt{\pi}\, h\,.$$

The Fourier representation of the δ function is

$$\delta(x) = (2\pi)^{-1} \int_{-\infty}^{\infty} \mathrm{e}^{\mathrm{i}kx}\, \mathrm{d}k\,.$$

(Here the equality is to be interpreted in the sense that the integral is equivalent to $\delta(x)$, when it occurs as the factor of an integrand.)

The following is a non-rigorous proof that

$$(2\pi)^{-1} \int_{-\infty}^{\infty} f(x)\, \mathrm{e}^{\mathrm{i}kx}\, \mathrm{d}k\, \mathrm{d}x = f(0)\,.$$

We write

$$(2\pi)^{-1}\int_{-\infty}^{\infty} e^{ikx}\,dk = (2\pi)^{-1}\lim_{K\to\infty}\int_{-K}^{K} e^{ikx}\,dk = \lim_{K\to\infty}\frac{K}{\pi}\frac{\sin(Kx)}{Kx}. \qquad (\beta)$$

Now

$$\int_{-\infty}^{\infty}\frac{\sin(Kx)}{\pi x}\,dx = 1$$

for any K, and therefore also in the limit $K\to\infty$. Moreover for $x = 0$ the expression on the right of equation (β) becomes infinitely large. For $x \neq 0$ its oscillations become infinitely fast, so that its contribution to the integral over x reduces to that of the points in neighbourhood of $x = 0$, and $f(x)$ may be replaced by $f(0)$ and taken out of the integral sign.

Likewise one can define a three-dimensional δ function

$$\delta(\boldsymbol{x}) = \delta(x_1)\,\delta(x_2)\,\delta(x_3)$$

with the Fourier representation

$$\delta(\boldsymbol{x}) = (2\pi)^{-3}\int e^{i\boldsymbol{k}\cdot\boldsymbol{x}}\,d^3k = (2\pi)^{-3}\int_{-\infty}^{\infty} e^{ik_1x_1}\,dk_1\int_{-\infty}^{\infty} e^{ik_2x_2}\,dk_2\int_{-\infty}^{\infty} e^{ik_3x_3}\,dk_3.$$

NOTE 3. Much use will be made of the formula

$$\delta(f(x)) = \sum_n \frac{\delta(x-x_n)}{|f'(x_n)|},$$

where $f(x_n) = 0,\quad f'(x_n) \neq 0$.

In order to prove this, the x axis is divided into N intervals by the points $a_1 = -\infty,\ \ b_1 = a_2,\ \ b_2, \dots b_{N-1} = a_N,\ \ b_N = \infty$. Each interval contains only one zero of $f(x)$.

For an arbitrary continuous function $F(x)$ we have

$$\int F(x)\,\delta(f(x))\,dx = \sum_n\int_{a_n}^{b_n} F(x)\,\delta(f(x))\,dx\,.$$

For a generic interval

$$\int_{a_n}^{b_n} F(x)\,\delta(f(x))\,\mathrm{d}x = \int_{\min}^{\max} F(x(y))\,\delta(y) \left|\frac{\mathrm{d}x}{\mathrm{d}y}\right| \mathrm{d}y\,,$$

where $y = f(x)$, $\max = \max[f(b_n), f(a_n)]$, $\min = \min[f(b_n), f(a_n)]$. Hence it follows that

$$\int F(x)\,\delta(f(x))\,\mathrm{d}x = \sum_n \frac{F(x_n)}{|f'(x_n)|}$$

(where x_n are the zero's of $y = f(x)$), showing that $\delta(f(x))$ has the same properties as the linear combination of Dirac functions

$$\sum_n \delta(x - x_n)\,|f'(x_n)|^{-1}\,.$$

For $f(x) = x^2 - a^2$ and $a > 0$

$$\delta(x^2 - a^2) = \frac{1}{2a}\{\delta(x - a) + \delta(x + a)\}\,.$$

NOTE 4. Using the formula given in the preceding note, it can be shown that the triple integral

$$(2\pi)^{-\frac{3}{2}} \int A(\boldsymbol{k})\,\mathrm{e}^{\mathrm{i}(\boldsymbol{k}\cdot\boldsymbol{x} - \omega t)}\,\mathrm{d}^3\mathrm{k}\,, \quad (\omega = c(\boldsymbol{k}^2 + \kappa^2)^{\frac{1}{2}})\,,$$

can be written in the form of a fourfold integral

$$\int A(\mathrm{k})\,\mathrm{e}^{\mathrm{i}(\mathrm{k},\,x)}\,\mathrm{d}^4k\,,$$

where $A(\mathrm{k}) = (2\pi)^{-\frac{3}{2}}\, 2\omega/c\, A(\boldsymbol{k})\, \eta(\mathrm{k})\, \delta(\mathrm{k}^2 + \kappa^2)$, $\eta(\mathrm{k}) = 1$ if $k_0 > 0$, $\eta(\mathrm{k}) = 0$ for $k_0 < 0$ and $\mathrm{d}^4k = \mathrm{d}^3\mathrm{k}\,\mathrm{d}k_0$.

In fact

$$\delta(k_0^2 - \omega^2/c^2) = c/2\omega\,\{\delta(k_0 - \omega/c) + \delta(k_0 + \omega/c)\},\ \text{for } \omega > 0\,.$$

Thus

$$\int A(\boldsymbol{k})\,\mathrm{e}^{\mathrm{i}(\boldsymbol{k}\cdot\boldsymbol{x} - \omega t)}\,\mathrm{d}^3\mathrm{k} = \int A(\boldsymbol{k})\,2\omega/c\,\delta(k_0^2 - \omega^2/c^2)\,\eta(\mathrm{k})\,\mathrm{e}^{\mathrm{i}(\mathrm{k},\,x)}\,\mathrm{d}^4k\,.$$

The function $\eta(\mathrm{k})$ eliminates the negative frequencies which would result from $\delta(k_0 + \omega/c)$. Because of the relativistic dispersive law (3) one has

$$k_0^2 - \omega^2/c^2 = k_0^2 - \boldsymbol{k}^2 - \kappa^2 = -\,\mathrm{k}^2 - \kappa^2\,.$$

Remembering that $\delta(-x) = \delta(x)$ and putting $\mathsf{A}(\mathsf{k}) = (2\pi)^{-\frac{3}{2}}\, 2\omega/c\; A(\boldsymbol{k})\, \eta(\mathsf{k})\, \delta(\mathsf{k}^2 + \kappa^2)$ it follows that

$$(2\pi)^{-\frac{3}{2}} \int A(\boldsymbol{k})\, \mathrm{e}^{\mathrm{i}(\boldsymbol{k}\cdot\boldsymbol{x}-\omega t)}\, \mathrm{d}^3\mathrm{k} = \int \mathsf{A}(\mathsf{k})\, \mathrm{e}^{\mathrm{i}(\mathsf{k},\mathsf{x})}\, \mathrm{d}^4\mathsf{k}\,.$$

The exponential, and the argument of the δ function, are invariant, as k_1, k_2, k_3, k_4 transform like the components of a four-vector. The δ function limits the domain of integration to time-like four-vectors whose square is $-\kappa^2$. Since the sign of the fourth component of a time-like vector is invariant, the function $\eta(\mathsf{k})$ of such a four-vector is invariant. The volume element $\mathrm{d}^4\mathsf{k}$ is also invariant under Lorentz transformations.

Now suppose that, for a certain wave packet, the modulus of $A(\boldsymbol{k}) = |A(\boldsymbol{k})|\, \mathrm{e}^{\mathrm{i}\phi(\boldsymbol{k})}$, ($\phi$ real), has a maximum for a certain value of $\boldsymbol{k}$, and is appreciably different from zero only in the finite domain $(k_i - \eta, k_i + \eta)$ $(i = 1, 2, 3)$ with $\eta \ll |\boldsymbol{k}|$. In the neighbourhood of the maximum the wave function has the form

$$\psi(\boldsymbol{x}, t) \simeq (2\pi)^{-\frac{3}{2}}\, |A(\boldsymbol{k})|\, \mathrm{e}^{\mathrm{i}(\boldsymbol{k}\cdot\boldsymbol{x}-\omega t+\phi)} \iiint_{-\eta}^{+\eta} \mathrm{e}^{\mathrm{i}\boldsymbol{\eta}'\cdot[\nabla_{k'}(\boldsymbol{k}'\cdot\boldsymbol{x}-\omega' t+\phi(\boldsymbol{k}'))]_{k'=k}}\, \mathrm{d}^3\boldsymbol{\eta}'$$

$$= 8(2\pi)^{-\frac{3}{2}}\, |A(\boldsymbol{k})|\, \mathrm{e}^{\mathrm{i}(\boldsymbol{k}\cdot\boldsymbol{x}-\omega t+\phi)} \prod_{i=1}^{3} \frac{\sin\left[\eta \dfrac{\partial}{\partial k_i}(\boldsymbol{k}\cdot\boldsymbol{x} - \omega t + \phi)\right]}{\left(x_i - \dfrac{\partial\omega}{\partial k_i} t + \dfrac{\partial\phi}{\partial k_i}\right)},$$

where the symbol ∇_k denotes a vector with components

$$\nabla_k \equiv \left(\frac{\partial}{\partial k_1}, \frac{\partial}{\partial k_2}, \frac{\partial}{\partial k_3}\right).$$

Therefore the amplitude of the wave function is appreciable only at the value $\boldsymbol{k}$ for which $|A(\boldsymbol{k}')|$ has a maximum and at the value of $\boldsymbol{x}$ for which

$$\{\nabla_{k'}[\boldsymbol{k}'\cdot\boldsymbol{x} - \omega(\boldsymbol{k}')t + \phi(\boldsymbol{k}')]\}_{k'=k} = 0\,,$$

i.e.

$$\boldsymbol{x} = t\nabla_k\omega - \nabla_k\phi\,.$$

This last equation defines, as a function of time, the position of the *centre* of the wave packet, which moves with the velocity

$$v_{\mathrm{g}} = \frac{c^2\boldsymbol{k}}{\omega}\,. \tag{7}$$

This is called the *group velocity*. This expression is acceptable since $|\boldsymbol{v}_g| < c$ in accordance with the postulate of relativity theory.

In the classical limit when diffraction effects may be neglected, a particle, even if subject to forces, can be represented by a wave packet of finite extension which moves in space with a group velocity determined at any time by the wave vector for which $|A(\boldsymbol{k}, t)|$ is a maximum. The momentum $\boldsymbol{p} = \hbar\boldsymbol{k}$ and the energy $E = \hbar\omega$ associated with this wavevector may be identified with the momentum and the energy of the particle. From equation (7) one sees that $\boldsymbol{v}_g$ is the velocity of a particle of momentum $\boldsymbol{p}$ and energy E

$$\boldsymbol{v}_g = \frac{c^2\boldsymbol{p}}{E}\,.$$

Notice that this argument may be reversed. If one requires that the group velocity $\boldsymbol{v}_g = \nabla_{\hbar k}\hbar\omega$ must coincide with the velocity of a particle $\boldsymbol{v} = \nabla_p E$, one has

$$\boldsymbol{p} = \hbar\boldsymbol{k} + \boldsymbol{c}\,,$$

where $\boldsymbol{c}$ is independent of $\boldsymbol{k}$. Now a relation of the type $p_i = \hbar k_i + c_i$ $(i = 1, 2, 3)$ in a certain system of reference would take the form $p_i' = \hbar k_i' + c_i'$ $(i = 1, 2, 3)$ with respect to another one, obtained by rotating the first, c_i' being a linear homogeneous combination of the c_i's.

Since the relation between $\boldsymbol{p}$ and $\boldsymbol{k}$ must be invariant under rotations we must require that $c_i = c_i'$, and this is only possible if $c_i = 0$.

This last approach was followed by de Broglie to establish the relationship between momentum and wavevector. The approach is independent of relativity, and is in the spirit of the correspondence principle. Moreover it may be adapted to cases in which the particle is subject to forces.

3. Charged particles in an electromagnetic field

We review some of the basic equations of relativistic electrodynamics, which will frequently be used in the following. It is well known that Maxwell's equations may be written in tensor form

$$\frac{\partial F_{\mu\nu}}{\partial x_\nu} = \frac{1}{c}\,j_\mu\,,$$
$$\frac{\partial \bar{F}_{\mu\nu}}{\partial x_\nu} = 0\,, \tag{8}$$

where

$$F_{\mu\nu} = \begin{pmatrix} 0 & H_3 & -H_2 & -\mathrm{i}\,E_1 \\ -H_3 & 0 & H_1 & -\mathrm{i}\,E_2 \\ H_2 & -H_1 & 0 & -\mathrm{i}\,E_3 \\ \mathrm{i}\,E_1 & \mathrm{i}\,E_2 & \mathrm{i}\,E_3 & 0 \end{pmatrix}, \quad \overline{F}_{\mu\nu} = \tfrac{1}{2}\,\varepsilon_{\mu\nu\varrho\sigma}\,F_{\varrho\sigma}\,.$$

Here $\varepsilon_{\mu\nu\varrho\sigma}$ is the four-dimensional Ricci-Levi Civita tensor, antisymmetric in all its indices $\varepsilon_{\mu\nu\varrho\sigma} = -\varepsilon_{\mu\nu\sigma\varrho} = \varepsilon_{\mu\sigma\nu\varrho} = -\varepsilon_{\sigma\mu\nu\varrho} \ldots$ and with $\varepsilon_{1234} = 1$ (see, however, Chapter VII, §1).

It may be shown that a necessary and sufficient condition for the second of equations (8) to be satisfied is that $F_{\mu\nu}$ may be written as a four-dimensional *curl* *

$$F_{\mu\nu} = \frac{\partial A_\nu}{\partial x_\mu} - \frac{\partial A_\mu}{\partial x_\nu}\,. \tag{9}$$

From the tensor character of $F_{\mu\nu}$ it follows that the four-potential A_μ is a four-vector. The first of equations (8) becomes

$$\Box\, A_\mu - \frac{\partial}{\partial x_\mu}\left(\frac{\partial A_\nu}{\partial x_\nu}\right) = -\frac{1}{c}\,j_\mu\,,$$

where (10)

$$\Box \equiv \Delta - \frac{1}{c^2}\frac{\partial^2}{\partial t^2} = \frac{\partial^2}{\partial x_\mu\,\partial x_\mu}$$

is the d' Alembert operator.

If the *Lorentz condition*

$$\frac{\partial A_\mu}{\partial x_\mu} = 0$$

is imposed, eq. (10) becomes

$$\Box\, A_\mu = -\frac{1}{c}\,j_\mu\,. \tag{10'}$$

For a given $F_{\mu\nu}$ the four-potential A_μ is not uniquely determined by eq. (9). In fact the electromagnetic tensor is invariant under the *gauge transformation*

* The vector potential of a uniform magnetic field $\boldsymbol{H}$ is given by $\boldsymbol{A} = \frac{1}{2}\boldsymbol{H} \times \boldsymbol{x}$, $\boldsymbol{x} \equiv (x, y, z)$.

of the *second kind*

$$A'_\mu = A_\mu + \frac{\partial \Lambda}{\partial x_\mu}, \quad \Lambda = \Lambda(x).$$

Only the field strengths have a direct physical meaning, and they do not change under a gauge transformation. Any relation involving electromagnetic quantities must be gauge invariant, if it is to have a physical meaning. From an arbitrary four-potential, one satisfying the Lorentz condition can be obtained by making a gauge transformation with Λ satisfying

$$\Box \Lambda = -\frac{\partial A_\mu}{\partial x_\mu}.$$

In the case of eqs. (10′), the only permissible gauge transformations are those for which

$$\Box \Lambda = 0$$

(*special gauge transformations*).

The Lagrangian of a particle of charge $-e$ in a static electromagnetic field is

$$L = eA_0 - \frac{e}{c} \boldsymbol{A} \cdot \boldsymbol{v} - mc^2 \left(1 - \frac{v^2}{c^2}\right)^{\frac{1}{2}}.$$

Thus the canonical momentum is

$$p_i = \frac{\partial L}{\partial v_i} = m\gamma v_i - \frac{e}{c} A_i, \quad \left(\gamma^{-2} = 1 - \frac{v^2}{c^2}\right),$$

and the Hamiltonian

$$H = -eA_0 + c\left[\left(\boldsymbol{p} + \frac{e}{c}\boldsymbol{A}\right)^2 + m^2c^2\right]^{\frac{1}{2}} = m\gamma c^2 - eA_0.$$

Hence the equations of motion for a particle of energy E

$$v_i = \frac{c^2\left(p_i + \frac{e}{c} A_i\right)}{E + eA_0}, \quad (i = 1, 2, 3),$$

$$\dot{p}_i = e\left(\frac{\partial A_0}{\partial x_i} - \frac{v_k}{c}\frac{\partial A_k}{\partial x_i}\right).$$

The last equation is not invariant under the *static* gauge transformation $\boldsymbol{A} \to \boldsymbol{A} + \nabla_x \Lambda(\boldsymbol{x})$, but it becomes so if $e\dot{A}_i/c$ is added to both sides. In fact one gets

$$\frac{\mathrm{d}}{\mathrm{d}t}\left(p_i + \frac{e}{c} A_i\right) = \frac{\mathrm{d}}{\mathrm{d}t}(m\gamma v_i) = e\frac{\partial A_0}{\partial x_i} - \frac{e}{c} v_k F_{ik} ,$$

where we took into account that $\boldsymbol{A}$ is an implicit function of t through $\boldsymbol{x} = \boldsymbol{x}(t)$. In order to give a more significant form to the right-hand side, we consider the three-dimensional Ricci-Levi Civita tensor ε_{ikl}, by means of which the vector product may be written in the form

$$(\boldsymbol{a} \times \boldsymbol{b})_i = \varepsilon_{ijk}\, a_j\, b_k \,.$$

Therefore

$$(\boldsymbol{v} \times \boldsymbol{H})_i = \varepsilon_{ikl}\, v_k\, \varepsilon_{lmn} \frac{\partial A_n}{\partial x_m} \,.$$

Using the identity

$$\varepsilon_{ikl}\, \varepsilon_{lmn} = \delta_{im}\, \delta_{kn} - \delta_{in}\, \delta_{km} ,$$

where δ_{ab} is the Kronecker symbol, one finally obtains

$$(\boldsymbol{v} \times \boldsymbol{H})_i = v_k F_{ik} ,$$

and therefore

$$\frac{\mathrm{d}}{\mathrm{d}t}\left(\boldsymbol{p} + \frac{e}{c}\boldsymbol{A}\right) = -e\left(\boldsymbol{E} + \frac{\boldsymbol{v}}{c} \times \boldsymbol{H}\right),$$

which is the Lorentz force.

In order that the equation

$$\boldsymbol{v} = \frac{c^2\left(\boldsymbol{p} + \frac{e}{c}\boldsymbol{A}\right)}{E + eA_0}$$

may be invariant – as is physically necessary – under the gauge transformation $\boldsymbol{A} \to \boldsymbol{A} + \nabla_x \Lambda$, the latter must be accompanied by the momentum transformation

$$\boldsymbol{p} \to \boldsymbol{p} - \frac{e}{c}\nabla_x \Lambda \,.$$

For a particle in an electrostatic field it is easy to establish de Broglie's relation. Assuming that $E = \hbar\omega$, then on requiring that $\boldsymbol{v} = \nabla_k\, \omega$, one gets

$$\hbar \boldsymbol{k} = \boldsymbol{p}$$

as for a free particle.

The wavelength $\lambda = 2\pi/|\boldsymbol{k}|$ is given by

$$\lambda = \frac{h}{|\boldsymbol{p}|} = \frac{hc}{\sqrt{(eA_0 + E)^2 - (mc^2)^2}}$$

and is a function of position.

For small velocities ($|\boldsymbol{p}| \ll mc$), energy and momentum of a free particle are given by the Taylor expansions of $m\gamma c^2$ and $m\gamma \boldsymbol{v}$ with respect to v/c terminated at the second term, $\boldsymbol{p} = m\boldsymbol{v}$ + *terms of third order in* v/c, $E \simeq mc^2 (1 + \frac{1}{2} \boldsymbol{v}^2/c^2)$. Correspondingly we have

$$\omega = \frac{E}{\hbar} = \frac{mc^2}{\hbar} + \frac{\hbar \boldsymbol{k}^2}{2m}, \quad \boldsymbol{v} = \frac{\hbar \boldsymbol{k}}{m}$$

and

$$\lambda = \frac{h}{m\,|\boldsymbol{v}|}\,.$$

The last of these equations is precisely the relation between wavelength and velocity as was found by de Broglie.

Notice that the term $mc^2/\hbar$ in the expression for ω gives the same periodic factor $\exp\,[-\,\mathrm{i}mc^2/\hbar]$ for all plane waves, and, therefore, also for an arbitrary wave function $\psi(\boldsymbol{x}, t)$. Thus one may redefine the zero point of both energy and angular frequency

$$\omega \to \omega - c\kappa\,, \quad E \to E - mc^2\,.$$

For an electron in an electrostatic field

$$E = -\,eA_0 + \frac{\boldsymbol{p}^2}{2m_\mathrm{e}}, \quad \omega = -\,\frac{eA_0}{\hbar} + \frac{\hbar \boldsymbol{k}^2}{2m_\mathrm{e}}, \quad \lambda = \frac{h}{\sqrt{2m_\mathrm{e}(E + eA_0)}}\,.$$

These non-relativistic expressions may be used in discussing experiments on diffraction of electrons (Davisson and Germer (1927), reflection on mono-crystals; G. P. Thomson (1928) and Rupp (1928), Debye-Scherrer rings from diffraction by crystal powders).

It is convenient to measure the energy of electrons in electron volts and the wavelength in Ångstroms. Writing

$$\lambda = \frac{h}{\sqrt{2m_\mathrm{e}\,E}} = \frac{h}{m_\mathrm{e}c}\left(\frac{m_\mathrm{e}c^2}{2E}\right)^{\frac{1}{2}}$$

and remembering that the electron *Compton wavelength*, i.e. the change of wavelength of a photon scattered 90° by an electron at rest, is *

$$h/m_e c = 0.0242 \text{ Å}$$

and the rest energy of an electron is

$$m_e c^2 = 0.51 \text{ MeV}$$

we have

$$\lambda = 12.2 (E_{eV})^{-\frac{1}{2}} \text{ Å} ,$$

where E_{eV} is the kinetic energy of the electron measured in eV.

4. Principle of least action and Fermat's principle

In giving the first formulation of wave mechanics, de Broglie made use of the analogy between the principle of *least action* of mechanics and *Fermat*'s principle of geometrical optics. The first principle states that, of all motions $\boldsymbol{x} = \boldsymbol{x}(t)$ leading from the position $P_1 = P(t_1)$ to $P_2 = P(t_2)$, the one actually realized makes the *action*, i.e. the integral

$$A = \int_{t_1}^{t_2} L(\boldsymbol{x}, \dot{\boldsymbol{x}}, t) \, dt ,$$

stationary (*Hamilton*'s principle). That is, for variations $\boldsymbol{x}(t) \to \boldsymbol{x}(t) + \delta\boldsymbol{x}(t)$ subject to the condition $\delta P(t_1) = \delta P(t_2) = 0$, we must have

$$\delta \int_{t_1}^{t_2} L(\boldsymbol{x}, \dot{\boldsymbol{x}}, t) \, dt = \int_{t_1}^{t_2} \left(\frac{\partial L}{\partial \dot{x}_i} \delta \dot{x}_i + \frac{\partial L}{\partial x_i} \delta x_i \right) dt =$$

$$= \int_{t_1}^{t_2} \left(\frac{\partial L}{\partial x_i} - \frac{d}{dt} \frac{\partial L}{\partial \dot{x}_i} \right) \delta x_i \, dt = 0 ,$$

* Some useful data ($\lambda_c = h/mc$):

$\lambda_c = 0.024 \times 10^{-8}$ cm for an electron
$\lambda_c = 1.32 \times 10^{-13}$ cm for a hydrogen atom
$\lambda_c = 2.2 \times 10^{-25}$ cm for a colloidal particle ($m = 10^{-12}$ g)
$\lambda_c = 2.2 \times 10^{-37}$ cm for a body of mass 1 g.

Since the kinetic energy of a particle in thermal equilibrium is $E = \frac{3}{2}kT = 0.62 \times 10^{-13}$ erg at $T = 300$ °K, the de Broglie wavelength is

$\lambda = 1.8 \times 10^{-6}$ Å for a colloidal particle
$\lambda = 1.4$ Å for a hydrogen atom
$\lambda = 62$ Å for an electron.

It is also useful to note that the characteristic time of atomic phenomena is

$$h/m_e c^2 = 0.8 \times 10^{-20} \text{ sec.}$$

and thus *Lagrange*'s equations hold:

$$\frac{\mathrm{d}}{\mathrm{d}t}\frac{\partial L}{\partial \dot{x}_i} - \frac{\partial L}{\partial x_i} = 0\,, \quad (i = 1, 2, 3)\,.$$

The principle of least action can also be formulated by restricting the variations $\boldsymbol{x}(t) \to \boldsymbol{x}(t) + \delta\boldsymbol{x}$ to those which do not change the energy (*Maupertuis*' principle). Since $L = \sum_k p_k\, \dot{q}_k - H$, in this case we have

$$\delta \int_{P_1}^{P_2} \boldsymbol{p} \cdot \mathrm{d}\boldsymbol{x} = 0\,, \quad \delta P_1 = \delta P_2 = 0\,,$$

(where $\boldsymbol{p}$ must be regarded as a function of $\boldsymbol{x}$), with the subsidiary condition

$$\delta H = 0 \quad \text{or} \quad H = \text{const}\,.$$

In this form the principle of least action is analogous to Fermat's principle of geometrical optics

$$\delta \int_{P_1}^{P_2} \boldsymbol{k} \cdot \mathrm{d}\boldsymbol{x} = 0\,, \quad \delta P_1 = \delta P_2 = 0\,,$$

where $\boldsymbol{k} = 2\pi\nu n_\nu(x)\boldsymbol{k}/c\,|\boldsymbol{k}|$, n_ν being the refractive index for light of frequency ν (isotropic medium). This principle determines a ray from P_1 to P_2 as a line for which the optical path is stationary.

Now classical mechanics is the limit of wave mechanics, just as geometrical optics is the limit of wave optics, when

$$|\nabla_x \lambda| \ll 1\,.$$

A ray may be interpreted as the classical trajectory along which a particle moves with velocity equal to the group velocity of the associated wave packet.

Using these ideas de Broglie interpreted Sommerfeld's condition as the condition for a standing wave to exist around the nucleus (Fig. 2).
In order that the wave be stationary one must have

$$\oint \boldsymbol{k} \cdot \mathrm{d}\boldsymbol{x} = 2\pi n\,,$$

where n is an integer. Since $\boldsymbol{p} = \hbar\boldsymbol{k}$ the Sommerfeld condition follows at once.

It is well known that, for small quantum numbers, one cannot speak, even approximately, of trajectories, so that Fig. 2 represents an actual physical situation only if one is dealing with high quantum numbers. In this case the classical picture is a good approximation (correspondence principle); for low quantum numbers the torus invades the whole space.

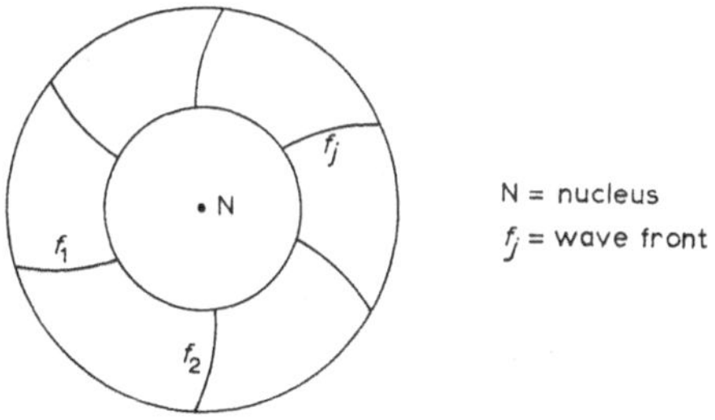

Fig. 2

5. Heisenberg relations

Let us assume that the intensity of a wave within a volume element d^3x is given by

$$\mathfrak{N}(x)\, d^3x = |\psi(\boldsymbol{x}, t)|^2\, d^3x\,,$$

i.e. by what, in non-relativistic wave mechanics, is interpreted as a probability density. However in this section we shall obtain formulae which would remain valid if the intensity were defined as any bilinear form of ψ^* and ψ and their derivatives. Denoting by N the total intensity we have

$$N = \int \psi^*\psi\, d^3x = (2\pi)^{-3} \int A^*(\boldsymbol{k}')\, A(\boldsymbol{k})\, e^{i(\omega'-\omega)t}\, e^{i(\boldsymbol{k}-\boldsymbol{k}')\cdot \boldsymbol{x}}\, d^3x\, d^3k'\, d^3k\,,$$

and, employing the properties of the δ function,

$$N = \int |A(\boldsymbol{k})|^2\, d^3k\,, \tag{11}$$

which expresses the property of completeness of the plane wave expansion (5) for wave functions involving positive frequencies only.

Since $A(\boldsymbol{k})$, for a free particle, is independent of time, the total intensity of a wave packet remains constant. From (11) one sees that $|A|^2$ may be interpreted as a density in $\boldsymbol{k}$ space, so that

$$\mathfrak{N}(\boldsymbol{k})\, d^3k = |A(\boldsymbol{k})|^2\, d^3k$$

represents the fraction of the total intensity transported by the wave for which k_i lies between k_i and $k_i + dk_i$, $(i = 1, 2, 3)$.

This suggests that the *centre* of the wave packet is defined as the mean value of $\boldsymbol{x}$

$$\langle \boldsymbol{x} \rangle = N^{-1} \int \boldsymbol{x}\, \psi^* \psi \, \mathrm{d}^3\mathrm{x} \, .$$

It is easy to show that $\langle \boldsymbol{x} \rangle$ can be expressed as an integral in $\boldsymbol{k}$ space as we have done with N,

$$N \langle \boldsymbol{x} \rangle = \int \left[A^*(\boldsymbol{k})\, \mathrm{e}^{\mathrm{i}\omega t} \right] \mathrm{i} \nabla_k \left[A(\boldsymbol{k})\, \mathrm{e}^{-\mathrm{i}\omega t} \right] \mathrm{d}^3\mathrm{k} \, . \tag{12}$$

NOTE 5. In establishing eq. (12) and similar equations later, it is useful to notice that the derivatives of the δ function can be expressed in the following way:

$$\frac{\partial^n}{\partial x^n} \delta(x - x_0) = (2\pi)^{-1} \int (\mathrm{i}k)^n \, \mathrm{e}^{\mathrm{i}k\,(x - x_0)} \, \mathrm{d}k \, .$$

Now, since

$$\int f(x) \frac{\partial}{\partial x} \delta(x - x_0) \, \mathrm{d}x = - \left(\frac{\partial f}{\partial x} \right)_{x = x_0} ,$$

for an arbitrary function $F(\boldsymbol{k})$ which can be developed in a power series of $\boldsymbol{k}$

$$F(\boldsymbol{k}) = \sum_{M=0}^{\infty} \sum_{\substack{lmn \\ l+m+n=M}} a_{lmn} \, k_1^l \, k_2^m \, k_3^n \, ,$$

one has

$$(2\pi)^{-3} \int f(\boldsymbol{x}) \, F(\boldsymbol{k}) \, \mathrm{e}^{\mathrm{i}\boldsymbol{k} \cdot (\boldsymbol{x} - \boldsymbol{x}_0)} \, \mathrm{d}^3\mathrm{k} \, \mathrm{d}^3\mathrm{x} =$$

$$= \int f(\boldsymbol{x}) \sum_{M=0}^{\infty} \sum_{\substack{lmn \\ l+m+n=M}} a_{lmn} \, \mathrm{i}^{-(l+m+n)} \frac{\partial^l}{\partial x_1^l} \frac{\partial^m}{\partial x_2^m} \frac{\partial^n}{\partial x_3^n} \delta(\boldsymbol{x} - \boldsymbol{x}_0) \, \mathrm{d}^3\mathrm{x} =$$

$$= \left[\sum_{M=0}^{\infty} \sum_{\substack{lmn \\ l+m+n=M}} a_{lmn} \, \mathrm{i}^{(l+m+n)} \frac{\partial^l}{\partial x_1^l} \frac{\partial^m}{\partial x_2^m} \frac{\partial^n}{\partial x_3^n} f(\boldsymbol{x}) \right]_{x = x_0} .$$

Introducing the wave function $\varphi(\boldsymbol{k}, t)$ in $\boldsymbol{k}$ space defined by (6) of § 2, we have

$$N \langle \boldsymbol{x} \rangle = \int \varphi^*(\boldsymbol{k}, t) \, \mathrm{i} \nabla_k \, \varphi(\boldsymbol{k}, t) \, \mathrm{d}^3\mathrm{k} \, .$$

If (12) is worked out further, one obtains

$$N \langle \boldsymbol{x} \rangle = \int A^* \, \mathrm{i} \nabla_k A \, \mathrm{d}^3\mathrm{k} + t \int |A|^2 \, \nabla_k \, \omega \, \mathrm{d}^3\mathrm{k} \, .$$

Since N and the two integrals are constant in time, the centre of the packet moves with uniform velocity which can be interpreted as the average group velocity.

The mean value of $\boldsymbol{k}$, called the *spectral centre*, is, on the other hand, time independent, and can be expressed as an integral in $\boldsymbol{x}$ space

$$N\langle \boldsymbol{k}\rangle = \int A^*(\boldsymbol{k})\, A(\boldsymbol{k})\, \boldsymbol{k}\, \mathrm{d}^3\mathrm{k} = -\int \psi^*\, \mathrm{i}\, \nabla_x\, \psi\, \mathrm{d}^3\mathrm{x}\,.$$

Generalizing the above formulae, it is easy to show that the *mean value* of any function of $\boldsymbol{x}$ is given by

$$N\langle F(\boldsymbol{x})\rangle = \int \varphi^*\, F(\mathrm{i}\,\nabla_k)\, \varphi\, \mathrm{d}^3\mathrm{k}\,,$$

and that of a function of $\boldsymbol{k}$ by

$$N\langle F(\boldsymbol{k})\rangle = \int \psi^*\, F(-\mathrm{i}\,\nabla_x)\, \psi\, \mathrm{d}^3\mathrm{x}\,.$$

For example one has

$$N\langle k_i^2\rangle = -\int \psi^*\, \frac{\partial^2}{\partial x_i^2}\, \psi\, \mathrm{d}^3\mathrm{x}$$

and, integrating by parts (henceforth, in performing integrations by parts, we shall always assume that the wave function is zero outside a finite domain) one obtains

$$N\langle k_i^2\rangle = \int \frac{\partial\psi^*}{\partial x_i}\frac{\partial\psi}{\partial x_i}\, \mathrm{d}^3\mathrm{x} \tag{13}$$

(no sum over i).
Similarly

$$N\langle x_i^2\rangle = \int \frac{\partial\varphi^*}{\partial k_i}\frac{\partial\varphi}{\partial k_i}\, \mathrm{d}^3\mathrm{k} \tag{13'}$$

(no sum over i).

So far we have considered the wave packet at a certain time. Let us now fix a point $\boldsymbol{x}$ in space. We denote by $\psi(t)$ the wave function at the given point, and by $a(\omega)$ its Fourier transform

$$\psi(t) = (2\pi)^{-\frac{1}{2}} \int_{-\infty}^{+\infty} a(\omega)\, \mathrm{e}^{-\mathrm{i}\omega t}\, \mathrm{d}\omega\,.$$

By comparison with previous formulae, and introducing polar coordinates, one sees that

$$\begin{aligned}\psi(t) &= (2\pi)^{-\frac{3}{2}} \int A(\boldsymbol{k})\, \mathrm{e}^{\mathrm{i}\boldsymbol{k}\cdot\boldsymbol{x}}\, \mathrm{e}^{-\mathrm{i}\omega t}\, \mathrm{d}^3\mathrm{k} \\ &= (2\pi)^{-\frac{3}{2}} \int A(\boldsymbol{k})\, \mathrm{e}^{\mathrm{i}\boldsymbol{k}\cdot\boldsymbol{x}}\, k^2\, \frac{\mathrm{d}\,|\boldsymbol{k}|}{\mathrm{d}\omega}\, \mathrm{e}^{-\mathrm{i}\omega t}\, \mathrm{d}\Omega_k\, \mathrm{d}\omega\end{aligned}$$

(Ω_k = solid angle in $\boldsymbol{k}$ space). Therefore

$$a(\omega) = \begin{cases} 0 & \text{for } \omega < c\kappa\,, \\ \dfrac{k^2}{2\pi}\dfrac{\mathrm{d}\,|\boldsymbol{k}|}{\mathrm{d}\omega}\displaystyle\int A(\boldsymbol{k})\,\mathrm{e}^{\mathrm{i}\boldsymbol{k}\cdot\boldsymbol{x}}\,\mathrm{d}\Omega_k & \text{for } \omega > c\kappa\,. \end{cases}$$

We can then define

$$N\,\langle t\rangle = \int t\psi^*\psi\,\mathrm{d}t\,,$$
$$N\,\langle\omega\rangle = \int \omega\, a^* a\,\mathrm{d}\omega\,,$$
$$N = \int |\psi|^2\,\mathrm{d}t = \int |a|^2\,\mathrm{d}\omega\,.$$

We may now discuss the question of the relation between the wave packet extension in $\boldsymbol{x}$ space and the corresponding extension in $\boldsymbol{k}$ space. It is easy to understand that such extensions are mutually restrictive. Consider a one-dimensional wave packet consisting of plane waves whose wavelengths λ lie in the neighbourhood of a particular wavelength λ_0. Then, if the wave packet has a spatial extension Δx, it contains $n = \Delta x/\lambda_0$ complete oscillations for waves of wavelength λ_0. Since, however, its components must interfere destructively outside the interval Δx, there must be waves present in it with wavelength λ such that

$$\frac{\Delta x}{\lambda} \gtrsim n + 1\,.$$

Therefore $\Delta\lambda = \lambda - \lambda_0$ is the smallest interval of wavelength by which the packet can be realized. Thus, neglecting terms of second order,

$$\frac{\Delta x}{\lambda} - \frac{\Delta x}{\lambda_0} \simeq -\frac{\Delta x \Delta\lambda}{\lambda_0^2} \geqslant 1 \quad \text{or} \quad \Delta x \Delta k \geqslant 2\pi\,, \tag{14}$$

which represents an approximate formulation of the *uncertainty* principle. Remembering that $\boldsymbol{p} = \hbar\boldsymbol{k}$, (14) can be written in the form

$$\Delta x\,\Delta p \geqslant h\,.$$

The above considerations may be formulated quantitatively by defining the *mean square deviation*

$$(\Delta x_i)^2 = \langle(x_i - \langle x_i\rangle)^2\rangle = \langle x_i^2\rangle - \langle x_i\rangle^2\,, \quad (i = 1, 2, 3)\,,$$

and similarly

$$(\Delta k_i)^2 = \langle k_i^2\rangle - \langle k_i\rangle^2\,, \quad (i = 1, 2, 3)\,,$$
$$(\Delta t)^2 = \langle t^2\rangle - \langle t\rangle^2\,,$$
$$(\Delta\omega)^2 = \langle\omega^2\rangle - \langle\omega\rangle^2\,.$$

Introducing the new variables

$$\boldsymbol{K} = \boldsymbol{k} - \langle \boldsymbol{k} \rangle , \quad \boldsymbol{X} = \boldsymbol{x} - \langle \boldsymbol{x} \rangle ,$$

and putting (for simplicity we take $t = 0$)

$$\Psi(\boldsymbol{X}) = \psi(\boldsymbol{x})\, \mathrm{e}^{-\mathrm{i}\, \langle k \rangle \cdot x} , \quad \Phi(\boldsymbol{K}) = \varphi(\boldsymbol{k})\, \mathrm{e}^{\mathrm{i}\, \langle x \rangle \cdot (k - \langle k \rangle)} ,$$

we can write

$$\begin{aligned} N(\Delta x_i)^2 &= \int X_i^2\, |\Psi(\boldsymbol{X})|^2\, \mathrm{d}^3\mathrm{X} , \\ N(\Delta k_i)^2 &= \int K_i^2\, |\Phi(\boldsymbol{K})|^2\, \mathrm{d}^3\mathrm{K} , \end{aligned} \quad (i = 1, 2, 3) ,$$

where

$$N = \int |\Psi|^2\, \mathrm{d}^3\mathrm{X} = \int |\Phi|^2\, \mathrm{d}^3\mathrm{K} .$$

Notice that Ψ and Φ are each other's Fourier transforms, i.e. they are in the same mutual relation as ψ and φ.

An analogous procedure can be used for $(\Delta t)^2$ and $(\Delta\omega)^2$, correspondingly defining $\Psi(T)$, $A(\Omega)$ and so on.

A rigorous proof of uncertainty relations can now be given. For this purpose we consider the quadratic form in Ψ and $\partial\Psi/\partial X_i$

$$D_i = \left| \frac{X_i}{2\, \langle X_i^2 \rangle}\, \Psi + \frac{\partial \Psi}{\partial X_i} \right|^2 \geqslant 0 ,$$

namely

$$D_i = \frac{X_i^2 - 2\, \langle X_i^2 \rangle}{4\, \langle X_i^2 \rangle^2}\, \Psi^*\Psi + \frac{1}{2} \frac{\partial}{\partial X_i} \left(\frac{X_i}{\langle X_i^2 \rangle}\, \Psi^*\Psi \right) + \frac{\partial \Psi^*}{\partial X_i} \frac{\partial \Psi}{\partial X_i} \geqslant 0 ,$$

(no sum convention of equal indices in this section). On integrating we have

$$0 \leqslant \int D_i\, \mathrm{d}^3\mathrm{X} = - \frac{N}{4\, \langle X_i^2 \rangle} + N\, \langle K_i^2 \rangle ,$$

from which

$$(\Delta x_i)^2\, (\Delta k_i)^2 \geqslant \tfrac{1}{4} .$$

Analogously, starting from the quadratic form

$$D_0 = \left| \frac{T}{2\, \langle T^2 \rangle}\, \Psi(T) + \frac{\partial \Psi}{\partial T} \right|^2 \geqslant 0 ,$$

one obtains

$$(\Delta t)^2\, (\Delta\omega)^2 \geqslant \tfrac{1}{4}, \quad \text{or} \quad (\Delta E)^2\, (\Delta t)^2 \geqslant \tfrac{1}{4}\, \hbar^2 .$$

The product $(\Delta x_i)^2 (\Delta k_i)^2$ is a minimum when $D_i = 0$, i.e. when the wave function satisfies the differential equation

$$\frac{\partial \Psi}{\partial X_i} = - \frac{X_i}{2 (\Delta x_i)^2}\, \Psi .$$

Using the method of separation of variables, the solution is

$$\Psi = F(X_j) \exp\left[-\frac{X_i^2}{4(\Delta x_i)^2} \right],$$

$F(X_j)$ being a constant of integration independent of X_i.

In order that $\Delta x_i \Delta k_i$ be a minimum for all three values of i, $(i = 1, 2, 3)$, one must have, going back to the old variables,

$$\psi(\boldsymbol{x}) = \alpha \exp\left[-\sum_i \frac{(x_i - \langle x_i \rangle)^2}{4(\Delta x_i)^2} + \mathrm{i} \langle \boldsymbol{k} \rangle \cdot \boldsymbol{x} \right]. \tag{15}$$

Because of the symmetry between $\boldsymbol{x}$ and $\boldsymbol{k}$, the wave function of this packet in $\boldsymbol{k}$ space must be of the type

$$\varphi(\boldsymbol{k}) = \beta \exp\left[-\sum_i \frac{(k_i - \langle k_i \rangle)^2}{4(\Delta k_i)^2} - \mathrm{i} \langle \boldsymbol{x} \rangle \cdot (\boldsymbol{k} - \langle \boldsymbol{k} \rangle) \right]. \tag{16}$$

In eqs. (15) and (16), α and β are constants determined by normalization. The apparent asymmetry of equations (15) and (16) is due to the difference between the transformations $\psi \to \Psi$, $\varphi \to \Phi$.

Notice that $(\Delta k_i)^2$, like $\langle k_i \rangle$, is independent of the time. On the other hand $(\Delta x_i)^2$ depends quadratically on the time. Now, if at a time $t = 0$ $(\Delta x_i)^2 (\Delta k_i)^2$ is a minimum for the wave packet, it will not be so at a subsequent instant.

NOTE 6. Determination of α and β in eqs. (15) and (16).
Since $N = \int \psi^* \psi \, \mathrm{d}^3x$, one must have

$$N = |\alpha|^2 \int \exp\left[-\sum_i \frac{(x_i - \langle x_i \rangle)^2}{2(\Delta x_i)^2} \right] \mathrm{d}^3x .$$

Using Gauss' integral

$$\int_{-\infty}^{\infty} \mathrm{e}^{-a^2 x^2} = \frac{\sqrt{\pi}}{a},$$

$$\int \exp\left[-\frac{(x_i - \langle x_i \rangle)^2}{2(\Delta x_i)^2} \right] \mathrm{d}x_i = \sqrt{2}\, \Delta x_i \int \mathrm{e}^{-\xi_i^2} \, \mathrm{d}\xi_i = (2\pi)^{\frac{1}{2}} \Delta x_i .$$

Therefore

$$|\alpha|^{-2} = (2\pi)^{\frac{3}{2}}\,\Delta x_1\,\Delta x_2\,\Delta x_3\,N^{-1}\,.$$

Similarly

$$|\beta|^{-2} = (2\pi)^{\frac{3}{2}}\,\Delta k_1\,\Delta k_2\,\Delta k_3\,N^{-1}\,.$$

NOTE 7. Using the expression for the time dependence of $(\Delta x_i)^2(\Delta k_i)^2$, one can derive, in yet another way, the wave function in $\boldsymbol{k}$ space of a wave packet for which $(\Delta x_i)^2(\Delta k_i)^2 = \frac{1}{4}$ at $t = 0$.
Using (13′) one has

$$\langle x_i^2\rangle = N^{-1}\int\left[\frac{\partial}{\partial k_i}(A^*\,\mathrm{e}^{\mathrm{i}\omega t})\right]\left[\frac{\partial}{\partial k_i}(A\,\mathrm{e}^{-\mathrm{i}\omega t})\right]\mathrm{d}^3\mathrm{k}$$

$$= N^{-1}\left[\int\left|\frac{\partial A}{\partial k_i}\right|^2\mathrm{d}^3\mathrm{k} + \mathrm{i}t\int\frac{\partial\omega}{\partial k_i}\left(A^*\frac{\partial A}{\partial k_i} - \frac{\partial A^*}{\partial k_i}A\right)\mathrm{d}^3\mathrm{k}\,+\right.$$

$$\left. + \,t^2\int\left(\frac{\partial\omega}{\partial k_i}\right)^2|A|^2\,\mathrm{d}^3\mathrm{k}\right],$$

$$\langle x_i\rangle^2 = N^{-2}\left[\mathrm{i}\int A^*\frac{\partial A}{\partial k_i}\,\mathrm{d}^3\mathrm{k} + t\int\frac{\partial\omega}{\partial k_i}|A|^2\,\mathrm{d}^3\mathrm{k}\right]^2$$

$$= N^{-2}\left[-\left(\int A^*\frac{\partial A}{\partial k_i}\,\mathrm{d}^3\mathrm{k}\right)^2 + 2\mathrm{i}t\left(\int A^*\frac{\partial A}{\partial k_i}\,\mathrm{d}^3\mathrm{k}\right)\times\right.$$

$$\left.\times\left(\int\frac{\partial\omega}{\partial k_i}|A|^2\,\mathrm{d}^3\mathrm{k}\right) + t^2\left(\int\frac{\partial\omega}{\partial k_i}|A|^2\,\mathrm{d}^3\mathrm{k}\right)^2\right].$$

Since $(\Delta k_i)^2$ is constant, the condition for the minimum of $(\Delta x_i)^2.$ $(\Delta k_i)^2$ is that the term linear in t in $(\Delta x_i)^2 = \langle x_i^2\rangle - \langle x_i\rangle^2$ be zero. Hence the integral equation for A is

$$\left[\int|A|^2\,\mathrm{d}^3\mathrm{k}\right]\left[\int\frac{\partial\omega}{\partial k_i}\left(A^*\frac{\partial A}{\partial k_i} - \frac{\partial A^*}{\partial k_i}A\right)\mathrm{d}^3\mathrm{k}\right] -$$

$$-\,2\left(\int A^*\frac{\partial A}{\partial k_i}\,\mathrm{d}^3\mathrm{k}\right)\left(\int\frac{\partial\omega}{\partial k_i}|A|^2\,\mathrm{d}^3\mathrm{k}\right) = 0\,.$$

It is easy to show that

$$A = C \exp\left[-\sum_i \frac{(k_i - \langle k_i \rangle)^2}{4(\Delta k_i)^2} - \mathrm{i}\,\langle \boldsymbol{x} \rangle \cdot (\boldsymbol{k} - \langle \boldsymbol{k} \rangle)\right]$$

(C = constant) is a solution of this equation.

PART I

PARTICLES OF SPIN ZERO

CHAPTER I

WAVE EQUATION

1. Dispersive law and wave equation

In writing the expansion

$$\psi(\boldsymbol{x}, t) = (2\pi)^{-\frac{3}{2}} \int A(\boldsymbol{k})\, e^{i(\boldsymbol{k}\cdot\boldsymbol{x}-\omega t)}\, d^3k\,,$$

we have, so to say, given an integral formulation of the *principle of superposition*. By establishing the wave equation, we will give the differential formulation of this principle.

First we notice that

$$\Delta^n\psi(\boldsymbol{x}, t) = (2\pi)^{-\frac{3}{2}} \int A(\boldsymbol{k})(-\,\boldsymbol{k}^2)^n\, e^{i(\boldsymbol{k}\cdot\boldsymbol{x}-\omega t)}\, d^3k\,,$$

where Δ is the Laplace operator.

Remembering that $\omega = \omega(\boldsymbol{k}^2)$, we can define the operator

$$\omega(-\,\Delta) = \sum_{n=0}^{\infty} a_n(-\,\Delta)^n\,.$$

Here the coefficients a_n are equal to those of the expansion of $\omega(\boldsymbol{k}^2)$ in a power series of $\boldsymbol{k}^2$. Thus we have:

$$i\frac{\partial}{\partial t}\psi(\boldsymbol{x}, t) = (2\pi)^{-\frac{3}{2}} \int A(\boldsymbol{k})\,\omega(\boldsymbol{k}^2)\, e^{i(\boldsymbol{k}\cdot\boldsymbol{x}-\omega t)}\, d^3k =$$

$$= (2\pi)^{-\frac{3}{2}} \int A(\boldsymbol{k})\,\omega(-\,\Delta)\, e^{i(\boldsymbol{k}\cdot\boldsymbol{x}-\omega t)}\, d^3k = \omega(-\,\Delta)\,\psi(\boldsymbol{x}, t)\,.$$

A generic wave function therefore satisfies the differential equation

$$\left[i\frac{\partial}{\partial t} - \omega(-\,\Delta)\right]\psi(\boldsymbol{x}, t) = 0\,.$$

Thus, to every dispersion function $\omega = \omega(\boldsymbol{k}^2)$, there corresponds a definite *wave equation*.

The dispersion function of non-relativistic wave mechanics

$$\omega = \frac{\hbar \boldsymbol{k}^2}{2m} + \frac{mc^2}{\hbar}$$

leads to the *Schroedinger equation* (with a term for the rest energy, which is usually omitted)

$$\hbar \mathrm{i} \frac{\partial}{\partial t} \psi = - \frac{\hbar^2}{2m} \Delta \psi + mc^2 \psi. \tag{1}$$

The operator acting on ψ in this equation may be formally obtained by performing the substitution

$$E \to \hbar \mathrm{i} \frac{\partial}{\partial t}, \quad \boldsymbol{p} \to \frac{\hbar}{\mathrm{i}} \nabla_x$$

in the expression for the energy.

The more familiar form of the Schroedinger equation does not contain the term for the rest energy, since it is derived from the dispersion function $\omega = \hbar \boldsymbol{k}^2/2m$. Its solutions differ from those of (1) by the phase factor $\exp\left[- \mathrm{i}\, mc^2 t/\hbar\right]$.

From the relativistic dispersion function

$$\omega = c(\boldsymbol{k}^2 + \kappa^2)^{\frac{1}{2}}$$

follows the wave equation

$$\left[\mathrm{i} \frac{\partial}{\partial t} - c(- \Delta + \kappa^2)^{\frac{1}{2}} \right] \psi = 0\,. \tag{2}$$

This, like the Schroedinger equation, is of the first order in the time. We could then ask why nobody has ever thought of choosing this equation as the basis of relativistic wave mechanics. Obviously, the corresponding theory would be relativistically invariant (see *Note 4* in the Introduction). The only reason is that this equation, in contrast to the non-relativistic equation, is not linear in Δ; therefore its extension to the case where forces are present proves to be impracticable.

Therefore, one adopts the *Klein-Gordon equation*

$$\left[\Delta - \frac{1}{c^2} \frac{\partial^2}{\partial t^2} - \kappa^2 \right] \psi(\boldsymbol{x}, t) = 0\,, \tag{3}$$

obtained by multiplying (2) from the left by the operator

$$\left[\mathrm{i} \frac{\partial}{\partial t} + c(- \Delta + \kappa^2)^{\frac{1}{2}} \right].$$

Formally, (3) may be obtained by the replacement

$$p_\mu \to -\mathrm{i}\hbar \frac{\partial}{\partial x_\mu},$$

in the equation

$$p_\mu p_\mu + m^2c^2 = 0\,.$$

Notice that all solutions of (2) are solutions of (3), but not vice versa. In fact, while eq. (2) has only positive frequency solutions, eq. (3) has negative frequency solutions as well. For example, any plane wave $\psi(\boldsymbol{x}, t) = \exp[\mathrm{i}\,\boldsymbol{k}\cdot\boldsymbol{x} - \mathrm{i}\,\omega t]$ is a solution of (3) if $\omega^2 = c^2(\boldsymbol{k}^2 + \kappa^2)$. This last equation has a positive root ($\omega > 0$) and a negative one ($\omega < 0$) corresponding to the two possible values of the energy of a particle of momentum $\boldsymbol{p}$. As we shall see later, negative frequencies are a source of complications. In the case of a free particle whose energy is initially positive, these difficulties do not arise, but they become serious when a particle is subject to forces which may cause transitions from positive to negative energy states.

The Klein-Gordon equation may be written in the covariant form

$$(\Box - \kappa^2)\,\psi = 0\,. \tag{4}$$

The relativistic wave mechanics of particles of spin zero employs a one-component wave function, which is a scalar with respect to Lorentz transformations, and satisfies eq. (4) in the absence of forces. For particles of spin $\frac{1}{2}$, 1 etc. it is necessary to use wave functions with more than one component.

In the early days of wave mechanics, attempts were made to describe electrons by means of a scalar function satisfying the Klein-Gordon equation. The discovery of spin showed the inadequacy of such a theory, employed nowadays to describe *pions* (π^+, π^- and π^0 mesons, having masses approximately 273 m_e), which are spin zero particles.

2. Green's function

One of the basic differences between the Schroedinger and the Klein-Gordon equation is the different order of the time derivative.

The Schroedinger equation is of the first order in the time, so that, if the function $\psi(\boldsymbol{x}, t)$ is known throughout space at a certain time, it is determined completely at all previous and subsequent times. The construction of

$\psi(\boldsymbol{x}, t)$ from $\psi(\boldsymbol{x}', t')$, $(t' \neq t)$, is carried out by means of the *Green's function* *

$$G_S(\boldsymbol{x} - \boldsymbol{x}', t - t') = (2\pi)^{-3} \int e^{i\boldsymbol{k}\cdot(\boldsymbol{x}-\boldsymbol{x}') - i(\hbar k^2/2m)(t-t')} \, d^3k \,. \tag{5}$$

This obeys the equation

$$\left(\hbar i \frac{\partial}{\partial t} + \frac{\hbar^2}{2m} \Delta\right) G_S(\boldsymbol{x} - \boldsymbol{x}', t - t') = 0 \,, \quad (t \neq t') \,, \tag{6}$$

and satisfies the initial condition

$$G_S(\boldsymbol{x} - \boldsymbol{x}', 0) = \delta(\boldsymbol{x} - \boldsymbol{x}') \,. \tag{7}$$

If $\psi(\boldsymbol{x}', t')$ is known,

$$\psi(\boldsymbol{x}, t) = \int G_S(\boldsymbol{x} - \boldsymbol{x}', t - t') \, \psi(\boldsymbol{x}', t') \, d^3x' \tag{8}$$

is a solution of the Schroedinger equation by eq. (6), and satisfies the initial condition

$$\psi(\boldsymbol{x}, t) = \psi(\boldsymbol{x}, t') \,, \quad (t = t') \,,$$

by eq. (7).

NOTE 1. For the convenience of the reader, the method of constructing Green's functions is briefly recalled.

If the functions $\varphi_n(x)$, orthogonal and normalized, i.e. satisfying the relations

$$\int_a^b \varphi_m^*(x) \, \varphi_n(x) \, dx = \delta_{mn} \,,$$

form a complete set relative to the interval (a, b) of the x axis, completeness is expressed by the relation

$$\sum_m \varphi_m(x) \, \varphi_m^*(x') = \delta(x - x') \,. \tag{α}$$

In fact, if the function $\psi(x)$ can be expanded in terms of φ_n in the given interval as

$$\psi(x) = \sum_m a_m \, \varphi_m(x) \,,$$

* An introduction to Green's functions can be found e.g. in H. Margenau and G. M. Murphy, *The Mathematics of Physics and Chemistry* (Van Nostrand, 1956), p. 534.

the coefficients are

$$a_n = \int_a^b \varphi_n^*(x)\, \psi(x)\, dx\,,$$

as is seen by multiplication of the expansion by $\varphi_n^*(x)$ and subsequent integration. On the other hand, inserting the last expression in the expansion one has

$$\psi(x) = \int_a^b \psi(x') \sum_m \varphi_m(x)\, \varphi_m^*(x')\, dx'\,,$$

and, in order that this be identically true, the above completeness relation must hold.

Using these properties, it is easy to construct the Green's function for the one-dimensional Schroedinger equation in the interval $0 \leqslant x \leqslant a$ with the boundary conditions $\psi(0) = \psi(a) = 0$.

From the expansion

$$\psi(x) = \sum_{n=1}^{\infty} A_n \sin(n\pi x/a)\,, \qquad A_n = 2a^{-1} \int_0^a \psi(x) \sin(n\pi x/a)\, dx$$

($\psi(x)$ satisfying the boundary conditions $\psi(0) = \psi(a) = 0$) it follows that the completeness of the system of functions $\sin(n\pi x/a)$ in the interval $(0, a)$ is expressed by

$$\delta(x - x') = 2a^{-1} \sum_{n=1}^{\infty} \sin(n\pi x/a) \sin(n\pi x'/a),\ 0 < |x - x'| < a\,.$$

Thus

$$G_S(x - x', t - t') = 2a^{-1} \sum_{n=1}^{\infty} \sin(n\pi x/a) \sin(n\pi x'/a)\, e^{-i\omega_n (t-t')}\,,$$

where

$$\omega_n = \frac{\hbar\, n^2 \pi^2}{2ma^2}\,.$$

$G_S(x - x', t - t')$ is clearly a solution of the Schroedinger equation for $t \neq t'$, and $G_S(x - x', 0) = \delta(x - x')$.

The completeness of the set of exponentials $(2\pi)^{-\frac{3}{2}}\, e^{i\boldsymbol{k}\cdot\boldsymbol{x}}$ is expressed by

$$\delta(\boldsymbol{x} - \boldsymbol{x}') = (2\pi)^{-3} \int e^{i\boldsymbol{k}\cdot\boldsymbol{x} - i\boldsymbol{k}\cdot\boldsymbol{x}'}\, d^3k\,.$$

(Since the functions of the set are defined in all space, $\int d^3k$ replaces the sum of eq. (α).)

The Green's function (5) is obtained on transforming the integral to the right of the last equation by inserting factors which cancel out for $t = t'$, and make it a solution of the Schroedinger equation for $t \neq t'$:

$$G_S = (2\pi)^{-3} \int \exp i[\boldsymbol{k}\cdot(\boldsymbol{x} - \boldsymbol{x}') - (\hbar\, \boldsymbol{k}^2/2m)(t - t')]\, d^3k .$$

A compact expression for G_S is easily found. For $t > t'$ we put

$$k_i = \left[\frac{2m}{\hbar(t - t')}\right]^{\frac{1}{2}} \xi_i + \left[\frac{m}{\hbar(t - t')}\right](x_i - x_i') .$$

Thus we have

$$G_S(\boldsymbol{x} - \boldsymbol{x}', t - t') = (2\pi)^{-3} \left[\frac{2m}{\hbar(t - t')}\right]^{\frac{3}{2}} \times$$

$$\times \int \exp i[-\xi^2 + m(\boldsymbol{x} - \boldsymbol{x}')^2/2\hbar(t - t')]\, d^3\xi .$$

Since

$$\int_{-\infty}^{\infty} e^{-i\xi_i^2}\, d\xi_i = \sqrt{\pi}\, e^{-\frac{1}{4}i\pi} ,$$

we finally get

$$G_S = \left[\frac{m}{h(t - t')}\right]^{\frac{3}{2}} e^{-\frac{3}{4}i\pi}\, e^{im(x-x')^2/2\hbar(t-t')} . \qquad (\beta)$$

NOTE 2. As an example of use of Green's functions, we determine the time development of the one-dimensional wave packet which, at $t = 0$, is given by

$$\psi(x', 0) = C\, e^{-(x'/2\Delta_0 x)^2 + imvx'/\hbar}$$

($(\Delta_0 x)^2$ is the mean square deviation at $t = 0$ and C is a constant).

Using the analogue of (β) for the one-dimensional case

$$\psi(x, t) = C\left(\frac{m}{ht}\right)^{\frac{1}{2}} e^{-\frac{1}{4}i\pi}\, e^{imx^2/2\hbar t} \times$$

$$\times \int_{-\infty}^{+\infty} \exp\left[\left(\frac{im}{2\hbar t} - \frac{1}{(2\Delta_0 x)^2}\right) x'^2 + \frac{im}{\hbar t}(vt - x)\, x'\right] dx' .$$

Now the integral above is of the type

$$\int_{-\infty}^{+\infty} e^{-ax^2 + 2ibx}\, dx = (\pi/a)^{\frac{1}{2}}\, e^{-b^2/a} ,$$

and so

$$\psi(x, t) = C\left(1 + \frac{2i\,\hbar t}{m(2\Delta_0 x)^2}\right)^{-\frac{1}{2}} e^{imx^2/2\hbar t} \times$$

$$\times \exp\left[-(x-vt)^2\left(1+\frac{im(2\Delta_0 x)^2}{2\hbar t}\right)\Big/(2\Delta_0 x)^2\left(1+\frac{4\hbar^2 t^2}{m^2(2\Delta_0 x)^4}\right)\right].$$

Putting

$$(2\Delta x)^2 = (2\Delta_0 x)^2\left[1+\left(\frac{2\hbar t}{m(2\Delta_0 x)^2}\right)^2\right], \qquad (\gamma)$$

one has

$$|\psi(x, t)|^2 = |C|^2\frac{\Delta_0 x}{\Delta x}\, e^{-(x-vt)^2/2(\Delta x)^2}.$$

This shows that the centre of the packet moves with velocity v and its width changes according to (γ). In order that $\Delta x = 2\Delta_0 x$ we must have $t = 2\sqrt{3}\, m(\Delta_0 x)^2/\hbar$. For $m = 1.7 \times 10^{-24}$ g (hydrogen atom) and $\Delta_0 x = 10^{-8}$ cm we have $t = 5.6 \times 10^{-13}$ sec. On the other hand for $m = 10^{-3}$ g and $\Delta_0 x = 10^{-3}$ cm we have $t \simeq 3.3 \times 10^{18}$ sec $\simeq \frac{1}{3} \times 10^{11}$ years.

The Klein-Gordon equation is, instead, of the second order in the time. Thus a solution is fully determined, if one knows its values, and those of its first time derivative, in all space at a particular time. More generally, if σ is a space-like surface in Minkowski space (that is a surface such that any of its points is outside the light cone having its vertex at any other point) then a knowledge of ψ and of the derivative of ψ along the normal to σ at all points of σ determines ψ at all later and earlier instants.

Instead of (8) we have for the Klein-Gordon equation

$$\psi(\boldsymbol{x}, t) =$$

$$\int\left[\psi(\boldsymbol{x}', t')\frac{1}{c}\frac{\partial}{\partial t'}G_{KG}(\boldsymbol{x}, t; \boldsymbol{x}', t') - \frac{1}{c}\frac{\partial\psi(\boldsymbol{x}', t')}{\partial t'}G_{KG}(\boldsymbol{x}, t; \boldsymbol{x}', t')\right]d^3x'.$$

Here G_{KG} is the *relativistic Green's function* having the properties

a) $G_{KG}(\boldsymbol{x}, t; \boldsymbol{x}', t') = 0$, for $t = t'$,

b) $$\left[\frac{1}{c}\frac{\partial}{\partial t'}G_{KG}(\boldsymbol{x}, t; \boldsymbol{x}', t')\right]_{t=t'} = -\left[\frac{1}{c}\frac{\partial}{\partial t}G_{KG}(\boldsymbol{x}, t; \boldsymbol{x}', t')\right]_{t=t'} = \delta(\boldsymbol{x}-\boldsymbol{x}'),$$

$$c)\ (\Box - \kappa^2)\, G_{KG}(\boldsymbol{x}, t; \boldsymbol{x}', t') = 0\,, \qquad t \neq t'\,,$$

$$d)\ \left[\frac{\partial^2}{\partial t\, \partial t'}\, G_{KG}(\boldsymbol{x}, t; \boldsymbol{x}', t')\right]_{t=t'} = 0\,.$$

This Green's function is obtained by a procedure analogous to that used for G_S:

$$G_{KG}(\boldsymbol{x}, t; \boldsymbol{x}', t') = G_{KG}(\boldsymbol{x} - \boldsymbol{x}', t - t') =$$

$$= -\tfrac{1}{2}(2\pi)^{-3}\, \mathrm{i}c \int \left(\mathrm{e}^{\mathrm{i}\boldsymbol{k}\cdot(\boldsymbol{x}-\boldsymbol{x}') - \mathrm{i}\omega(t-t')} - \mathrm{e}^{\mathrm{i}\boldsymbol{k}\cdot(\boldsymbol{x}'-\boldsymbol{x}) - \mathrm{i}\omega(t'-t)} \right) \frac{\mathrm{d}^3 k}{\omega(\boldsymbol{k}^2)}$$

$$= (2\pi)^{-3}\, c \int \sin\left[(t' - t)\, \omega(\boldsymbol{k}^2)\right] \mathrm{e}^{\mathrm{i}\boldsymbol{k}\cdot(\boldsymbol{x}-\boldsymbol{x}')} \frac{\mathrm{d}^3 k}{\omega(\boldsymbol{k}^2)}\,. \qquad (9)$$

Notice that in G_{KG} there appears a term $\mathrm{e}^{\mathrm{i}\boldsymbol{k}\cdot(\boldsymbol{x}'-\boldsymbol{x}) - \mathrm{i}\omega(t'-t)}$ corresponding to negative frequencies.

3. Transformations of the wave function

First notice that the operator of the Klein-Gordon equation is invariant under Lorentz transformations, space rotations and four-dimensional translations (i.e. under the transformations of the *inhomogeneous Lorentz group*; for the time being we do not consider space and time inversions, which will be dealt with in Chap. VII).

In fact under the above transformations

$$\frac{\partial^2}{\partial x_\mu\, \partial x_\mu} - \left(\frac{mc}{\hbar}\right)^2 = \frac{\partial^2}{\partial x'_\mu\, \partial x'_\mu} - \left(\frac{mc}{\hbar}\right)^2.$$

Moreover a typical plane wave solution

$$\psi(x) = \mathrm{e}^{\mathrm{i}k_\mu x_\mu}\,,$$

with $k_0 = \omega/c = (\boldsymbol{k}^2 + \kappa^2)^{\frac{1}{2}}$, is a scalar under a Lorentz transformation, i.e. we have

$$\psi(x) = \mathrm{e}^{\mathrm{i}k_\mu x_\mu} = \mathrm{e}^{\mathrm{i}k_\mu' x_\mu'} = \psi'(x')\,.$$

In the general case, where, instead of a plane wave, we have a wave packet represented by the wave function $\psi(\boldsymbol{x}, t) = \psi(x)$, the wave function is also a scalar under Lorentz transformations, space rotations and four-dimensional translations. The transformed wave function is

$$\psi'(x') = \psi(x)\,, \qquad (10)$$

obtained by expressing x as a function of x', ($x = x(x')$), in the wave function for the original frame.

NOTE 3. It is useful to explain more clearly the use of primes. Take a plane wave $\psi(x) = \mathrm{e}^{\mathrm{i}(k,x)}$. Its transformation is

$$\psi(x) = \mathrm{e}^{\mathrm{i}(k,x)} = \mathrm{e}^{\mathrm{i}(k,x(x'))} = \mathrm{e}^{\mathrm{i}(k(k'),x(x'))} = \mathrm{e}^{\mathrm{i}(k',x')} = \psi'(x') .$$

The difference between $\psi'(x') = \psi(x)$ and $\psi'(x)$ must be carefully noted. For the plane wave considered

$$\psi'(\xi) = \mathrm{e}^{\mathrm{i}(k',\xi)} , \quad \text{whereas} \quad \psi(\xi) = \mathrm{e}^{\mathrm{i}(k,\xi)} ,$$

i.e. the *form* of the function ψ is not invariant. $\psi(x)$ and $\psi'(x')$ refer to the *same* space-time point, $\psi(x)$ and $\psi'(x)$ to two different points having the coordinates x_1, x_2, x_3, x_4 with respect to the old and the new frame of reference, respectively.

Let us now see how the wave function in $\boldsymbol{k}$ space is transformed under Lorentz transformations and space rotations. For this purpose consider the Fourier integral

$$\psi(x) = \int A(k)\, \mathrm{e}^{\mathrm{i}(k,x(x'))}\, \mathrm{d}^4k .$$

Here

$$A(k) = 2(2\pi)^{-\frac{3}{2}} A(\boldsymbol{k})\, \eta(k)\, \delta(k^2 + \kappa^2)\, \omega/c , \tag{11}$$

as has been shown in the Introduction (*Note 4*). Making the substitution $k = k(k')$, that is transforming the variables of integration in the same way as x, since $\mathrm{d}^4k = \mathrm{d}^4k'$, one has

$$\psi(x) = \int A(k(k'))\, \mathrm{e}^{\mathrm{i}(k(k'),x(x'))}\, \mathrm{d}^4k' = \int A(k(k'))\, \mathrm{e}^{\mathrm{i}(k',x')}\, \mathrm{d}^4k' . \tag{12}$$

Now the plane wave expansion of $\psi'(x')$ is given by the Fourier integral

$$\psi'(x') = \int A'(k')\, \mathrm{e}^{\mathrm{i}(k',x')}\, \mathrm{d}^4k' . \tag{13}$$

Thus comparison of (12) with (13), taking (10) into account, gives

$$A'(k') = A(k) .$$

Therefore, whereas $A(k)$ defined by eq. (11) is a scalar under Lorentz transformations, $A(\boldsymbol{k})$ is not ($A(\boldsymbol{k})$ and $A(k)$ differ not only by invariant factors, but also by the factor ω).

NOTE 4. $A(k)$ is not a scalar under four-dimensional translations

$$x_\mu = x'_\mu + a_\mu , \quad (\mu = 1, 2, 3, 4) .$$

In fact

$$\psi(x) = \psi(x' + a) = \int A(k)\, e^{i(k,a)}\, e^{i(k,x')}\, d^4k\,.$$

Now

$$\psi'(x') = \int A'(k')\, e^{i(k',x')}\, d^4k'\,.$$

Therefore, since $\psi'(x') = \psi(x)$ and $k' = k$, we have

$$A'(k') = e^{i(k,a)}\, A(k)\,,$$

that is, $A'(k')$ differs from $A(k)$ by a phase factor. However,

$$|A'(k')|^2 = |A(k)|^2,$$

as it must obviously be for physical reasons.

Let us now go back to the Schroedinger equation. It might seem natural to expect the differential operator $\hbar i\, \partial/\partial t + \hbar^2 \Delta/2m$ to be invariant under a Galilei transformation

$$\boldsymbol{x}' = \boldsymbol{x} - \boldsymbol{v}t\,, \quad t = t'\,. \tag{14}$$

As is well known, this is the limit of a Lorentz transformation for $c \to \infty$.

Actually, from (14) it follows that *

$$\frac{\partial}{\partial t'} = \frac{\partial}{\partial t} + \boldsymbol{v}\cdot\boldsymbol{\nabla}_x\,, \quad \boldsymbol{\nabla}_x = \boldsymbol{\nabla}_{x'}\,,$$

and, therefore,

$$\left(\hbar i \frac{\partial}{\partial t'} + \frac{\hbar^2}{2m}\Delta'\right) = \left(\hbar i \frac{\partial}{\partial t} + \frac{\hbar^2}{2m}\Delta + \hbar i\, \boldsymbol{v}\cdot\boldsymbol{\nabla}_x\right).$$

Thus the Schroedinger operator is not invariant under a Galilei transformation. Nor is the phase $i(\boldsymbol{p}\cdot\boldsymbol{x} - Et)/\hbar$ of a plane wave invariant. In fact, for the transformation (14), the non-relativistic momentum and energy are transformed according to

$$\boldsymbol{p}' = \boldsymbol{p} - m\boldsymbol{v}\,, \quad E' = E + \tfrac{1}{2}mv^2 - \boldsymbol{v}\cdot\boldsymbol{p}\,,$$

so that

$$\boldsymbol{p}'\cdot\boldsymbol{x}' - E't' = (\boldsymbol{p}\cdot\boldsymbol{x} - Et) + (\tfrac{1}{2}mv^2 t - m\boldsymbol{v}\cdot\boldsymbol{x})\,. \tag{15}$$

* We recall that, when a coordinate transformation

$$y_i = f_i(x_1 \ldots x_n),\ (i = 1, 2, \ldots n)\,,$$

is made, the differential operators are transformed according to

$$\frac{\partial}{\partial y_i} = \sum_k \frac{\partial x_k}{\partial y_i}\frac{\partial}{\partial x_k}\,.$$

Here $\partial x_k/\partial y_i$ is the derivative with respect to y_i of the function $x_k = x_k(y_1, y_2 \ldots y_n)$ obtained by expressing the x's in terms of the y's.

NOTE 5. A Lorentz transformation reduces to a Galilei transformation when the relative velocity $v \ll c$. On the other hand, for $\boldsymbol{u} = c^2\boldsymbol{p}/E \ll c$ the momentum of the particle tends to the classical expression $m\boldsymbol{u}$, and the kinetic energy to $\frac{1}{2}m\boldsymbol{u}^2$. Why then is $\boldsymbol{p} \cdot \boldsymbol{x} - Et$ not invariant under Galilei transformations for $\boldsymbol{p} = m\boldsymbol{u}$ and $E = \frac{1}{2}m\boldsymbol{u}^2 + mc^2$, whereas it is invariant under Lorentz transformations for $\boldsymbol{p} = m\gamma\boldsymbol{u}$ ($\gamma^{-2} = 1 - u^2/c^2$) and $E = m\gamma c^2$? The reason is that

$$\tfrac{1}{2}\, m\boldsymbol{u}'^2\, t = \Big(\lim_{c\to\infty} (E' - mc^2)\Big)\Big(\lim_{c\to\infty} t'\Big) =$$

$$= \lim_{c\to\infty} \left[(E' - mc^2)t'\right] \neq \lim_{c\to\infty}\left[m\gamma c^2 t - m\gamma\boldsymbol{u}\cdot\boldsymbol{x} + m\gamma'\boldsymbol{u}'\cdot\boldsymbol{x}' - mc^2 t\right]$$

$$= \tfrac{1}{2}\, m\boldsymbol{u}^2 t - m\boldsymbol{u}\cdot\boldsymbol{x} + m\boldsymbol{u}'\cdot\boldsymbol{x}'\,,$$

as is easily found by calculation.

Since in eq. (15) $\frac{1}{2}\,m\boldsymbol{v}^2 t - m\boldsymbol{v}\cdot\boldsymbol{x}$ is independent of $\boldsymbol{p}$ and E, according to the superposition principle we may conclude that the transformation law for a generic wave function is

$$\psi'(\boldsymbol{x}', t') = e^{i(\frac{1}{2}mv^2 t - m\boldsymbol{v}\cdot\boldsymbol{x})/\hbar}\,\psi(\boldsymbol{x}, t)\,. \tag{16}$$

It is easily verified that from

$$\left(\hbar i\frac{\partial}{\partial t'} + \frac{\hbar^2}{2m}\Delta'\right)\psi'(\boldsymbol{x}', t') = 0$$

it follows that

$$\left(\hbar i\frac{\partial}{\partial t} + \frac{\hbar^2}{2m}\Delta\right)\psi(\boldsymbol{x}, t) = 0\,.$$

Notice that the probability density $\psi^*\psi$ is invariant under a change of phase $\psi \to \psi\, e^{if}$ and, therefore, also under (16), which is indeed a phase transformation:

$$|\psi(\boldsymbol{x}, t)|^2 = |\psi'(\boldsymbol{x}', t')|^2\,.$$

Let us now deduce the transformation law of the wave function in $\boldsymbol{k}$ space. One has

$$A'(\boldsymbol{k}')\,e^{-i\omega' t'} = (2\pi)^{-\frac{3}{2}}\int \psi'(\boldsymbol{x}', t')\,e^{-i\boldsymbol{x}'\cdot\boldsymbol{k}'}\,d^3x' =$$

$$= e^{i(\frac{1}{2}mv^2 + \hbar\boldsymbol{k}'\cdot\boldsymbol{v})t/\hbar}\,(2\pi)^{-\frac{3}{2}}\int \psi(\boldsymbol{x}, t)\,e^{-i(\boldsymbol{k}' + m\boldsymbol{v}/\hbar)\cdot\boldsymbol{x}}\,d^3x =$$

$$= e^{i(\frac{1}{2}mv^2 + \hbar\boldsymbol{k}'\cdot\boldsymbol{v})\,t/\hbar}\,A(\boldsymbol{k}' + m\boldsymbol{v}/\hbar)\,e^{-i\,\omega(|\boldsymbol{k}' + m\boldsymbol{v}/\hbar|)\,t}\,,$$

where we introduced the new integration variables $\boldsymbol{x} = \boldsymbol{x}' + \boldsymbol{v}t$ and made use of eq. (16). Now if we put $\boldsymbol{k}' = \boldsymbol{k} - m\boldsymbol{v}/\hbar$ – which is the law according

to which $\boldsymbol{k}$ must transform in order that de Broglie's relations be invariant under Galilei transformations – we obtain

$$A'(\boldsymbol{k}')\, \mathrm{e}^{-\mathrm{i}\omega' t'} = \mathrm{e}^{\mathrm{i}(\boldsymbol{k}\cdot\boldsymbol{v} - mv^2/2\hbar)t}\, A(\boldsymbol{k})\, \mathrm{e}^{-\mathrm{i}\,\omega t}\,.$$

Therefore the probability density in $\boldsymbol{k}$ space is also invariant

$$|A(\boldsymbol{k})|^2 = |A'(\boldsymbol{k}')|^2\,.$$

Obviously, the Schroedinger operator is invariant under space rotations and space-time translations. Under such transformations, therefore, the wave function behaves as a scalar

$$\psi'(\boldsymbol{x}', t') = \psi(\boldsymbol{x}, t)\,.$$

Chapter II

LAGRANGIAN FORMALISM

1. Variational principle

The Klein-Gordon and the Schroedinger equation may be deduced from a variational principle. This consists in assuming that the *action integral* *

$$A = \int_{V(x)} \mathfrak{L}\left(\psi(x), \psi^*(x), \frac{\partial\psi(x)}{\partial x_\mu}, \frac{\partial\psi^*(x)}{\partial x_\mu}\right) \mathrm{d}^4 x$$

is stationary for variations of the type $\psi^*(x) \to \psi^*(x) + \delta\psi^*(x)$, $\delta\psi^*(x)$ being an arbitrary infinitesimal function, differentiable and satisfying the only condition: $\delta\psi^* = 0$ on the surface S enclosing the integration region $V(x)$. If, on the other hand, A is assumed to be stationary with respect to variations $\psi(x) \to \psi(x) + \delta\psi(x)$ with $\delta\psi = 0$ on S, one obtains the complex conjugate wave equation. The function

$$\mathfrak{L}(x) = \mathfrak{L}\left(\psi(x), \psi^*(x), \frac{\partial\psi(x)}{\partial x_\mu}, \frac{\partial\psi^*(x)}{\partial x_\mu}\right)$$

is called the *Lagrangian density.*

Note 1. The methods used here are borrowed from the mechanics of continuous systems. The technique employed may be illustrated by presenting the Lagrangian formulation of the motion of a vibrating string. The kinetic energy is

$$T = \tfrac{1}{2}\mu \int_a^b \left(\frac{\partial y}{\partial t}\right)^2 \mathrm{d}x$$

---

* Terminology: *action* = space-time integral of Lagrangian density over all space and a certain time interval;
action integral = integral of Lagrangian density over arbitrary space-time region.

(a, b are the fixed ends of the string, which is assumed to have a mass density independent of x) and the potential energy is

$$V = \tfrac{1}{2} K \int_a^b \left(\frac{\partial y}{\partial x}\right)^2 \mathrm{d}x \,.$$

The Lagrangian is

$$L = \int_a^b \mathfrak{L}\, \mathrm{d}x = \tfrac{1}{2} \int_a^b \left[\mu \left(\frac{\partial y}{\partial t}\right)^2 - K \left(\frac{\partial y}{\partial x}\right)^2 \right] \mathrm{d}x \,.$$

The equation of motion of the vibrating string follows from the assumption that the action

$$A = \int_{t_1}^{t_2} L\, \mathrm{d}t$$

is stationary for variations $y(x, t) \rightarrow y(x, t) + \delta y(x, t)$ such that

$$\delta y(a, t) = \delta y(b, t) = 0 \,, \quad \delta y(x, t_1) = \delta y(x, t_2) = 0 \,.$$

Thus

$$\delta A = \int_{t_1}^{t_2} \mathrm{d}t \int_a^b \left(\mu \frac{\partial y}{\partial t} \frac{\partial \delta y}{\partial t} - K \frac{\partial y}{\partial x} \frac{\partial \delta y}{\partial x} \right) \mathrm{d}x = 0 \,.$$

Partial integration gives

$$\int_{t_1}^{t_2} \mathrm{d}t \int_a^b \delta y \left(\mu \frac{\partial^2 y}{\partial t^2} - K \frac{\partial^2 y}{\partial x^2} \right) \mathrm{d}x = 0$$

and, since δy is arbitrary, we must have

$$\mu \frac{\partial^2 y}{\partial t^2} = K \frac{\partial^2 y}{\partial x^2} \,.$$

NOTE 2. The method used for the vibrating string is easily extended to cases where the "excitation" is a function of $\boldsymbol{x}$ and t, $y(\boldsymbol{x}, t)$, and/or where there is more than one excitation function. For instance, the Lagrangian density for the Klein-Gordon and the Schroedinger equation will involve the two functions $\psi(x)$ and $\psi^*(x)$, regarded as independent, which, roughly speaking, amounts to attaching two

degrees of freedom to each position in space. The necessity of using two functions ψ and ψ^* to describe a charged particle of spin zero will become clear when considering the interaction with an electromagnetic field. In wave mechanics for particles of non-zero spin more than two independent functions occur.

NOTE 3. The Lagrangian densities $\mathfrak{L} = \mathfrak{L}(x)$ appearing in this book will depend on certain functions u_i, $(i = 1, 2, \dots n)$, and their first derivatives $\partial u_i/\partial x_\mu$,

$$\mathfrak{L}(x) = \mathfrak{L}\left(u_i(x), \frac{\partial u_i(x)}{\partial x_\mu}\right).$$

The resulting equations of motion will be, at most, of the second order. The variational principles will be of the type

$$\delta A = \delta \int\limits_{V(x)} \mathfrak{L}\, d^4x = \int\limits_{V(x)} \delta\mathfrak{L}\, d^4x = 0\,,$$

$\delta\mathfrak{L}$ being the variation of the Lagrangian density under an infinitesimal variation of one of the functions u_i, $u_i \to u_i + \delta u_i$ with $\delta u_i = 0$ on the boundary of $V(x)$, but otherwise quite arbitrary. Then

$$\delta A = \int \left[\frac{\partial \mathfrak{L}}{\partial u_i}\,\delta u_i + \sum_\mu \left(\partial\mathfrak{L}/\partial\left(\frac{\partial u_i}{\partial x_\mu}\right)\right)\delta\,\frac{\partial u_i}{\partial x_\mu}\right] d^4x = 0\,.$$

Because

$$\delta\,\frac{\partial u_i}{\partial x_\mu} = \frac{\partial}{\partial x_\mu}\,\delta u_i\,,$$

integration by parts gives

$$\delta A = \int\limits_{V(x)} \delta u_i \left[\frac{\partial\mathfrak{L}}{\partial u_i} - \sum_\mu \frac{\partial}{\partial x_\mu}\left(\partial\mathfrak{L}/\partial\,\frac{\partial u_i}{\partial x_\mu}\right)\right] d^4x = 0\,.$$

Since δu_i is arbitrary, this leads to the Euler equations

$$\frac{\partial}{\partial x_\mu}\,\frac{\partial\mathfrak{L}}{\partial\left(\dfrac{\partial u_i}{\partial x_\mu}\right)} = \frac{\partial\mathfrak{L}}{\partial u_i}\,, \quad (i = 1, 2, \dots n)\,. \tag{1}$$

These are as many of these equations as there are independent functions u_i. (For further details see, for instance, N. N. Bogoliubov and D. V. Shirkov, *Introduction to the Theory of Quantized Fields* (Interscience, New York, 1959) Chap. I.)

For the Klein-Gordon equation the Lagrangian density is

$$\mathfrak{L} = -\frac{\hbar^2}{2m}\left(\frac{\partial\psi^*}{\partial x_\mu}\frac{\partial\psi}{\partial x_\mu} + \kappa^2\psi^*\psi\right). \tag{2}$$

From the stationary property of the action integral for arbitrary variations $\psi^* \to \psi^* + \delta\psi^*$ ($\delta\psi^*$ vanishing at the boundary), following the method outlined in *Note 3*, we obtain the Klein-Gordon equation. Similarly variations of the type $\psi \to \psi + \delta\psi$ lead to the complex conjugate equation.

The Lagrangian density and the action integral for the Klein-Gordon equation are relativistically invariant.

For the Schroedinger equation the Lagrangian density is

$$\mathfrak{L} = -\frac{\hbar^2}{2m}\nabla_x\psi^*\cdot\nabla_x\psi + \frac{\hbar}{2\mathrm{i}}\left(\frac{\partial\psi^*}{\partial t}\psi - \psi^*\frac{\partial\psi}{\partial t}\right). \tag{3}$$

The equation obtained by varying ψ^* is

$$-\frac{\hbar^2}{2m}\Delta\psi = \hbar\mathrm{i}\frac{\partial\psi}{\partial t},$$

and that obtained by varying ψ is the complex conjugate

$$-\frac{\hbar^2}{2m}\Delta\psi^* = -\hbar\mathrm{i}\frac{\partial\psi^*}{\partial t}.$$

Obviously, the Schroedinger equation for ψ^* is different from that for ψ. On writing

$$\psi = \mathrm{Re}\,\psi + \mathrm{i}\,\mathrm{Im}\,\psi$$

one finds that the real and the imaginary part of ψ obey the system of differential equations

$$-\frac{\hbar^2}{2m}\Delta(\mathrm{Re}\,\psi) = -\hbar\frac{\partial}{\partial t}(\mathrm{Im}\,\psi),$$

$$-\frac{\hbar^2}{2m}\Delta(\mathrm{Im}\,\psi) = \hbar\frac{\partial}{\partial t}(\mathrm{Re}\,\psi).$$

Since $\mathrm{Re}\,\psi$ and $\mathrm{Im}\,\psi$ are mutually related, it will, in general, not be possible, according to the Schroedinger wave mechanics, to describe physical situations by means of a single real wave function.

In the relativistic case, however, ψ and ψ^* obey the same equation since the Klein-Gordon operator $(\partial^2/\partial x_\mu\partial x_\mu) - \kappa^2$, in contrast to $(\hbar^2/2m)\,\Delta + \hbar\mathrm{i}\,\partial/\partial t$, does not involve the imaginary unit. It might seem, therefore, that relativistic wave mechanics could be formulated in terms of only one single real function, in analogy with electromagnetism, where the basic quantities $\boldsymbol{E}$ and $\boldsymbol{H}$ are real. However, a relativistic wave mechanics built on such an assumption would be incapable of describing charged particles in electromagnetic fields (see Chap. V, § 1).

The Lagrangian densities (2) and (3), from which the Klein-Gordon and the Schroedinger equation are deduced, have invariance properties which lead to some very important consequences. Thus the relativistic Lagrangian density is invariant under the transformations of the inhomogeneous Lorentz group, namely under linear transformations

$$x'_\mu = a_\mu + a_{\mu\nu}\, x_\nu \,,$$

where a_μ are constants characterizing space (a_1, a_2, a_3) and time (a_4) translations, while $a_{\mu\nu}$ are the coefficients of a homogeneous Lorentz transformation leaving x^2 invariant (rotation in space-time). Space rotations, under which $\boldsymbol{x}^2$ is invariant and the time unchanged, are a special case of the homogeneous Lorentz transformations. The Lagrangian density of the Schroedinger equation is invariant under space and time translations. However, it is not invariant under a general space-time rotation, but only under space rotations.

2. Noether's theorem

Noether * showed that, if the action integral

$$A = \int_{V(x)} \mathfrak{L}(x)\, \mathrm{d}^4x$$

of a continuous system – which, for the sake of brevity, we shall refer to as a "field" – is invariant under a coordinate transformation depending on n parameters, there exist n constants of motion. This means that n space integrals exist, involving the wave function and its space and time derivatives, which remain constant in time.

Let us consider an infinitesimal coordinate transformation

$$x'_\mu = x_\mu + \delta x_\mu \,. \tag{4}$$

We denote by $\psi'_i(x')$ and $\partial\psi'_i(x')/\partial x'_\mu$ the wave functions which correspond to $\psi_i(x)$ and $\partial\psi_i(x)/\partial x_\mu$ in the new system of coordinates, namely the trans-

* E. Noether, Nachr. d. Kgl. Ges. d. Wiss. Göttingen (1918), p. 235.

formed functions of ψ_i and $\partial\psi_i/\partial x_\mu$. For the time being we will not make any assumption upon the transformation law $\psi_i(x) \to \psi'_i(x')$.

Now, when we say that the action integral is invariant under a given transformation, we mean that the following equation is satisfied *

$$\int_{V'(x')} \mathfrak{L}'(x')\,\mathrm{d}^4x' = \int_{V(x)} \mathfrak{L}(x)\,\mathrm{d}^4x\,. \tag{5}$$

Here $V'(x')$ is defined as follows. If the four-dimensional region $V(x)$ is enclosed by the surface $f(x) = 0$, $V'(x')$ is the region of the space x' enclosed by the surface $f'(x') = f(x(x')) = 0$. On the other hand

$$\mathfrak{L}'(x') \equiv \mathfrak{L}\left(\psi'_i(x'),\ \frac{\partial\psi'_i(x')}{\partial x'_\mu}\right),$$

i.e. $\mathfrak{L}'(x')$ is the same function of $\psi'_i(x')$ and $\partial\psi'_i(x')/\partial x'_\mu$ as $\mathfrak{L}(x)$ is of $\psi_i(x)$ and $\partial\psi_i(x)/\partial x_\mu$.

When (5) holds, the equations of motion are also invariant under the given transformation. In fact the variational principle, when applied to the action integral

$$A' = \int_{V'(x')} \mathfrak{L}\left(\psi'_i(x'),\ \frac{\partial\psi'_i(x')}{\partial x'_\mu}\right)\mathrm{d}^4x'\,,$$

leads to equations of motion

$$\frac{\partial}{\partial x'_\mu}\frac{\partial\mathfrak{L}'(x')}{\partial\left(\dfrac{\partial\psi'_i(x')}{\partial x'_\mu}\right)} = \frac{\partial\mathfrak{L}'(x')}{\partial\psi'_i(x')}\,,$$

which have the same form as those in the original system

$$\frac{\partial}{\partial x_\mu}\frac{\partial\mathfrak{L}(x)}{\partial\left(\dfrac{\partial\psi_i(x)}{\partial x_\mu}\right)} = \frac{\partial\mathfrak{L}(x)}{\partial\psi_i(x)}\,,$$

the functional dependence of $\mathfrak{L}'(x')$ on $\psi'_i(x')$ and $\partial\psi'_i(x')/\partial x'_\mu$ and of $\mathfrak{L}(x)$ on $\psi_i(x)$ and $\partial\psi_i(x)/\partial x_\mu$ being the same.

---

* This is not to be confused with the trivial identity

$$\int_{V'(x')} \mathfrak{L}\left[\psi_i(x(x')),\ \frac{\partial x_\nu'}{\partial x_\mu}\frac{\partial\psi_i(x(x'))}{\partial x_\nu'}\right]\left|\frac{\partial(x)}{\partial(x')}\right|\mathrm{d}^4x' = \int_{V(x)} \mathfrak{L}\left(\psi_i(x),\ \frac{\partial\psi_i(x)}{\partial x_\mu}\right)\mathrm{d}^4x\,.$$

Notice that (5) does not, in general, imply the numerical equality $\mathfrak{L}'(x') = \mathfrak{L}(x)$. In fact the integral on the left-hand side of (5) may be written

$$\int_{V(x)} \mathfrak{L}'(x') \left| \frac{\partial(x')}{\partial(x)} \right| \mathrm{d}^4 x$$

and the Jacobian $J = \partial(x')/\partial(x)$ is, in general, different from unity.

Having so dealt with some of the implications of the action invariance, we shall now prove Noether's theorem. To this end, we write (5) in the more explicit form

$$0 = \int_{V(x)} \left[(\mathfrak{L}'(x') - \mathfrak{L}(x)) |J| + \mathfrak{L}(x)(|J| - 1) \right] \mathrm{d}^4 x \,.$$

Now for an infinitesimal transformation of the type (4), we have *

$$J \simeq 1 + \partial \delta x_\mu / \partial x_\mu \,, \tag{6}$$

(sum convention of equal indices).

Thus, if infinitesimal terms of order higher than the first are neglected, eq. (5) becomes

$$0 = \int_{V(x)} (\delta \mathfrak{L} + \mathfrak{L}\, \partial \delta x_\mu / \partial x_\mu)\, \mathrm{d}^4 x \,, \tag{7}$$

with $\delta \mathfrak{L} = \mathfrak{L}'(x') - \mathfrak{L}(x)$.

Now we put

$$\delta \psi_i = \psi_i'(x') - \psi_i(x) \simeq \bar{\delta} \psi_i + \frac{\partial \psi_i(x)}{\partial x_\nu} \delta x_\nu \,,$$

thus distinguishing the variation in form ($\bar{\delta} \psi_i = \psi_i'(x') - \psi_i(x') \simeq \psi_i'(x) - - \psi_i(x)$) from the variation due to the change of the argument in the trans-

* The explicit form of the Jacobian is

$$\frac{\partial(x')}{\partial(x)} = \begin{vmatrix} 1 + \dfrac{\partial \delta x_1}{\partial x_1} & \dfrac{\partial \delta x_1}{\partial x_2} & \dfrac{\partial \delta x_1}{\partial x_3} & \dfrac{\partial \delta x_1}{\partial x_4} \\ \dfrac{\partial \delta x_2}{\partial x_1} & 1 + \dfrac{\partial \delta x_2}{\partial x_2} & \dfrac{\partial \delta x_2}{\partial x_3} & \dfrac{\partial \delta x_2}{\partial x_4} \\ \cdot & \cdot & \cdot & \cdot \\ \dfrac{\partial \delta x_4}{\partial x_1} & \dfrac{\partial \delta x_4}{\partial x_2} & \dfrac{\partial \delta x_4}{\partial x_3} & 1 + \dfrac{\partial \delta x_4}{\partial x_4} \end{vmatrix} = A_{11} + \frac{\partial \delta x_1}{\partial x_1} A_{11} + \text{terms of second order,}$$

(A_{ik} is the co-factor of the element of the i-th row and k-th column). Performing an analogous expansion of A_{11}, then of A_{22} etc., always neglecting all infinitesimal quantities of order higher than the first, we obtain (6).

formation of $\psi_i(x)$. Similarly we have

$$\delta\mathfrak{L} = \mathfrak{L}'(x') - \mathfrak{L}(x') + \mathfrak{L}(x') - \mathfrak{L}(x) \simeq \bar{\delta}\mathfrak{L} + \frac{\partial\mathfrak{L}}{\partial x_\mu}\,\delta x_\mu\,.$$

(Here $\partial\mathfrak{L}/\partial x_\mu$ stands for the derivative with respect to x_μ of $\mathfrak{L}(x)$, regarded as an implicit function of x_μ through $\psi_i(x)$ and $\partial\psi_i(x)/\partial x_\mu$.) The explicit expression for $\bar{\delta}\mathfrak{L}$ is

$$\bar{\delta}\mathfrak{L} = \frac{\partial\mathfrak{L}}{\partial\psi_i}\,\bar{\delta}\psi_i + \frac{\partial\mathfrak{L}}{\partial\left(\dfrac{\partial\psi_i}{\partial x_\mu}\right)}\,\bar{\delta}\left(\frac{\partial\psi_i}{\partial x_\mu}\right).$$

Moreover we have

$$\bar{\delta}\,\frac{\partial\psi_i}{\partial x_\mu} = \frac{\partial\psi_i'(x')}{\partial x_\mu'} - \left(\frac{\partial\psi_i(x)}{\partial x_\mu}\right)_{x=x'} = \frac{\partial\psi_i'(x')}{\partial x_\mu'} - \frac{\partial\psi_i(x')}{\partial x_\mu'} = \frac{\partial}{\partial x_\mu'}\,\bar{\delta}\psi_i =$$

$$= \frac{\partial\bar{\delta}\psi_i}{\partial x_\mu} + \text{terms of second order}\,.$$

Then, using the equations of motion

$$\frac{\partial}{\partial x_\mu}\left[\partial\mathfrak{L}/\partial\left(\frac{\partial\psi_i}{\partial x_\mu}\right)\right] = \frac{\partial\mathfrak{L}}{\partial\psi_i}\,,$$

the variation $\bar{\delta}\mathfrak{L}$ can be written in the form

$$\bar{\delta}\mathfrak{L} = \frac{\partial}{\partial x_\mu}\left[\frac{\partial\mathfrak{L}}{\partial\left(\dfrac{\partial\psi_i}{\partial x_\mu}\right)}\,\bar{\delta}\psi_i\right].$$

Therefore eq. (7) becomes

$$0 = \int\limits_{V(x)} \frac{\partial}{\partial x_\mu}\left[\frac{\partial\mathfrak{L}}{\partial\left(\dfrac{\partial\psi_i}{\partial x_\mu}\right)}\,\bar{\delta}\psi_i + \mathfrak{L}\,\delta x_\mu\right] \mathrm{d}^4x\,. \tag{8}$$

Let us suppose that the coordinate transformation (4) depends on n parameters $\alpha_1, \alpha_2, \ldots \alpha_n$ in the following way

$$\delta x_\mu = A_{\mu j}(x)\,\delta\alpha_j\,, \tag{9}$$

and that the wave functions are transformed according to

$$\psi_i'(x') = \psi_i(x) + \delta\psi_i\,, \quad \delta\psi_i = \Psi_{ij}\,\delta\alpha_j\,. \tag{10}$$

For (9) and (10), (8) reads

$$0 = \int\limits_{V(x)} \delta\alpha_j \frac{\partial}{\partial x_\mu} \left[\frac{\partial \mathfrak{L}}{\partial \left(\dfrac{\partial \psi_i}{\partial x_\mu} \right)} \left(\Psi_{ij} - \frac{\partial \psi_i}{\partial x_\nu} A_{\nu j} \right) + \mathfrak{L}\, A_{\mu j} \right] \mathrm{d}^4 x \,.$$

Since the $\delta\alpha_j$ are arbitrary, this equation is equivalent to the vanishing of n four-divergences

$$\frac{\partial \Theta_\mu^{(j)}}{\partial x_\mu} = 0\,, \quad (j = 1, 2, \ldots n)\,, \tag{11}$$

where

$$\Theta_\mu^{(j)} = \frac{\partial \mathfrak{L}}{\partial \left(\dfrac{\partial \psi_i}{\partial x_\mu} \right)} \left(\Psi_{ij} - \frac{\partial \psi_i}{\partial x_\nu} A_{\nu j} \right) + \mathfrak{L}\, A_{\mu j}\,. \tag{12}$$

As is well known, (11) is the differential expression of a *conservation law.* The corresponding integral form of such a law is obtained on integrating over the space-time region comprised between the planes $x_4 = \mathrm{i}ct_1$ and $x_4 = \mathrm{i}ct_2$. Using the four-dimensional divergence theorem, one finds

$$\left(\int \Theta_4^{(j)}\, \mathrm{d}^3\mathrm{x}\right)_{t=t_1} = \left(\int \Theta_4^{(j)}\, \mathrm{d}^3\mathrm{x}\right)_{t=t_2},$$

namely

$$\int \Theta_4^{(j)}\, \mathrm{d}^3\mathrm{x} = \text{const}\,, \quad (j = 1, 2, \ldots n)\,. \tag{13}$$

NOTE 4. As an example of the formalism developed, we determine the equation of motion corresponding to the Lagrangian density

$$\mathfrak{L} = -\frac{\hbar^2}{2m} \left(\frac{\partial \psi}{\partial x_\mu} \frac{\partial \psi}{\partial x_\mu} + \tfrac{1}{2} f \psi^r \right). \tag{α}$$

In particular we want to find the values of the exponent r in the Lagrangian density, and of s in the *scale transformation*

$$\begin{aligned} x' &= \lambda x\,, \quad (\text{i.e. } \boldsymbol{x}' = \lambda \boldsymbol{x}\,,\ x_0' = \lambda x_0)\,, \\ \psi'(x') &= \lambda^s \psi(x)\,, \end{aligned} \tag{β}$$

such that the integral of the Lagrangian density (α), extended over an arbitrary four-dimensional region, be invariant.

Substituting the given Lagrangian density in the Euler-Lagrange equation (1), we obtain

$$\Box \psi = \tfrac{1}{4} r f \psi^{r-1}\,.$$

On the other hand, if we impose the condition

$$\mathfrak{L}(x)\,\mathrm{d}^4x = \mathfrak{L}'(x')\,\mathrm{d}^4x'\,,$$

since, according to (β), $\mathrm{d}^4x' = \lambda^4\,\mathrm{d}^4x$, we see that it must be

$$\mathfrak{L}'(x') = \lambda^{-4}\,\mathfrak{L}(x)\,. \tag{γ}$$

Now making the transformation (β) on $\mathfrak{L}(x)$ we get

$$\mathfrak{L}'(x') = -\frac{\hbar^2}{2m}\left(\lambda^{2(s-1)}\frac{\partial\psi}{\partial x_\mu}\frac{\partial\psi}{\partial x_\mu} + \tfrac{1}{2}f\lambda^{rs}\psi^r\right).$$

In order that (γ) be satisfied we must have

$$r = 4\,,\quad s = -1\,,$$

so that the equation of motion is

$$\Box\,\psi - f\psi^3 = 0\,.$$

The above invariance of the action leads to a conservation law. Let us consider an infinitesimal transformation (β)

$$\lambda = 1 + \varepsilon\,,\quad x'_\mu = x_\mu + \varepsilon x_\mu\,,\quad \psi'(x') = \psi(x) - \varepsilon\psi(x)\,.$$

Using the formulae of § 2 we have

$$A_\mu = x_\mu\,,\quad \Psi = -\psi\,.$$

Therefore, because of (12) and (13) the constant of motion is

$$\int \Omega_0\,\mathrm{d}^3\mathrm{x} = \mathrm{const}$$

with

$$\Omega_0 = \frac{mc}{\hbar^2}\left[\frac{\partial\mathfrak{L}}{\partial\dot\psi}\left(\frac{\partial\psi}{\partial x_\mu}x_\mu + \psi\right) - \mathfrak{L}t\right] =$$

$$= \frac{\partial\psi}{\partial x_0}\left(\frac{\partial\psi}{\partial x_\mu}x_\mu + \psi\right) + \tfrac{1}{2}x_0\left(\frac{\partial\psi}{\partial x_\mu}\frac{\partial\psi}{\partial x_\mu} + \tfrac{1}{2}f\psi^4\right).$$

3. Energy-momentum tensor

The invariance of the action integral with respect to space and time translations leads to the vanishing of the divergence of the energy-momentum tensor.

The action integral relative to the Klein-Gordon equation is indeed invariant under a translation a of the origin of the reference system

$$x'_\mu = x_\mu + \delta x_\mu\,, \quad \delta x_\mu = -\,a_\mu\,.$$

In the notation of the previous section

$$A_{\mu\nu} = -\,\delta_{\mu\nu}\,, \quad \delta\psi = \psi'(\mathrm{x}') - \psi(\mathrm{x}) = 0\,.$$

Therefore eqs. (11) and (12) give in this case

$$\frac{\partial T_{\mu\nu}}{\partial x_\mu} = 0\,, \tag{14}$$

where

$$T_{\mu\nu} = \frac{\hbar^2}{2m}\left[\frac{\partial\psi^*}{\partial x_\mu}\frac{\partial\psi}{\partial x_\nu} + \frac{\partial\psi^*}{\partial x_\nu}\frac{\partial\psi}{\partial x_\mu} - \left(\frac{\partial\psi^*}{\partial x_\lambda}\frac{\partial\psi}{\partial x_\lambda} + \kappa^2\,\psi^*\psi\right)\delta_{\mu\nu}\right]$$

is the *energy-momentum tensor*. Notice that this tensor is *symmetric*.

The constant of motion following from the invariance of the action under time translations is the energy

$$E = -\int T_{44}\,\mathrm{d}^3\mathrm{x} = \frac{\hbar^2}{2m}\int\left(\nabla_x\psi^*\cdot\nabla_x\psi + \frac{1}{c^2}\frac{\partial\psi^*}{\partial t}\frac{\partial\psi}{\partial t} + \kappa^2\,\psi^*\psi\right)\mathrm{d}^3\mathrm{x}\,.$$

This may be cast into the form

$$E = \frac{\hbar\mathrm{i}}{2mc^2}\int\left[\psi^*\left(\hbar\mathrm{i}\frac{\partial}{\partial t}\right)\frac{\partial\psi}{\partial t} - \frac{\partial\psi^*}{\partial t}\left(\hbar\mathrm{i}\frac{\partial}{\partial t}\right)\psi\right]\mathrm{d}^3\mathrm{x}\,. \tag{15}$$

On the other hand, the constant of motion corresponding to a space translation along the i-th axis is the i-th component of the momentum

$$P_i = \frac{1}{\mathrm{i}c}\int T_{4i}\,\mathrm{d}^3\mathrm{x} = -\frac{\hbar^2}{2mc^2}\int\left(\frac{\partial\psi^*}{\partial t}\frac{\partial\psi}{\partial x_i} + \frac{\partial\psi^*}{\partial x_i}\frac{\partial\psi}{\partial t}\right)\mathrm{d}^3\mathrm{x}\,.$$

By partial integration we find

$$P_i = \frac{\hbar\mathrm{i}}{2mc^2}\int\left[\psi^*\left(\frac{\hbar}{\mathrm{i}}\frac{\partial}{\partial x_i}\right)\frac{\partial\psi}{\partial t} - \frac{\partial\psi^*}{\partial t}\left(\frac{\hbar}{\mathrm{i}}\frac{\partial}{\partial x_i}\right)\psi\right]\mathrm{d}^3\mathrm{x}\,. \tag{16}$$

The equations (15) and (16) may be condensed to *

$$P_\mu = \frac{\hbar i}{2mc^2} \int \left[\psi^* \left(\frac{\hbar}{i} \frac{\partial}{\partial x_\mu} \right) \frac{\partial \psi}{\partial t} - \frac{\partial \psi^*}{\partial t} \left(\frac{\hbar}{i} \frac{\partial}{\partial x_\mu} \right) \psi \right] d^3x \, .$$

The action integral relative to the Schroedinger equation is also invariant under space and time translations. This leads to (14) with

$$T_{\mu\nu} = \frac{\hbar^2}{2m} \left(\frac{\partial \psi^*}{\partial x_\mu} \frac{\partial \psi}{\partial x_\nu} + \frac{\partial \psi^*}{\partial x_\nu} \frac{\partial \psi}{\partial x_\mu} \right) (1 - \delta_{\mu 4}) + \tfrac{1}{2} \hbar c \left[\psi^* \frac{\partial \psi}{\partial x_\nu} - \frac{\partial \psi^*}{\partial x_\nu} \psi \right] \delta_{\mu 4} +$$

$$+ \left[- \frac{\hbar^2}{2m} \frac{\partial \psi^*}{\partial x_k} \frac{\partial \psi}{\partial x_k} - \tfrac{1}{2} \hbar c \left(\psi^* \frac{\partial \psi}{\partial x_4} - \frac{\partial \psi^*}{\partial x_4} \psi \right) \right] \delta_{\mu\nu}$$

(no sum over μ). From this we have

$$E = - \int T_{44} \, d^3x = \frac{\hbar^2}{2m} \int \nabla_x \psi^* \cdot \nabla_x \psi \, d^3x = \int \psi^* \left(- \frac{\hbar^2}{2m} \Delta \right) \psi \, d^3x \, ,$$

or, remembering the Schroedinger equation,

$$E = \int \psi^* \hbar i \frac{\partial}{\partial t} \psi \, d^3x \, . \tag{17}$$

Similarly

$$P_i = \frac{1}{ic} \int T_{4i} \, d^3x = \frac{\hbar}{2i} \int \left(\psi^* \frac{\partial \psi}{\partial x_i} - \frac{\partial \psi^*}{\partial x_i} \psi \right) d^3x = \int \psi^* \frac{\hbar}{i} \frac{\partial}{\partial x_i} \psi \, d^3x . \tag{18}$$

We notice that $T_{4i} \neq T_{i4}$, so that the energy-momentum tensor of the non-relativistic theory is *not symmetric*. Its space components, however, are symmetric,

$$T_{ik} = T_{ki} \, .$$

NOTE 5. We discuss the invariance properties of the action integral for an infinite vibrating string of linear density μ, independent of x, and tension K.

---

* We shall see later that the *mean* energy and momentum of a spin zero particle with the wave function ψ are

$$E = - K \int T_{44} \, d^3x, \quad P_i = \frac{K}{ic} \int T_{4i} \, d^3x \, ,$$

where

$$K^{-1} = \frac{\hbar i}{2mc^2} \int \left(\psi^* \frac{\partial \psi}{\partial t} - \frac{\partial \psi^*}{\partial t} \psi \right) d^3x \quad \text{(independent of time)} \, .$$

One sees at once that the Lagrangian density

$$\mathfrak{L} = \tfrac{1}{2}\,\mu\left(\frac{\partial y}{\partial t}\right)^2 - \tfrac{1}{2}\,K\left(\frac{\partial y}{\partial x}\right)^2$$

and the action integral are invariant under time translations and space translations along the x axis. Application of Noether's theorem provides the constants of motion

$$E = \tfrac{1}{2}\int\left[\mu\left(\frac{\partial y}{\partial t}\right)^2 + K\left(\frac{\partial y}{\partial x}\right)^2\right]\mathrm{d}x$$

and

$$P = -\,\mu\int\frac{\partial y}{\partial t}\frac{\partial y}{\partial x}\,\mathrm{d}x\,.$$

NOTE 6. The Lagrangian density

$$\mathfrak{L} = -\,\frac{\hbar^2}{2m}\left(\frac{\partial\psi^*}{\partial x_\mu}\frac{\partial\psi}{\partial x_\mu} + \kappa^2\,\psi^*\psi\right) +$$

$$+\,\tfrac{1}{2}\,M\dot{\boldsymbol{\xi}}^2\delta(\boldsymbol{x}-\boldsymbol{\xi}) + f\,\psi^*\psi\;u(\boldsymbol{x}-\boldsymbol{\xi})\,,\quad \left(\kappa = \frac{mc}{\hbar}\right),$$

represents matter waves interacting with a (classical) particle of mass M, position $\boldsymbol{\xi}$ and velocity $|\dot{\boldsymbol{\xi}}| \ll c$ (f is a coupling constant). It is invariant under space and time translations.

Requiring that

$\delta A = \int\delta\mathfrak{L}\,\mathrm{d}^4x = 0$, for $\psi^* \to \psi^* + \delta\psi^*$, ($\delta\psi^* = 0$ on the boundary),

gives

$$(\Box - \kappa^2)\,\psi(\boldsymbol{x}, t) = -\,2mf\,\psi(\boldsymbol{x}, t)\,u(\boldsymbol{x}-\boldsymbol{\xi})/\hbar^2\,.$$

On the other hand, varying $\boldsymbol{\xi}(t)$

$$\boldsymbol{\xi}(t) \to \boldsymbol{\xi}(t) + \delta\boldsymbol{\xi}(t) \quad\text{with}\quad \delta\boldsymbol{\xi}(+\infty) = \delta\boldsymbol{\xi}(-\infty) = 0\,,$$

we obtain

$$M\,\frac{\mathrm{d}^2\boldsymbol{\xi}(t)}{\mathrm{d}t^2} = f\int\left\{u(\boldsymbol{x}-\boldsymbol{\xi}(t))\,\nabla_x\left[\psi^*(\boldsymbol{x}, t)\,\psi(\boldsymbol{x}, t)\right]\right\}\mathrm{d}^3\mathrm{x}\,.$$

From Noether's theorem follows the conservation of the energy

$$E = \frac{\hbar^2}{2m} \int \left(\nabla_x \psi^* \cdot \nabla_x \psi + \frac{1}{c^2} \frac{\partial \psi^*}{\partial t} \frac{\partial \psi}{\partial t} + \kappa^2 \psi^* \psi \right) \mathrm{d}^3\mathrm{x} + \\ + \tfrac{1}{2} M \dot{\xi}^2 - f \int \psi^* \psi\, u(\boldsymbol{x} - \xi)\, \mathrm{d}^3\mathrm{x}$$

and of the momentum

$$\boldsymbol{P} = - \frac{\hbar^2}{2mc^2} \int \left(\frac{\partial \psi^*}{\partial t} \nabla_x \psi + \nabla_x \psi^* \frac{\partial \psi}{\partial t} \right) \mathrm{d}^3\mathrm{x} + M \dot{\xi}$$

of the system (matter waves + particle of mass M). In the limit $f \to 0$ the energy reduces to the sum of the energies of free matter waves and of the free particle.

4. Angular momentum

As has been noted in § 1, both the relativistic and non-relativistic Lagrangian densities, and obviously also the corresponding action integrals, are invariant under space rotations. From this invariance three constants of motion will be deduced, in correspondence to the three parameters characterizing a rotation. These are the components of the angular momentum along the three coordinate axes.

Denoting by $\delta\boldsymbol{\alpha}$ the vector whose direction is that of the axis of rotation of the coordinate system and whose modulus is equal to the angle of the (counter-clockwise) rotation, the transformation of the coordinates of a fixed point takes the form

$$x_i' = x_i - \varepsilon_{ijk}\, \delta\alpha_j\, x_k\,, \quad (i = 1, 2, 3)\,, \quad x_4' = x_4\,.$$

Thus

$$A_{ij} = \varepsilon_{ikj}\, x_k\,, \quad A_{4j} = A_{j4} = A_{44} = 0\,.$$

Hence

$$\frac{\partial \mathfrak{M}_{\mu j}}{\partial x_\mu} = 0\,, \quad (j = 1, 2, 3)\,,$$

with

$$\mathfrak{M}_{\mu j} = \frac{\partial \mathfrak{L}}{\partial \left(\dfrac{\partial \psi}{\partial x_\mu} \right)} \varepsilon_{jkl}\, x_k \frac{\partial \psi}{\partial x_l} + \frac{\partial \mathfrak{L}}{\partial \left(\dfrac{\partial \psi^*}{\partial x_\mu} \right)} \varepsilon_{jkl}\, x_k \frac{\partial \psi^*}{\partial x_l} + \mathfrak{L}\, \varepsilon_{\mu kj}\, x_k\,, \; (\varepsilon_{4kj} = 0)\,.$$

Then, for the Klein-Gordon equation the components of the angular momentum are

$$M_j = -\frac{1}{\mathrm{i}c} \int \mathfrak{M}_{4j}\, \mathrm{d}^3\mathrm{x} =$$

$$= -\frac{\hbar \mathrm{i}}{2mc^2} \int \left[\frac{\partial \psi^*}{\partial t} \left(\varepsilon_{jkl}\, x_k \frac{\hbar}{\mathrm{i}} \frac{\partial}{\partial x_l} \right) \psi + \varepsilon_{jkl}\, x_k \frac{\hbar}{\mathrm{i}} \frac{\partial \psi^*}{\partial x_l} \frac{\partial \psi}{\partial t} \right] \mathrm{d}^3\mathrm{x}\,.$$

On integrating by parts we have

$$M_j = \frac{\hbar \mathrm{i}}{2mc^2} \int \left[\psi^* \left(\boldsymbol{x} \times \frac{\hbar}{\mathrm{i}} \boldsymbol{\nabla}_x \right)_j \frac{\partial \psi}{\partial t} - \frac{\partial \psi^*}{\partial t} \left(\boldsymbol{x} \times \frac{\hbar}{\mathrm{i}} \boldsymbol{\nabla}_x \right)_j \psi \right] \mathrm{d}^3\mathrm{x}\,. \qquad (19)$$

This shows that the Klein-Gordon equation describes spinless particles. In fact $\boldsymbol{M}$ coincides with the orbital angular momentum $\boldsymbol{L}$ involving the differential operator

$$\boldsymbol{x} \times \boldsymbol{p} = \boldsymbol{x} \times \frac{\hbar}{\mathrm{i}} \boldsymbol{\nabla}_x\,.$$

In a similar way from the Lagrangian density for the Schroedinger equation one obtains the constants of motion

$$M_j = -\frac{1}{\mathrm{i}c} \int \mathfrak{M}_{4j}\, \mathrm{d}^3\mathrm{x} = \int \psi^* \left(\boldsymbol{x} \times \frac{\hbar}{\mathrm{i}} \boldsymbol{\nabla}_x \right)_j \psi\, \mathrm{d}^3\mathrm{x}\,, \quad (j = 1, 2, 3)\,. \qquad (20)$$

NOTE 7. We shall now deduce the constants of motion corresponding to the invariance of the action integral with respect to an infinitesimal Lorentz transformation

$$x'_\mu = x_\mu + \delta x_\mu\,, \quad \delta x_\mu = \varepsilon_{\mu\nu}\, x_\nu\,, \quad \varepsilon_{\mu\nu} = -\, \varepsilon_{\nu\mu}\,.$$

The parameters characterizing this transformation are the six $\varepsilon_{\mu\nu}$ ($\mu \neq \nu$). The two indices μ and ν play here the rôle of the one single index labelling the transformation parameters in § 2.

Now we can write

$$\delta x_\mu = \varepsilon_{\lambda\nu}\, x_\nu\, \delta_{\lambda\mu} = \tfrac{1}{2}\, \varepsilon_{\lambda\nu} (x_\nu\, \delta_{\lambda\mu} - x_\lambda\, \delta_{\nu\mu})$$

and so

$$A_{\mu(\lambda\nu)} = \tfrac{1}{2}(x_\nu\, \delta_{\lambda\mu} - x_\lambda\, \delta_{\nu\mu})\,.$$

For the sake of generality we do not, for the moment, discard the possibility that $\delta\psi_i(x)$ may be different from zero, and write

$$\delta\psi_i(x) = \psi'_i(x') - \psi_i(x) = \tfrac{1}{2}\, \Psi_{i(\lambda\nu)}\, \varepsilon_{\lambda\nu}\,.$$

Then, from Noether's theorem, we have

$$\frac{\partial \mathfrak{H}_{\mu\nu\lambda}}{\partial x_\mu} = 0\,, \tag{21}$$

with

$$\mathfrak{H}_{\mu\nu\lambda} = x_\nu \left[-\frac{\partial \mathfrak{L}}{\partial\left(\dfrac{\partial\psi_i}{\partial x_\mu}\right)} \frac{\partial \psi_i}{\partial x_\lambda} + \mathfrak{L}\,\delta_{\lambda\mu} \right] - \tag{22}$$

$$- x_\lambda \left[-\frac{\partial \mathfrak{L}}{\partial\left(\dfrac{\partial\psi_i}{\partial x_\mu}\right)} \frac{\partial \psi_i}{\partial x_\nu} + \mathfrak{L}\,\delta_{\nu\mu} \right] + \frac{\partial \mathfrak{L}}{\partial\left(\dfrac{\partial\psi_i}{\partial x_\mu}\right)} \Psi_{i(\lambda\nu)}\,,$$

which can be written in the form

$$\mathfrak{H}_{\mu\nu\lambda} = x_\nu\, T_{\mu\lambda} - x_\lambda\, T_{\mu\nu} + \frac{\partial \mathfrak{L}}{\partial\left(\dfrac{\partial\psi_i}{\partial x_\mu}\right)} \Psi_{i(\lambda\nu)}\,, \tag{23}$$

($T_{\mu\nu}$ is the energy-momentum tensor).

The last term in eq. (23) can be ascribed to the intrinsic angular momentum (or spin) of the "field". For fields which are scalars under Lorentz transformations ($\Psi_{i(\lambda\nu)} = 0$), such an intrinsic angular momentum does not exist.

The constants of motion following from eq. (21) in the case of a spin zero field are

$$M_{\nu\lambda} = \frac{1}{\mathrm{i}c} \int \mathfrak{H}_{4\nu\lambda}\, \mathrm{d}^3\mathrm{x} = \frac{1}{\mathrm{i}c} \int (x_\nu\, T_{4\lambda} - x_\lambda\, T_{4\nu})\, \mathrm{d}^3\mathrm{x}\,.$$

Those with $4 \neq \lambda \neq \nu \neq 4$ can be identified with the components of the angular momentum deduced above.

For $\lambda = 4$, $\nu = 1, 2, 3$ one has

$$\int \mathrm{d}^3\mathrm{x}\, (x_\nu\, T_{44} - x_4\, T_{4\nu}) = \text{const}\,. \tag{24}$$

Since $-T_{44}$ is the energy density, the coordinates of the "centre" of the energy may be defined by

$$\boldsymbol{x}_{\mathrm{G}} = \frac{\int \boldsymbol{x}\, T_{44}\, \mathrm{d}^3\mathrm{x}}{\int T_{44}\, \mathrm{d}^3\mathrm{x}}\,.$$

On the other hand

$$P_k = -\,\mathrm{i}c^{-1} \int T_{4k}\, \mathrm{d}^3\mathrm{x}\,, \quad (k = 1, 2, 3)\,,$$

is the momentum.

Thus eq. (24) gives

$$\boldsymbol{x}_{\mathrm{G}} = \frac{c^2 \boldsymbol{P}}{E} t + \text{const} .$$

The "centre" of the energy of a spin zero particle moves with a uniform velocity.

NOTE 8. We have seen that the components of the angular momentum for the Klein-Gordon and the Schroedinger equation may be written in terms of the corresponding energy-momentum tensors as

$$M_j = \frac{1}{\mathrm{i}c} \varepsilon_{jlk} \int x_l \, T_{4k} \, \mathrm{d}^3\mathrm{x} .$$

Then

$$\frac{\mathrm{d}M_j}{\mathrm{d}t} = \varepsilon_{jik} \int x_i \frac{\partial T_{4k}}{\partial x_4} \mathrm{d}^3\mathrm{x} = \varepsilon_{jik} \int x_i \left(- \frac{\partial T_{lk}}{\partial x_l} \right) \mathrm{d}^3\mathrm{x} = \varepsilon_{jik} \int T_{lk} \, \delta_{il} \, \mathrm{d}^3\mathrm{x}$$

$$= \varepsilon_{jlk} \int T_{lk} \, \mathrm{d}^3\mathrm{x} .$$

This is zero if $T_{lk} = T_{kl}$, $(l, k = 1, 2, 3)$, which is true for both the relativistic and the non-relativistic theory. Note that for the latter $T_{ik} = T_{ki}$, but, in general, $T_{\mu\nu} \neq T_{\nu\mu}$, $(\nu, \mu = 1, 2, 3, 4)$.

Chapter III

PHYSICAL QUANTITIES AS MEAN VALUES

1. Schroedinger metric

As is well known, in non-relativistic wave mechanics energy, momentum and angular momentum may be represented as mean values of self-adjoint differential operators in a function space.

In this function space the product of two wave functions ψ_1 and ψ_2 is defined by

$$(\psi_1, \psi_2) = \int \psi_1^* \psi_2 \, d^3x = (\psi_2, \psi_1)^* ,$$

and the norm (ψ, ψ) of a wave function is positive definite ($(\psi, \psi) = 0$ only if $\psi = 0$). According to this *metric* the mean value of a generic operator A is given by *

$$\langle A \rangle = (\psi, A\psi) = \int \psi^* A\psi \, d^3x .$$

The operator adjoint of A, denoted by A^+, is defined by the relation

$$(\psi_1, A\psi_2) = (A^+ \psi_1, \psi_2) \tag{1}$$

or

$$\int \psi_1^* A\psi_2 \, d^3x = \int (A^+ \psi_1)^* \psi_2 \, d^3x .$$

An operator is said to be self-adjoint, if it is equal to its adjoint, that is

$$A = A^+ .$$

Now one sees from eqs. (17), (18) Chap. II, § 3 and eq. (20), § 4 that the energy, the momentum and the angular momentum can be considered as mean values of the operators $\hbar i \, \partial/\partial t = -(\hbar^2/2m) \Delta$, $-i\hbar \nabla_x$ and $-\boldsymbol{x} \times \hbar i \nabla_x$, respectively. These operators are all self-adjoint. For example

$$(\psi_1, \boldsymbol{p} \psi_2) = -i\hbar \int \psi_1^* \nabla_x \psi_2 \, d^3x = i\hbar \int (\nabla_x \psi_1^*) \psi_2 \, d^3x =$$
$$= \int (-i\hbar \nabla_x \psi_1)^* \psi_2 \, d^3x = (\boldsymbol{p} \psi_1, \psi_2) .$$

* Actually the mean value of A should be defined as

$$\langle A \rangle = (\psi, A\psi) / (\psi, \psi) .$$

However, the denominator is a constant of motion, and, for the sake of brevity, will be omitted.

Notice that, if $A = A^+$ and $\psi_1 = \psi_2$, eq. (1) gives

$$(\psi, A\psi) = (A\psi, \psi) = (\psi, A\psi)^* .$$

Thus the mean value of a self-adjoint operator is real.

2. Metric of Feshbach and Villars

The above considerations can be extended to the relativistic case, i.e. it is possible to interpret the basic mechanical quantities of the Klein-Gordon "field" as mean values of differential operators.

A way of extending the results of the previous section to the relativistic case is suggested by the considerable analogy between the formulae expressing the basic mechanical quantities in the two cases (see Chap. II, § 3, eqs. (15) and (17), (16) and (18), § 4, eqs. (19) and (20)). By a suitable definition of the metric in a function space of two-component wave functions, it will be possible to regard the relativistic energy, momentum and angular momentum as mean values of the same differential operators as in non-relativistic wave mechanics.

Following Feshbach and Villars * a system of two equations of the first order in the time for ψ and $\partial\psi/\partial x_4$ may be set up, which is equivalent to the Klein-Gordon equation.

We introduce the 2×2 matrices

$$\tau_1 = \begin{pmatrix} 0 & 1 \\ 1 & 0 \end{pmatrix}, \quad \tau_2 = \begin{pmatrix} 0 & -\mathrm{i} \\ \mathrm{i} & 0 \end{pmatrix}, \quad \tau_3 = \begin{pmatrix} 1 & 0 \\ 0 & -1 \end{pmatrix}, \quad I = \begin{pmatrix} 1 & 0 \\ 0 & 1 \end{pmatrix}$$

and the two-component wave function

$$\Psi = \begin{pmatrix} u \\ v \end{pmatrix}$$

with

$$u = \frac{1}{\sqrt{2}} \left(\psi - \frac{\hbar}{mc} \frac{\partial \psi}{\partial x_4} \right),$$

$$v = \frac{1}{\sqrt{2}} \left(\psi + \frac{\hbar}{mc} \frac{\partial \psi}{\partial x_4} \right).$$

* H. Feshbach and F. Villars, Revs. Mod. Phys. **30** (1958), 24. See also M. Taketani and S. Sakata, Proc. Phys. Math. Soc. (Japan) **22** (1940), 757.

Conversely ψ and $\partial\psi/\partial x_4$ are given in terms of u and v by

$$\psi = \frac{1}{\sqrt{2}}(u + v)\,,$$

$$-\frac{\hbar}{mc}\frac{\partial\psi}{\partial x_4} = \frac{1}{\sqrt{2}}(u - v)\,.$$

It is easy to verify that the equation

$$\hbar \mathrm{i}\frac{\partial}{\partial t}\Psi = H\,\Psi\,, \tag{2}$$

with

$$H = -\frac{\hbar^2}{2m}(\tau_3 + \mathrm{i}\,\tau_2)\,\Delta + mc^2\,\tau_3\,, \tag{3}$$

is equivalent to the Klein-Gordon equation. Notice that eq. (2), of the first order in the time, is formally identical with the Schroedinger equation.

NOTE 1. For the convenience of the reader we recall the basic rules of matrix algebra.

Let $A = \|a_{ik}\|$ and $B = \|b_{ik}\|$ be two matrices having p rows and q columns and p' rows and q' columns respectively. The sum of such matrices, for $p = p'$, $q = q'$, is obtained by summing the corresponding elements

$$C = A + B = \|a_{ik} + b_{ik}\| = \|c_{ik}\|\,.$$

The product of a number (possibly complex) by a matrix is obtained by multiplying each element of the matrix by the number. For example

$$\tau_3 + \mathrm{i}\,\tau_2 = \begin{pmatrix}1 & 0\\ 0 & -1\end{pmatrix} + \begin{pmatrix}0 & 1\\ -1 & 0\end{pmatrix} = \begin{pmatrix}1 & 1\\ -1 & -1\end{pmatrix}.$$

The product of a $(p \times q)$ matrix A with a $(p' \times q')$ matrix B for $q = p'$ is

$$C = AB = \|c_{ik}\| = \Big\|\sum_j a_{ij}\,b_{jk}\Big\|\,.$$

In general $AB \neq BA$. If $BA = AB$, A and B are said to commute.

The unit matrix is $I = \|\delta_{mn}\|$. The inverse A^{-1} of a square $(p = q)$

matrix A is defined by the equation

$$AA^{-1} = A^{-1}A = I .$$

The Hermitian conjugate $A^\dagger$ of a matrix A is obtained by replacing each element by its complex conjugate and exchanging rows and columns. Thus

$$(A + B)^\dagger = A^\dagger + B^\dagger , \quad (AB)^\dagger = B^\dagger A^\dagger .$$

A matrix is Hermitian if $A = A^\dagger$. The sum of two Hermitian matrices is Hermitian. On the other hand, the product of two Hermitian matrices is Hermitian only if the matrices commute with each other.

A matrix U is said to be unitary if

$$UU^\dagger = I = U^\dagger U \quad \text{or} \quad U^\dagger = U^{-1} .$$

Notice finally that, in eq. (2), Ψ may be regarded as a 2×1 matrix and that the matrix $\tau_3 + i\tau_2$, occurring in the expression (3) of the Hamiltonian, is not Hermitian.

A metric in the space of the two-component wave functions is now introduced by defining the product of two such functions Ψ_1 and Ψ_2 as

$$(\Psi_1, \Psi_2) = \tfrac{1}{2} \int \Psi_1^\dagger \tau_3 \Psi_2 \, d^3x . \tag{4}$$

Here

$$\Psi^\dagger = (u^* \quad v^*) = \frac{1}{\sqrt{2}} \left(\psi^* + \frac{\hbar}{mc} \frac{\partial \psi^*}{\partial x_4} \quad \psi^* - \frac{\hbar}{mc} \frac{\partial \psi^*}{\partial x_4} \right) .$$

The mean value of an operator A is given by (see footnote on p. 56)

$$\langle A \rangle = (\Psi, A\Psi) = \tfrac{1}{2} \int \Psi^\dagger \tau_3 A \Psi \, d^3x .$$

If A is a differential operator, this equation can easily be written in terms of the components of Ψ

$$(\Psi, A\Psi) = \tfrac{1}{2} \int (u^* A u - v^* A v) \, d^3x .$$

Recalling the definition of u and v this becomes

$$\langle A \rangle = (\Psi, A \Psi) = \frac{\hbar i}{2mc^2} \int \left(\psi^* A \frac{\partial \psi}{\partial t} - \frac{\partial \psi^*}{\partial t} A \psi \right) d^3x . \tag{5}$$

According to the metric (4), the operator *adjoint* of A, denoted by A^+, is defined by the relation

$$(\Psi_1, A\Psi_2) = (A^+\Psi_1, \Psi_2) .$$

An operator is self-adjoint if

$$A^+ = A \,. \tag{6}$$

Note that, according to this definition, the adjoint of a 2×2 matrix M (involving neither $\boldsymbol{x}$ nor its derivatives) is

$$M^+ = \tau_3 \, M^\dagger \, \tau_3 \,.$$

Thus τ_3 is both self-adjoint and Hermitian, whereas τ_1 and τ_2, though Hermitian, are not self-adjoint. One has

$$\tau_1^+ = -\,\tau_1^\dagger = -\,\tau_1\,, \quad \tau_2^+ = -\,\tau_2^\dagger = -\,\tau_2\,, \quad \tau_3^+ = \tau_3^\dagger = \tau_3\,,$$

whereas

$$(\mathrm{i}\,\tau_1)^+ = \tau_3(\mathrm{i}\,\tau_1)^\dagger\,\tau_3 = -\,\mathrm{i}\,\tau_3\,\tau_1\,\tau_3 = \mathrm{i}\,\tau_1$$

and similarly

$$(\mathrm{i}\,\tau_2)^+ = \mathrm{i}\,\tau_2\,, \quad (\mathrm{i}\,\tau_3)^+ = -\,\mathrm{i}\,\tau_3\,.$$

An operator U is said to be unitary if

$$(U\Psi_1,\, U\Psi_2) = (\Psi_1,\, \Psi_2)$$

for any two wave functions, namely if

$$U^{-1} = U^+\,.$$

Thus a 2×2 unitary matrix (involving neither $\boldsymbol{x}$ nor its derivatives), for which $U^{-1} = U^\dagger$, is in general *not* a unitary operator according to the Feshbach-Villars metric.

For $A = I$, eq. (5) reduces to the norm of Ψ

$$(\Psi,\, \Psi) = \frac{\hbar \mathrm{i}}{2mc^2} \int \left(\psi^* \frac{\partial \psi}{\partial t} - \frac{\partial \psi^*}{\partial t} \psi \right) \mathrm{d}^3 x\,.$$

This is not positive definite, as in the Schroedinger case. It is however so, if ψ is a stationary wave of positive frequency. In many cases $\hbar\mathrm{i}(\psi^* \, \partial\psi/\partial t - \partial\psi^*/\partial t\,\psi)/2mc^2$ may be interpreted as a "density" for one particle. Since it is a bilinear form of ψ^* and ψ and their first derivatives, many of the results obtained in the Introduction, § 5, can immediately be extended to relativistic wave functions.

According to eq. (5) we may write, in analogy with the non-relativistic case,

$$\boldsymbol{P} = \left(\Psi,\, \frac{\hbar}{\mathrm{i}} \nabla_x \Psi \right),$$

$$E = (\Psi,\, H\,\Psi)\,,$$

$$M = \left(\Psi, x \times \frac{\hbar}{\mathrm{i}} \nabla_x \Psi \right).$$

The operators $-\mathrm{i}\hbar \nabla_x$, H, $-\mathrm{i}\hbar\, x \times \nabla_x$, are self-adjoint according to the definition (6).

Notice that the energy

$$E = -\frac{\hbar^2}{2mc^2} \int \left(\psi^* \frac{\partial^2 \psi}{\partial t^2} - \frac{\partial \psi^*}{\partial t} \frac{\partial \psi}{\partial t} \right) \mathrm{d}^3\mathrm{x}$$

is positive definite for any ψ, irrespectively of the sign of the frequency of the plane wave components.

3. Position operator and Zitterbewegung

In coordinate space, the *position* is described by the operator x, which is merely the argument of the wave function. This is clearly self-adjoint, $(\Psi_1, x\,\Psi_2) = (x\,\Psi_1, \Psi_2)$. We shall now find the operator which, in k space, corresponds to x.

In the spirit of a one-particle theory, we confine ourselves to wave functions ψ with representation

$$\psi(x, \mathrm{t}) = (2\pi)^{-\frac{3}{2}} \int A(k)\, \mathrm{e}^{\mathrm{i}\,(k\cdot x - \omega t)}\, \mathrm{d}^3\mathrm{k}$$

in which only positive frequencies occur.

Now for a generic function $f(x)$ we have

$$(\Psi_1, f(x)\, \Psi_2) =$$

$$(\hbar/2mc^2)\,(2\pi)^{-3} \int A_1^*(k)\, A_2(k')\, \mathrm{e}^{\mathrm{i}\,(\omega-\omega')\,t} (\omega + \omega') f(x)\, \mathrm{e}^{\mathrm{i}\,(k'-k)\cdot x}\, \mathrm{d}^3\mathrm{x}\, \mathrm{d}^3\mathrm{k}\, \mathrm{d}^3\mathrm{k}' =$$

$$= (\hbar/2mc^2) \int A_1^*(k)\, A_2(k')\, \mathrm{e}^{\mathrm{i}\,(\omega-\omega')\,t}\, (\omega + \omega') f(-\mathrm{i}\, \nabla_{k'})\, \delta(k' - k)\, \mathrm{d}^3\mathrm{k}\, \mathrm{d}^3\mathrm{k}'.$$

By partial integration, and using the properties of the δ function, this may be written

$$(\Psi_1, f(x)\, \Psi_2) = (\hbar/2mc^2) \int \varphi_1^*(k, t) \{ f(\mathrm{i}\, \nabla_{k'}) [(\omega + \omega')\, \varphi_2(k', t)] \}_{k'=k}\, \mathrm{d}^3\mathrm{k}. \quad (7)$$

For $f(x) = 1$ this equation reduces to

$$(\Psi_1, \Psi_2) = (c\kappa)^{-1} \int \varphi_1^*\, \varphi_2\, \omega\, \mathrm{d}^3\mathrm{k} = (c\kappa)^{-1} \int A_1^*(k)\, A_2(k)\, \omega\, \mathrm{d}^3\mathrm{k},$$

showing that the product of two functions, defined according to the Feshbach-Villars metric, is independent of time, as can also be seen directly from the Klein-Gordon equation (see Chap. V, § 1). (This is true also for wave functions involving both positive and negative frequencies.)

Putting $f(\boldsymbol{x}) = \boldsymbol{x}$ and $\Psi_1 = \Psi_2$ in eq. (7), we obtain the mean value of $\boldsymbol{x}$ expressed as an integral in $\boldsymbol{k}$ space,

$$(\Psi, \boldsymbol{x}\,\Psi) = (c\,\kappa)^{-1} \int \varphi^* \,\omega\, \mathrm{i}\big(\nabla_k + \tfrac{1}{2}\,\boldsymbol{k}/(\boldsymbol{k}^2 + \kappa^2)\big)\, \varphi \,\mathrm{d}^3\mathrm{k}\,.$$

Thus, while in the non-relativistic theory the position is given, in $\boldsymbol{k}$ space, by the operator $\mathrm{i}\nabla_k$, in the relativistic theory it is given by

$$\mathrm{i}\left(\nabla_k + \frac{\boldsymbol{k}}{2(\boldsymbol{k}^2 + \kappa^2)}\right). \tag{8}$$

It is easy to verify that

$$\langle \boldsymbol{x} \rangle = \langle \nabla_k\, \omega \rangle\, t + \text{const}\,, \tag{9}$$

the mean value being evaluated according to the Feshbach-Villars metric, in agreement with the result obtained in the Introduction. Thus the mean group velocity $\langle \nabla_k\, \omega \rangle$ may be interpreted as the velocity of the particle described by the wave packet.

In the non-relativistic theory the eigenfunctions of $\boldsymbol{x}$ are δ functions,

$$\boldsymbol{x}\,\delta(\boldsymbol{x} - \boldsymbol{x}_0) = \boldsymbol{x}_0\,\delta(\boldsymbol{x} - \boldsymbol{x}_0)\,,$$

while the eigenfunctions of the operator $\mathrm{i}\nabla_k$ are the exponentials $\mathrm{e}^{-\mathrm{i}\boldsymbol{k}\cdot\boldsymbol{x}_0}$.

On the other hand, an eigenfunction of the operator $\mathrm{i}(\nabla_k + \frac{1}{2}\,\boldsymbol{k}/(\boldsymbol{k}^2 + \kappa^2))$ belonging to the eigenvalue $\boldsymbol{x}_0$ is

$$\varphi(\boldsymbol{k}) = \kappa^{\frac{1}{2}}\, \mathrm{e}^{-\mathrm{i}\boldsymbol{k}\cdot\boldsymbol{x}_0}\, (\boldsymbol{k}^2 + \kappa^2)^{-\frac{1}{4}}\,, \tag{10}$$

as may be easily verified. In the non-relativistic limit, i.e. $\kappa \to \infty$ and $\mathrm{i}(\nabla_k + \frac{1}{2}\,\boldsymbol{k}/(\boldsymbol{k}^2 + \kappa^2)) \to \mathrm{i}\nabla_k$, the function (10) reduces to $\mathrm{e}^{-\mathrm{i}\boldsymbol{k}\cdot\boldsymbol{x}_0}$.

Reverting now to coordinate space, the eigenfunction of the operator $\boldsymbol{x}$ corresponding to the eigenvalue $\boldsymbol{x}_0$ is

$$\psi(\boldsymbol{x}) = (2\pi)^{-\frac{3}{2}} \int \kappa^{\frac{1}{2}} (\boldsymbol{k}^2 + \kappa^2)^{-\frac{1}{4}}\, \mathrm{e}^{\mathrm{i}\boldsymbol{k}\cdot(\boldsymbol{x}-\boldsymbol{x}_0)}\, \mathrm{d}^3\mathrm{k}$$

or

$$\psi(\boldsymbol{x}) = -\,\pi\,\kappa^3\, \mathrm{e}^{\frac{1}{8}\mathrm{i}\pi}\, (2/\rho)^{\frac{5}{4}}\, H^{(1)}_{\frac{5}{4}}\,(\mathrm{i}\rho)/2^{\frac{3}{2}}\, \Gamma(\tfrac{1}{4})\,. \tag{11}$$

Here $\rho = r\,\kappa = |\boldsymbol{x} - \boldsymbol{x}_0|\,\kappa$, and Γ and H are the gamma and a Hankel function, respectively.

NOTE 2. To deduce eq. (11) we put $r = |\boldsymbol{x} - \boldsymbol{x}_0|$ and write the integral in polar coordinates $|\boldsymbol{k}| = u$, θ and ϕ, taking the direction of the vector $\boldsymbol{x} - \boldsymbol{x}_0$ as the axis $\theta = 0$.
Thus

$$\psi(\boldsymbol{x}) = (\kappa/2\pi)^{\frac{1}{2}} \int_0^\infty (u^2 + \kappa^2)^{-\frac{1}{4}}\, u^2\, \mathrm{d}u \int_{-1}^{1} \mathrm{e}^{\mathrm{i}ur\cos\theta}\, \mathrm{d}\cos\theta\,.$$

Integrating with respect to $\cos\theta$ and putting $\rho = r\kappa$, $u = x\kappa$ we obtain

$$\psi(\boldsymbol{x}) = (2\pi)^{-\frac{1}{2}} (\kappa^3/\mathrm{i}\rho) \int_0^\infty (x^2+1)^{-\frac{1}{4}} (\mathrm{e}^{\mathrm{i}x\rho} - \mathrm{e}^{-\mathrm{i}x\rho})\, x\, \mathrm{d}x$$

$$= -(2\pi)^{-\frac{1}{2}} (\kappa^3/\rho)\, \partial/\partial\rho \int_0^\infty (x^2+1)^{-\frac{1}{4}} (\mathrm{e}^{\mathrm{i}x\rho} + \mathrm{e}^{-\mathrm{i}x\rho})\, \mathrm{d}x$$

$$= -\sqrt{2}\,\pi^{-\frac{1}{2}} (\kappa^3/\rho)\, \partial/\partial\rho \int_0^\infty \cos(x\rho)\, (x^2+1)^{-\frac{1}{4}}\, \mathrm{d}x\,.$$

We now recall the integral representation of the Basset function of order ν

$$K_\nu = (2^\nu\, \Gamma(\nu+\tfrac{1}{2})/\rho^\nu\, \Gamma(\tfrac{1}{2})) \int_0^\infty \cos(\rho x)(x^2+1)^{-(\nu+\frac{1}{2})}\, \mathrm{d}x\,,$$

(see G. N. Watson, *Theory of Bessel Functions* (Cambridge University Press, 1952), p. 78 and 185). Here $\Gamma(z)$ is the gamma function, whose integral representation is

$$\Gamma(z) = \int_0^\infty \mathrm{e}^{-t}\, t^{z-1}\, \mathrm{d}t\,, \quad \text{for } \operatorname{Re} z > 0\,,$$

(see H. Margenau and G. M. Murphy, *The Mathematics of Physics and Chemistry* (Van Nostrand, New York, 1956), p. 95).

The K_ν function may also be given in terms of a Hankel function H_ν

$$K_\nu(\rho) = \tfrac{1}{2}\,\mathrm{i}\pi\, \mathrm{e}^{\frac{1}{2}\mathrm{i}\pi\nu}\, H_\nu^{(1)}(\mathrm{i}\rho) = \tfrac{1}{2}\,\mathrm{i}\pi\, \mathrm{e}^{-\frac{1}{2}\mathrm{i}\pi\nu}\, H_{-\nu}^{(1)}(\mathrm{i}\rho)\,, \qquad (\alpha)$$

(see Watson, *loc. cit.*, p. 78, eq. (8)).

Using the recurrence formula for the K_ν functions

$$\left(\frac{\mathrm{d}}{\rho\mathrm{d}\rho}\right)^m [\rho^n\, K_n(\rho)] = (-1)^m\, \rho^{n-m}\, K_{n-m}(\rho)\,,$$

(see Watson, *loc. cit.*, p. 79, eq. (5)), we have

$$\frac{\partial}{\partial\rho} \int_0^\infty \cos(\rho x)(x^2+1)^{-\frac{1}{4}}\, \mathrm{d}x = [2^{\frac{1}{4}}\sqrt{\pi}/\Gamma(\tfrac{1}{4})]\, \frac{\mathrm{d}}{\mathrm{d}\rho} (\rho^{-\frac{1}{4}}\, K_{-\frac{1}{4}}(\rho)) =$$

$$= -\sqrt{\pi}\,(2/\rho)^{\frac{1}{4}}\, K_{-\frac{5}{4}}(\rho)/\Gamma(\tfrac{1}{4})\,.$$

Thus by eq. (α) we obtain eq. (11).

Notice that the function (11) tends to zero as $r^{-\frac{7}{4}}\,e^{-\kappa r}$ when $r \to \infty$, and has a singularity of order $r^{-\frac{5}{2}}$ for $r \to 0$.

The physical meaning of the result obtained * is connected with the fact that, in relativistic wave mechanics, the position of a particle of spin zero (and, as we shall see later, also of a particle of spin $\frac{1}{2}$) cannot be *localized* with an accuracy greater than the Compton wavelength $\lambda\!\!\!^{-}_{c} = \kappa^{-1}$, if only positive frequencies are used to form the wave packet describing the particle.

In the Schroedinger non-relativistic wave mechanics, the localization of a particle in a certain region V at a particular time, e.g. $t = 0$, is expressed by making $\psi(\boldsymbol{x}, 0)$ vanish outside V. Then the density of physical quantities such as the energy $(-\,\psi^*\,\hbar^2\,\Delta\,\psi/2m)$, the momentum $(-\,i\hbar\,\psi^*\,\nabla_x\,\psi)$ and the probability of presence of the particle $(\psi^*\,\psi)$, involving only ψ and its space derivatives, are all zero outside V. Thus localization in an arbitrarily small region is possible.

In the case of the Klein-Gordon equation, if, for $t = 0$, one assumes that

$$\psi(\boldsymbol{x}, 0) = (2\pi)^{-\frac{3}{2}}\left[\int A(\boldsymbol{k})\,e^{i(\boldsymbol{k}\cdot\boldsymbol{x}-\omega t)}\,d^3k\right]_{t=0}$$

is zero outside a certain region V, there is no guarantee that the densities of physical quantities such as the energy etc. vanish outside V. In fact, in contrast to the Schroedinger case, such densities involve $(\partial\psi/\partial t)_{t=0}$ which, in general, is not zero outside V since its Fourier transform is $-\,i\omega\,A(\boldsymbol{k})$. Thus the time derivative may be a function with properties totally different from those of $\psi(\boldsymbol{x}, 0)$, whose Fourier transform is $A(\boldsymbol{k})$.

Only if the linear dimensions of the region V are all much greater than $\lambda\!\!\!^{-}_{c}$, $\psi(\boldsymbol{x}, 0)$ and $(\partial\psi/\partial t)_{t=0}$ may both be zero outside V. In this case, in fact, the wave packet may be constructed as a superposition of plane waves with wavevectors satisfying the condition $|\boldsymbol{k}| \ll \kappa$. Thus $\omega = c(\boldsymbol{k}^2 + \kappa^2)^{\frac{1}{2}}$ is practically independent of $\boldsymbol{k}$ and the Fourier transform of $\partial\psi/\partial t$, evaluated at $t = 0$, differs from that of $\psi(\boldsymbol{x})$, approximately, by a constant $(-\,i\omega \simeq -\,ic\,\kappa)$. From this it follows that the structure of the two functions, at $t = 0$, is essentially the same.

To construct a wave packet with ψ and $\partial\psi/\partial t$ both equal to zero outside a region of linear dimensions of the order of $\lambda\!\!\!^{-}_{c}$, negative frequencies must be used.

In fact, in this case

$$\psi(\boldsymbol{x}, t) = (2\pi)^{-\frac{3}{2}}\int \left(A(\boldsymbol{k})\,e^{i(\boldsymbol{k}\cdot\boldsymbol{x}-\omega t)} + B(\boldsymbol{k})\,e^{i(\boldsymbol{k}\cdot\boldsymbol{x}+\omega t)}\right)d^3k\,, \qquad (12)$$

* See T. D. Newton and E. P. Wigner, Revs. Mod. Phys. **21** (1949), 400.

where $\omega = c\,|(k^2+\kappa^2)^{\frac{1}{2}}|$. Obviously the Fourier transform at $t = 0$ is $A(\boldsymbol{k}) + B(\boldsymbol{k})$, while that of $(\partial\psi/\partial t)_{t=0}$ is $-\mathrm{i}\omega(A - B)$. Having now *two* functions of $\boldsymbol{k}$ at our disposal, we can contrive to make both ψ and $\partial\psi/\partial t$ vanish outside V.

The introduction of negative frequencies, however, implies that one no longer considers a one-particle scheme, but has rather stepped into the domain of quantum field theory, where negative frequencies are associated with antiparticles.

We now want to show that, if ψ involves both positive and negative frequencies, the mean value of the position, defined by eq. (5) (i.e. taking as "weight" the quantity $(\hbar\mathrm{i}/2mc^2)(\psi^*\,\partial\psi/\partial t - \partial\psi^*/\partial t\;\psi)$) is not a linear function of time, in contrast with eq. (9).

The time derivative of $\langle \boldsymbol{x}\rangle$ involves oscillating functions with frequencies $2\omega \geqslant 2c\kappa$ (*Zitterbewegung*). Indeed one has

$$\frac{\mathrm{d}\,\langle \boldsymbol{x}\rangle}{\mathrm{d}t} = \frac{\hbar\mathrm{i}}{2mc^2}\int\left(\psi^*\,\frac{\partial^2\psi}{\partial t^2} - \frac{\partial^2\psi^*}{\partial t^2}\,\psi\right)\boldsymbol{x}\,\mathrm{d}^3\mathrm{x}\,. \tag{13}$$

This may be put in the form *

$$\frac{\mathrm{d}\,\langle \boldsymbol{x}\rangle}{\mathrm{d}t} = \frac{\hbar}{2m\mathrm{i}}\int[\psi^*\,\nabla_x\,\psi - (\nabla_x\,\psi^*)\,\psi]\,\mathrm{d}^3\mathrm{x}\,. \tag{14}$$

Substituting in this equation ψ as given by the expansion (12), one has

$$\frac{\mathrm{d}\,\langle \boldsymbol{x}\rangle}{\mathrm{d}t} = \frac{\hbar}{m}\left[\int(|A|^2+|B|^2)\,\boldsymbol{k}\,\mathrm{d}^3\mathrm{k} + \int(A^*B\,\mathrm{e}^{2\mathrm{i}\omega t} + AB^*\,\mathrm{e}^{-2\mathrm{i}\omega t})\,\boldsymbol{k}\,\mathrm{d}^3\mathrm{k}\right]. \tag{15}$$

The first term on the right-hand side, written in the form

$$(\hbar/m)\int(|A|^2+|B|^2)\,\boldsymbol{k}\,\mathrm{d}^3\mathrm{k} = (\hbar/mc^2)\int(|A|^2+|B|^2)\,\omega\,\nabla_k\,\omega\,\mathrm{d}^3\mathrm{k}$$

may be recognized as the mean value of the group velocity. It is constant.

---

* Using the Klein-Gordon equation, we can express $\partial^2\psi/\partial t^2$ and $\partial^2\psi^*/\partial t^2$ in terms of $\Delta\psi$, $\Delta\psi^*$, ψ and ψ^*. On substituting in eq. (13) we obtain

$$\int[\psi^*(\partial^2\psi/\partial t^2) - (\partial^2\psi^*/\partial t^2)\,\psi]\,\boldsymbol{x}\mathrm{dx}^3 = c^2\int[\psi^*\Delta\psi - (\Delta\psi^*)\,\psi]\,\boldsymbol{x}\mathrm{d}^3\mathrm{x}\,.$$

After partial integration this becomes

$$-\,c^2\int[\psi^*\nabla_x\psi + \boldsymbol{x}(\nabla_x\psi^*\cdot\nabla_x\psi) - (\nabla_x\psi^*)\,\psi - \boldsymbol{x}(\nabla_x\psi^*\cdot\nabla_x\psi)]\,\mathrm{d}^3\mathrm{x}$$

leading to eq. (14).

The oscillating terms are missing only when the plane wave expansion of $\psi(x)$ involves solely frequencies of one sign, e.g. positive frequencies ($B(\boldsymbol{k}) = B^*(\boldsymbol{k}) = 0$), in which case the motion of the centre of the wave packet is uniform.

4. Non-relativistic limit

It is interesting to discuss the non-relativistic limit of the Feshbach-Villars metric.

Clearly if

$$(E_{\text{kin}}/mc^2) \ll 1 \tag{16}$$

one of the two components of Ψ is negligible as compared with the other. Therefore Ψ reduces essentially to a one-component function like the wave function of the non-relativistic theory. In fact

$$u = 2^{-\frac{1}{2}}\left(\psi - \frac{\hbar}{mc}\frac{\partial\psi}{\partial x_4}\right) = 2^{-\frac{1}{2}}(2\pi)^{-\frac{3}{2}}\int A(\boldsymbol{k})\, e^{i(k,x)}\,(2 + E_{\text{kin}}/mc^2)\, d^3k\,,$$

$$v = 2^{-\frac{1}{2}}\left(\psi + \frac{\hbar}{mc}\frac{\partial\psi}{\partial x_4}\right) = 2^{-\frac{1}{2}}(2\pi)^{-\frac{3}{2}}\int A(\boldsymbol{k})\, e^{i(k,x)}\,(-E_{\text{kin}}/mc^2)\, d^3k\,,$$

that is

$$v^*v \ll u^*u\,.$$

Thus, in the limit (16), for an arbitrary differential operator A one has

$$(\Psi, A\Psi) = \tfrac{1}{2}\int (u^*Au - v^*Av)\, d^3x \simeq \tfrac{1}{2}\int u^*Au\, d^3x\,,$$

and the mean value of such an operator is

$$\langle A\rangle \simeq \frac{\int u^*Au\, d^3x}{\int u^*u\, d^3x}\,. \tag{17}$$

Now if we confine ourselves to stationary solutions * we have

$$(\Psi, \Psi) \simeq \tfrac{1}{2}\int u^*u\, d^3x \simeq (1 + E_{\text{kin}}/mc^2)\int \psi^*\psi\, d^3x$$

and, for an arbitrary differential operator,

$$(\Psi, A\Psi) \simeq (1 + E_{\text{kin}}/mc^2)\int \psi^*A\psi\, d^3x\,.$$

Thus

$$\langle A\rangle = \int \psi^* A\psi\, d^3x / \int \psi^*\psi\, d^3x\,. \tag{18}$$

On the other hand, in the limit (16), in which eqs. (17) and (18) have been

* The extension to non-stationary wave functions is immediate.

derived, the Klein-Gordon equation reduces to the Schroedinger equation, and therefore a solution of the former becomes also a solution of the latter.

This is easily verified:

$$\begin{aligned} c^{-2}\,\partial^2\psi/\partial t^2 &= -\,(2\pi)^{-\frac{3}{2}} \int A(k)\,\mathrm{e}^{\mathrm{i}\,(k,x)}\,(m^2c^4 + 2E_{\mathrm{kin}}\,mc^2 + E_{\mathrm{kin}}^2)\,\mathrm{d}^3k/\hbar^2c^2 \\ &\simeq -\,(2\pi)^{-\frac{3}{2}} \int A(k)\,\mathrm{e}^{\mathrm{i}\,(k,x)}\,mc^2\,(2E - mc^2)\,\mathrm{d}^3k/\hbar^2c^2 \\ &= \left[\kappa^2 - (2\mathrm{i}m/\hbar)\,\partial/\partial t\right]\psi\,, \end{aligned}$$

whence

$$0 = \left(\Delta - \frac{1}{c^2}\frac{\partial^2}{\partial t^2} - \kappa^2\right)\psi \simeq \left[\Delta - 2\,\kappa^2 + \left(\frac{2m\mathrm{i}}{\hbar}\right)\frac{\partial}{\partial t}\right]\psi\,.$$

(The right side gives the Schroedinger equation with the term for the rest energy.)

5. Diagonalization of the Feshbach-Villars Hamiltonian

The second (first) component of a wave function Ψ of positive (negative) frequency tends to zero as $\boldsymbol{p} \to 0$. Then Ψ becomes essentially a one-component wave function.

It is possible to introduce a representation in which this feature is realized at all energies, so that the Feshbach-Villars wave function has only one non-vanishing component. This is achieved by a transformation * giving a new Hamiltonian which does not involve the *odd* matrices τ_1 and τ_2 ** which mix the two components of the wave function.

In fact, the wave functions

$$\Psi' = U\,\Psi\,,$$

with ***

$$U = \mathrm{e}^{\mathrm{i}S}\,,$$

$$S = \tfrac{1}{2}\,\tau_1\,\mathrm{tg}^{-1}\left(\frac{-\,\mathrm{i}\boldsymbol{p}^2}{\boldsymbol{p}^2 + 2m^2c^2}\right) = \frac{\tau_1}{4\mathrm{i}}\log\left(\frac{\boldsymbol{p}^2 + m^2c^2}{m^2c^2}\right) = S^+\,,$$

obey the equation

$$H'\,\Psi' = \mathrm{i}\hbar\,\frac{\partial \Psi'}{\partial t}\,,$$

* Analogous to the Foldy-Wouthuysen transformation for particles of spin $\frac{1}{2}$ (see Part II, Chap. V, § 6), see K.M. Case, Phys. Rev. **95** (1954), 1323.

** τ_3 and the unit matrix, which do not mix the components of the wave function, are *even* matrices. Note that τ_3 anticommutes with any odd matrix, and commutes with any even matrix.

*** Since $U^{-1} = U^+$, this is a unitary transformation according to the Feshbach-Villars metric.

where *

$$H' = UHU^{-1} = e^{2iS} H$$
$$= \frac{1}{2mc\sqrt{p^2 + m^2c^2}} (p^2 + 2m^2c^2 + p^2 \tau_1) H$$
$$= \tau_3 \, c \sqrt{p^2 + m^2c^2} \, .$$

Since the operator $c\sqrt{p^2 + m^2c^2}$ is positive, (positive/negative) frequency wave functions $\Psi'_\pm$ have only the (first/second) component different from zero.

We now separate the Klein-Gordon wave function ψ into its positive and negative frequency parts ψ_+ and ψ_-,

$$\psi(x) = \psi_+(x) + \psi_-(x) = \sum_i A_i \psi^i_+(\boldsymbol{x}) \, e^{-i\omega^i_+ t} + \sum_i B_i \psi^i_-(\boldsymbol{x}) \, e^{-i\omega^i_- t} , \tag{19}$$

with $\omega^i_+ > 0$, $\omega^i_- < 0$.

A simple calculation then shows that $\Psi' = U\Psi$ has the form

$$\Psi' = \begin{pmatrix} \varphi_+(x) \\ \varphi^*_-(x) \end{pmatrix} ,$$

with

$$\varphi_+ = \frac{1}{mc} [2mc\sqrt{p^2 + m^2c^2}]^{\frac{1}{2}} \psi_+ , \quad \varphi^*_- = \frac{1}{mc} [2mc\sqrt{p^2 + m^2c^2}]^{\frac{1}{2}} \psi_- .$$

Since

$$c\sqrt{p^2 + m^2c^2} \, \psi_\pm = \pm \, i\hbar \frac{\partial}{\partial t} \psi_\pm$$

one has

$$c\sqrt{p^2 + m^2c^2} \, \varphi_\pm = i\hbar \frac{\partial}{\partial t} \varphi_\pm \, .$$

Thus, for steady wave functions,

$$c\sqrt{p^2 + m^2c^2} \, \varphi^i_\pm(\boldsymbol{x}) = \pm \, \omega^i_\pm \, \varphi^i_\pm(\boldsymbol{x}) ,$$

where

* Since the Hamiltonian anticommutes with τ_1 ($\tau_3\tau_1 = -\tau_1\tau_3$, $\tau_2\tau_1 = -\tau_1\tau_2$), one has

$$HU^{-1} = UH .$$

On the other hand, putting $2iS = i\tau_1\alpha$, one has

$$e^{i\tau_1\alpha} = \cos\alpha + i\tau_1 \sin\alpha$$

as can be verified by expanding the two members in power series of α and remembering that $\tau_1{}^{2n} = 1$ and $\tau_1{}^{2n+1} = \tau_1$.

$$\varphi_+^i(x) = \frac{1}{mc}\,[2mc\,\sqrt{p^2+m^2c^2}]^{\frac{1}{2}}\,\psi_+^i(x)\,,$$

$$\varphi_-^i(x) = \frac{1}{mc}\,[2mc\,\sqrt{p^2+m^2c^2}]^{\frac{1}{2}}\,(\psi_-^i(x))^*\,.$$

Note that $\varphi_\pm^i(x)$ are eigenfunctions of the operator $c\sqrt{p^2+m^2c^2}$ belonging to *positive* eigenvalues $\pm\,\omega_\pm^i$.

The charge and the energy of a wave packet with the wave function (19) are

$$Q = (\Psi, \Psi) = \tfrac{1}{2}\int \Psi^\dagger\,\tau_3\,\Psi\,\mathrm{d}^3\mathrm{x} = \tfrac{1}{2}\int \Psi'^\dagger (U^{-1})^\dagger\,\tau_3\,U^{-1}\,\Psi'\,\mathrm{d}^3\mathrm{x} =$$
$$= \tfrac{1}{2}\int \Psi'^\dagger\,\tau_3\,\Psi'\,\mathrm{d}^3\mathrm{x} = \tfrac{1}{2}\int (\varphi_+^*\,\varphi_+ - \varphi_-^*\,\varphi_-)\,\mathrm{d}^3\mathrm{x}\,,$$

$$E = (\Psi, H\Psi) = \tfrac{1}{2}\int \Psi^\dagger\,\tau_3\,H\Psi\,\mathrm{d}^3\mathrm{x} = \tfrac{1}{2}\int \Psi'^\dagger\,\tau_3\,H'\,\Psi'\,\mathrm{d}^3\mathrm{x}$$
$$= \tfrac{1}{2}\int (\varphi_+^*\,c\sqrt{p^2+m^2c^2}\,\varphi_+ + \varphi_-^*\,c\sqrt{p^2+m^2c^2}\,\varphi_-)\,\mathrm{d}^3\mathrm{x}\,.$$

Assuming for simplicity the frequency spectrum $\omega_\pm^i$ to be discrete * and the wave functions $\varphi_\pm^i(x)$ to obey the orthonormality relations **

$$\int (\varphi_\pm^j(x))^*\,\varphi_\pm^i(x)\,\mathrm{d}^3\mathrm{x} = \delta_{ji}\,,$$

we have

$$Q = \tfrac{1}{2}\sum_i (|A_i|^2 - |B_i|^2)\,,$$

$$E = \tfrac{1}{2}\sum_i (\omega_+^i|A_i|^2 - \omega_-^i|B_i|^2) = \tfrac{1}{2}\sum_i (|\omega_+^i|\;|A_i|^2 + |\omega_-^i|\;|B_i|^2)\,.$$

The position and velocity operators, which, in the old representation, are x and $v = \dot{x}$, are given in the new representation by UxU^{-1} and $U\dot{x}U^{-1}$. These, when worked out, are seen to be fairly complicated expressions.

Conversely, one may define the operators *mean position* $x_{\mathrm{av.}}$ and *mean velocity* $v_{\mathrm{av.}}$, given, in the new representation, by

$$(x_{\mathrm{av.}})' = x\,, \quad (v_{\mathrm{av.}})' = \frac{\mathrm{i}}{\hbar}\,[H', x] = \tau_3\,\frac{cp}{\sqrt{p^2+m^2c^2}}\,,$$

and, in the old representation, by $U^{-1}\,x\,U$ and

* The particle is enclosed in a box of finite size.

** If $\omega_+^j \neq \omega_+^i$, φ_+^j and φ_+^i are orthogonal (Schroedinger metric!), being eigenfunctions of the self-adjoint operator $c\sqrt{p^2+m^2c^2}$ belonging to different eigenvalues. In the exceptional case when $\omega_+^i = \omega_+^j$ though $j \neq i$, we assume that φ_+^j and φ_-^i are eigenfunctions of some other self-adjoint operator (e.g. the momentum) belonging to different eigenvalues. Similar considerations can be made for φ_-^j and φ_-^i.

$$U^{-1} \frac{c\, \boldsymbol{p}\tau_3}{\sqrt{\boldsymbol{p}^2 + m^2c^2}}\, U\,.$$

Expanding $U^{-1}\, \boldsymbol{x}\, U$ in power series of $\boldsymbol{p}/mc$, we have

$$\boldsymbol{x}_{\text{av.}} = \boldsymbol{x} + \frac{\mathrm{i}\tau_1}{2\kappa} \frac{\boldsymbol{p}}{mc} + \dots$$

in the ordinary representation.
In contrast to the ordinary position and velocity operators, $\boldsymbol{x}_{\text{av.}}$ and $\boldsymbol{v}_{\text{av.}}$ are closely related to the position and velocity of a particle in classical mechanics.

In fact, for an arbitrary wave function one has

$$\langle \boldsymbol{v}_{\text{av.}} \rangle = \tfrac{1}{2} \int \Psi^\dagger\, \tau_3\, \boldsymbol{v}_{\text{av.}}\, \Psi\, \mathrm{d}^3\mathrm{x} = \tfrac{1}{2} \int \Psi^\dagger\, \tau_3\, U^{-1}\, \tau_3 \frac{c\boldsymbol{p}}{\sqrt{\boldsymbol{p}^2 + m^2c^2}}\, U\Psi\, \mathrm{d}^3\mathrm{x}$$

$$= \tfrac{1}{2} \int \Psi^\dagger\, \mathrm{e}^{2\mathrm{i}S} \frac{c\boldsymbol{p}}{\sqrt{\boldsymbol{p}^2 + m^2c^2}}\, \Psi\, \mathrm{d}^3\mathrm{x}$$

$$= \tfrac{1}{2} \int \left(\psi^* \frac{\boldsymbol{p}}{m} \psi + \frac{\partial \psi^*}{\partial t} \frac{\hbar \boldsymbol{p}}{c\kappa(\boldsymbol{p}^2 + m^2c^2)} \frac{\partial \psi}{\partial t} \right) \mathrm{d}^3\mathrm{x}\,.$$

Inserting in this expression the representation

$$\psi = (2\pi)^{-\frac{3}{2}} \int \left[A(\boldsymbol{k})\, \mathrm{e}^{\mathrm{i}\,(\mathrm{k},\mathrm{x})} + B(-\,\boldsymbol{k})\, \mathrm{e}^{-\mathrm{i}\,(\mathrm{k},\mathrm{x})} \right] \mathrm{d}^3\mathrm{k}\,,$$

we find

$$\langle \boldsymbol{v}_{\text{av.}} \rangle = \int \frac{c\boldsymbol{k}}{\kappa} \left(|A(\boldsymbol{k})|^2 - |B(-\,\boldsymbol{k})|^2 \right) \mathrm{d}^3\mathrm{k}\,.$$

Thus, unlike $\langle \boldsymbol{v} \rangle$ (see eq. (15)), $\langle \boldsymbol{v}_{\text{av.}} \rangle$ does not exhibit any Zitterbewegung. Note that $B(-\,\boldsymbol{k})$ corresponds to momentum $-\,\hbar\boldsymbol{k}$. Therefore the mean velocity of a plane wave component is parallel to the momentum for both positive and negative frequency.

CHAPTER IV

TRANSLATION AND ROTATION OPERATORS

1. Commutation relations

The expression

$$[a, b] = ab - ba$$

is called the *commutator* of the two quantities a and b.

Two operators are said to commute if their commutator vanishes. Two components of the momentum commute,

$$[p_m, p_n] = -\hbar^2(\partial^2/\partial x_m\,\partial x_n - \partial^2/\partial x_n\,\partial x_m) = 0\,.$$

Two coordinates clearly commute in coordinate space. One can verify that the same is true for the corresponding operators in momentum space,

$$[x_m, x_n] =$$
$$= -\hbar^2\left[\left(\partial/\partial p_m + \tfrac{1}{2}\,p_m/(\boldsymbol{p}^2 + m^2c^2)\right), \left(\partial/\partial p_n + \tfrac{1}{2}\,p_n/(\boldsymbol{p}^2 + m^2c^2)\right)\right] = 0\,.$$

We recall that, if the commutator of two operators a and b is a c-number or an operator not having the eigenvalue zero, no function exists which is a common eigenfunction of a with eigenvalue a' and of b with eigenvalue b'. For if it did exist, then

$$[a, b]f = (ab - ba)f = (b'a' - a'b')f = 0\,,$$

in contrast to the assumption that $[a, b]$ does not have the eigenvalue zero.

Now the commutators of the coordinate and momentum components are c-numbers in general different from zero,

$$[x_m, p_n] =$$
$$= i\hbar\left[\left(\partial/\partial p_m + \tfrac{1}{2}\,p_m/(\boldsymbol{p}^2 + m^2c^2)\right)p_n - p_n\left(\partial/\partial p_m + \tfrac{1}{2}\,p_m/(\boldsymbol{p}^2 + m^2c^2)\right)\right] =$$
$$= i\hbar\,\delta_{mn}\,.$$

Therefore, a common eigenfunction of momentum and position does not exist.

Using the above commutation relations, one may now derive those between the coordinates and the angular momentum components,

$$[L_i, x_j] = \varepsilon_{imn}[x_m p_n, x_j] = \varepsilon_{imn}\, x_m [p_n, x_j] = i\hbar\, \varepsilon_{ijm}\, x_m\,.$$

Similarly one has

$$[L_i, p_j] = i\hbar\, \varepsilon_{ijm}\, p_m\,.$$

The commutator of two components of the angular momentum is *

$$\begin{aligned}[L_i, L_j] &= \varepsilon_{jkl}[L_i, x_k]\, p_l + \varepsilon_{jkl}\, x_k [L_i, p_l] \\ &= i\hbar(\varepsilon_{jkl}\,\varepsilon_{ikm}\, x_m\, p_l - \varepsilon_{jkl}\,\varepsilon_{iml}\, x_k\, p_m) \\ &= i\hbar\, \varepsilon_{ijl}\, L_l\,. \end{aligned} \tag{1}$$

We now want to show that L_i, $(i = 1, 2, 3)$, commutes with $\boldsymbol{L}^2 = L_1^2 + L_2^2 + L_3^2$. In fact

$$[L_i, L_i^2 + L_k^2 + L_j^2] = i\hbar\, [\varepsilon_{ikm}(L_k L_m + L_m L_k) + \varepsilon_{ijn}(L_j L_n + L_n L_j)]$$

(the indices i, k, j are a permutation of 1, 2, 3, and no sum convention for i, k and j is used). The only terms different from zero are those for which $n = k$, $m = j$, and these cancel out. Therefore

$$[L_i, \boldsymbol{L}^2] = 0\,.$$

From the commutation relations (1) it follows that there are no common eigenfunctions of all the components of the angular momentum. The only exceptions are the functions of $r = |\boldsymbol{x}|$ alone, which are eigenfunctions of L_1, L_2 and L_3 belonging to the eigenvalue zero, as

$$(\boldsymbol{x} \times \nabla_x) f(r) = (\boldsymbol{x} \times \boldsymbol{x}/r)\, \mathrm{d}f/\mathrm{d}r = 0\,.$$

On the other hand one can construct common eigenfunctions of $\boldsymbol{L}^2$ and any component of the angular momentum. These are the *spherical harmonics*.

NOTE 1. We will now show that, from the commutation relation $[A, B] = C$ for the self-adjoint operators A and B, the Heisenberg uncertainty relation

$$(\Delta A)^2\, (\Delta B)^2 \geqslant -\tfrac{1}{4} \langle C \rangle^2$$

$\big((\Delta A)^2 = \langle (A - \langle A \rangle)^2 \rangle$, $(\Delta B)^2 = \langle (B - \langle B \rangle)^2 \rangle\big)$ follows at once.

Obviously, putting $a = A - \langle A \rangle$ and $b = B - \langle B \rangle$, we have

$$[a, b] = [A, B]\,.$$

a and b are also self-adjoint operators.

* Use the relation $\varepsilon_{jkl}\,\varepsilon_{ikm} = \delta_{ij}\,\delta_{lm} - \delta_{jm}\,\delta_{li}$.

Now, if the metric is positive definite, i.e. $(\Psi, \Psi) \geqslant 0$, the norm of the wave function $\Phi = (a - \mathrm{i}\lambda b)\,\Psi$ (λ = real number) is $\geqslant 0$. Thus, for an arbitrary Ψ, we have

$$(\Phi, \Phi) = \langle a^2 + \lambda^2 b^2 - \mathrm{i}\lambda(ab - ba)\rangle = \langle a^2\rangle + \lambda^2\langle b^2\rangle - \mathrm{i}\lambda\langle C\rangle \geqslant 0\,.$$

This expression, for

$$\lambda = \mathrm{i}\langle C\rangle/2\langle b^2\rangle\,,$$

attains its minimum value

$$\langle a^2\rangle + \langle C\rangle^2/4\langle b^2\rangle \geqslant 0\,.$$

Thus we have

$$\langle a^2\rangle\langle b^2\rangle \geqslant -\tfrac{1}{4}\langle C\rangle^2\,,$$

namely

$$(\Delta A)^2\,(\Delta B)^2 \geqslant -\tfrac{1}{4}\langle(AB - BA)\rangle^2\,.$$

For $A = x_i$, $B = p_i$ this gives

$$(\Delta x_i)^2(\Delta p_i)^2 \geqslant \tfrac{1}{4}\hbar^2\,.$$

2. Spherical harmonics

For a study of the spherical harmonics it is convenient to introduce polar coordinates (r, θ, ϕ), according to the transformation formulae

$$x_1 = r\sin\theta\cos\phi\,,\quad x_2 = r\sin\theta\sin\phi\,,\quad x_3 = r\cos\theta\,.$$

Putting

$$L_1 = \hbar l_1\,,\quad L_2 = \hbar l_2\,,\quad L_3 = \hbar l_3\,,$$

we have

$$l_3 = -\,\mathrm{i}(x_1\,\partial/\partial x_2 - x_2\,\partial/\partial x_1) = -\,\mathrm{i}\,\partial/\partial\phi\,,$$

$$l_1 \pm \mathrm{i}\,l_2 = \mathrm{e}^{\pm \mathrm{i}\phi}(\pm\,\partial/\partial\theta + \mathrm{i}\cot\theta\,\partial/\partial\phi)\,. \tag{2}$$

It is easy to verify that

$$(l_1 \pm \mathrm{i}l_2)\,(l_1 \mp \mathrm{i}l_2) = l_1^2 + l_2^2 \mp \mathrm{i}(l_1 l_2 - l_2 l_1) = l_1^2 + l_2^2 \pm l_3\,.$$

Thus one can write

$$\boldsymbol{l}^2 = (l_1 \pm \mathrm{i}l_2)\,(l_1 \mp \mathrm{i}l_2) + l_3^2 \mp l_3\,.$$

Using the above identities we find

$$\boldsymbol{l}^2 = -\left[\frac{1}{\sin^2\theta}\frac{\partial^2}{\partial\phi^2} + \frac{1}{\sin\theta}\frac{\partial}{\partial\theta}\left(\sin\theta\frac{\partial}{\partial\theta}\right)\right] = -\,\Lambda\,.$$

Here Λ denotes the angular part of the expression for the Laplace operator in polar coordinates

$$\Delta = \frac{\partial^2}{\partial r^2} + \frac{2}{r}\frac{\partial}{\partial r} + \frac{1}{r^2}\Lambda .$$

The spherical harmonics are the eigenfunctions of Λ, namely of $\boldsymbol{l}^2$, and of $l_3 = -\,\mathrm{i}\,\partial/\partial\phi$.

A spherical harmonic of order l (l integer $\geqslant 0$), denoted by Y_l, is defined * as an harmonic function U_l of degree l (i.e. a homogeneous polynomial of degree l in x_1, x_2 and x_3, satisfying the equation $\Delta U_l = 0$) divided by r^l,

$$Y_l = r^{-l}\, U_l .$$

It is easy to verify that Y_l is an eigenfunction of Λ belonging to the eigenvalue $\lambda = -\,l(l+1)$. In fact, writing Δ in polar coordinates and operating on U_l, we find

$$0 = \Delta U_l = \Delta(r^l Y_l) = l(l-1)\, r^{l-2}\, Y_l + 2l\, r^{l-2}\, Y_l + r^{l-2}\, \Lambda\, Y_l ,$$

namely

$$\Lambda\, Y_l = -\,l(l+1)\, Y_l .$$

Now, as a homogeneous polynomial of degree l, U_l may be written in the form

$$U_l = \sum c_{pq}(x_1 + \mathrm{i}x_2)^p\,(x_1 - \mathrm{i}x_2)^q\, x_3^{\,l-p-q} . \tag{3}$$

The equation $\Delta U_l = 0$ yields the recurrence formula for the coefficients **

$$c_{p+1,\,q+1} = -\,(l-p-q)\,(l-p-q-1)\, c_{pq}/4(p+1)\,(q+1) . \tag{4}$$

Obviously there are $2l+1$ arbitrary constants in the definition (3) of U_l, for instance the coefficients $c_{p-q,0}\,(-\,l < p-q < l)$. Therefore we put

$$U_l = \sum_{m=-l}^{l} U_l^{(m)} ,$$

where

$$U_l^{(m)} = \sum_{q=0}^{q_{\max}} c_{m+q,\,q}\,(x_1 + \mathrm{i}x_2)^{m+q}(x_1 - \mathrm{i}x_2)^q\, x_3^{\,l-m-2q} .$$

---

* See, for instance, B. L. van der Waerden, *Die Gruppentheoretische Methode in der Quantenmechanik* (Springer, Berlin, 1932), p. 12.

** This may be easily shown by making the change of variables $\xi = x_1 + \mathrm{i}x_2$, $\eta = x_1 - \mathrm{i}x_2$, $\zeta = x_3$.

Then

$$\Delta = 4\,\partial^2/\partial\xi\partial\eta + \partial^2/\partial\zeta^2 ,$$

and the equation $\Delta U_l = 0$ gives

$$0 = \Delta U_l = \Sigma\,\bigl(4pqc_{pq}\xi^{p-1}\eta^{q-1}\zeta^{l-p-q} + (l-p-q)(l-p-q-1)\, c_{pq}\zeta^{l-p-q-2}\xi^p\eta^q\bigr) .$$

Since the three variables ξ, η and ζ are independent, the coefficients of terms obtained by grouping all equal powers of ξ, η and ζ must be zero. The recurrence formula is thus obtained.

Here $m = p - q$ is an integer comprised between $-l$ and l, $q_{\max} = \frac{1}{2}(l - m)$ or $(l - m - 1)/2$ according as $l - m$ is an even or an odd number. Clearly the function $U_l^{(m)}$ is uniquely determined once $c_{m,0}$ is given, since the recurrence formula (4) determines all other coefficients.

On expressing $U_l^{(m)}$ in polar coordinates, one notices that its dependence on ϕ is of the type $e^{im\phi}$. In fact

$$U_l^{(m)} = r^l \, e^{im\phi} \, (\sin\theta)^m \sum_{q=0}^{q_{\max}} c_{m+q,\,q} \, (\cos\theta)^{l-m-2q} \, (1 - \cos^2\theta)^q \, .$$

Thus we can write

$$U_l^{(m)} = r^l \, e^{im\phi} \, (\sin\theta)^m \, \mathfrak{Y}_l^{(m)}(\theta) \, .$$

It is easy to verify that $\mathfrak{Y}_l^{(m)}(\theta)$ is essentially the m-th derivative of a Legendre polynomial of degree l in the argument $\cos\theta$. Thus $\mathfrak{Y}_l^{(m)}$ may be expressed in terms of the associated Legendre polynomials $P_l^{(m)}$.

The $2l + 1$ polynomials $U_l^{(m)}$ are independent solutions of the equation $\Delta U = 0$. Therefore, they may be chosen as a base for the representation of any harmonic polynomial of degree l. The function $r^{-l} \, U_l^{(m)}$ is a spherical harmonic $Y_l^{(m)}$.

This does not uniquely determine $Y_l^{(m)}$, since $U_l^{(m)}$ involves an arbitrary constant. We adopt the definition *

$$Y_l^{(m)}(\theta, \phi) =$$

$$\frac{(-1)^{l+m}}{(2l)!!} \left[\frac{(2l+1)(l-m)!}{4\pi(l+m)!} \right]^{\frac{1}{2}} (\sin\theta)^m \frac{d^{l+m}}{(d\cos\theta)^{l+m}} (\sin\theta)^{2l} \, e^{im\phi} \, , \quad (5)$$

($(2l)!! = 2^l l!$), for which

$$Y_l^{(m)*} = (-1)^m \, Y_l^{(-m)} \, .$$

The functions of order $m = 0$ depend solely on θ and are essentially Legendre polynomials,

$$Y_l^{(0)}(\theta) = [(2l+1)/4\pi]^{\frac{1}{2}} \, P_l(\cos\theta) \, .$$

Those of order $m \neq 0$ involve the associated Legendre polynomials

$$P_l^{(m)}(\cos\theta) = \frac{(-1)^{l+m}}{(2l)!!} (\sin\theta)^m \frac{d^{l+m}}{(d\cos\theta)^{l+m}} (\sin\theta)^{2l} \, .$$

NOTE 1. The functions (5) differ from those in the article by Bethe

* See J. M. Blatt and V. F. Weisskopf, *Theoretical Nuclear Physics* (Wiley, New York, 1952), p. 783.

and Salpeter (Enc. of Phys., Vol. **35** (Springer, Berlin, 1957), p. 430) by a factor $(-1)^m$, which has been introduced in the definition of $P_l^{(m)}(\cos\theta)$. This difference is not essential. On the other hand, while in some textbooks the spherical harmonics for $m < 0$ are defined by the formula

$$Y_l^{(m)} = (-1)^m \left[(2l+1)(l-m)!/4\pi(l+m)!\right]^{\frac{1}{2}} P_l^{(|m|)} e^{im\phi},$$

the definition (5) holds both for positive and negative m.

Obviously, the functions (5) are eigenfunctions of $\boldsymbol{l}^2$ and l_3,

$$\boldsymbol{l}^2\, Y_l^{(m)} = l(l+1)\, Y_l^{(m)}, \quad l_3\, Y_l^{(m)} = m\, Y_l^{(m)},$$

and have the following properties:

$$\begin{aligned} (L_1 + iL_2)\, Y_l^{(m_l)} &= \hbar \left[(l-m_l)(l+m_l+1)\right]^{\frac{1}{2}}\, Y_l^{(m_l+1)}, \\ (L_1 - iL_2)\, Y_l^{(m_l)} &= \hbar \left[(l+m_l)(l-m_l+1)\right]^{\frac{1}{2}}\, Y_l^{(m_l-1)}. \end{aligned} \tag{6}$$

These formulae are also meaningful for $m_l = \pm l$ if we put $Y_l^{(l+1)} = 0$, $Y_l^{(-l-1)} = 0$.

It can be shown that the spherical harmonics are orthonormal, that is

$$\int Y_l^{(m)*}\, Y_{l'}^{(m')}\, d\Omega = \delta_{ll'}\, \delta_{mm'}.$$

They also satisfy the completeness relation

$$\sum_{l,\,m} Y_l^{(m)*}(\theta, \phi)\, Y_l^{(m)}(\theta', \phi') = \delta(\theta - \theta', \phi - \phi'),$$

where the function $\delta(\theta - \theta', \phi - \phi')$ has the property

$$\int F(\theta', \phi')\, \delta(\theta - \theta', \phi - \phi')\, d\Omega' = F(\theta, \phi),$$

($F(\theta, \phi)$ = arbitrary continuous function).

In $\boldsymbol{k}$ space the eigenfunctions of $\boldsymbol{L}^2$ and L_3 are the functions $Y_l^{(m)}(\boldsymbol{k})$, namely the spherical functions whose arguments are the polar angles of $\boldsymbol{k}$.

3. Translation and rotation operators

It has been seen in Chap. I that the Schroedinger and Klein-Gordon wave functions are transformed, under translations and space rotations, according to

$$\psi'(x') = \psi(x).$$

Now we shall derive formulae expressing $\psi'(x)$ in terms of $\psi(x)$, that is the transformation formulae for the *form* of the wave function.

For both translations and space rotations, there exist operators T such that

$$\psi'(x') = T\,\psi(x') . \tag{7}$$

Since $\psi'(x') = \psi(x)$, this is equivalent to

$$\psi(x') = T^{-1}\,\psi(x) .$$

We shall first deal with infinitesimal translations and rotations

$$\boldsymbol{x}' = \boldsymbol{x} + \delta\boldsymbol{x} .$$

For an arbitrary function $\psi(\boldsymbol{x})$ we have

$$\psi(\boldsymbol{x}') = \psi(\boldsymbol{x} + \delta\boldsymbol{x}) \simeq (1 + \delta\boldsymbol{x} \cdot \nabla_x)\,\psi(\boldsymbol{x}) . \tag{8}$$

Then, in the case of an infinitesimal translation ($\delta\boldsymbol{x} = -\,\delta\boldsymbol{a} =$ constant vector), eq. (8) gives

$$\psi(\boldsymbol{x}') \simeq (1 - \mathrm{i}\,\delta\boldsymbol{a} \cdot \boldsymbol{p}/\hbar)\,\psi(\boldsymbol{x}) ,$$

and

$$\psi'(\boldsymbol{x}) \simeq (1 + \mathrm{i}\,\delta\boldsymbol{a} \cdot \boldsymbol{p}/\hbar)\,\psi(\boldsymbol{x}) .$$

Therefore, $\boldsymbol{p} = -\,\mathrm{i}\hbar\,\nabla_x$ may be interpreted as the *generator* of infinitesimal translations.

Similarly, if $\delta\boldsymbol{\phi}$ is the vector characterizing an infinitesimal counter-clockwise rotation of the coordinate axes through an angle $\delta\phi$ around its direction, one has

$$\delta\boldsymbol{x} = \boldsymbol{x} \times \delta\boldsymbol{\phi}$$

for the change undergone by the coordinates of an arbitrary point. Thus the function $\psi(\boldsymbol{x})$ is transformed according to

$$\psi'(\boldsymbol{x}) = (1 + \mathrm{i}\,\delta\boldsymbol{\phi} \cdot \boldsymbol{L}/\hbar)\,\psi(\boldsymbol{x}) ,$$

from which one sees that $\boldsymbol{L} = \boldsymbol{x} \times \boldsymbol{p} = -\,\mathrm{i}\hbar\boldsymbol{x} \times \nabla_x$ is the generator of infinitesimal space rotations.

We shall now determine the operators which generate finite translations and rotations.

Since

$$\nabla_a\,\psi'(\boldsymbol{x}) = \mathrm{i}\,\frac{\boldsymbol{p}}{\hbar}\,\psi(\boldsymbol{x}) \quad \text{(translations)} ,$$

$$\frac{\partial\psi'(\boldsymbol{x})}{\partial\phi} = \mathrm{i}\,\frac{\boldsymbol{n} \cdot \boldsymbol{L}}{\hbar}\,\psi(\boldsymbol{x}) \quad \text{(rotations)} ,$$

integration with respect to $\boldsymbol{a}$ and ϕ gives

$$\psi'(\boldsymbol{x}) = e^{i\boldsymbol{a}\cdot\boldsymbol{p}/\hbar}\,\psi(\boldsymbol{x})$$

and

$$\psi'(\boldsymbol{x}) = e^{i\phi\boldsymbol{n}\cdot\boldsymbol{L}/\hbar}\,\psi(\boldsymbol{x}),$$

respectively.

Therefore, the translation and rotation operators are, respectively,

$$T_a = e^{i\boldsymbol{a}\cdot\boldsymbol{p}/\hbar}$$

and

$$T_{n,\phi} = e^{i\phi\boldsymbol{n}\cdot\boldsymbol{L}/\hbar}. \tag{9}$$

Here $\boldsymbol{a}$ is the translation vector, $\boldsymbol{n}$ the unit vector giving the direction of the axis of rotation, and ϕ the angle of rotation.

The operators (9) are clearly unitary, i.e. $T^+ = T^{-1}$.

A plane wave $e^{i\boldsymbol{k}\cdot\boldsymbol{x}}$ is an eigenfunction of the translation operator belonging to the eigenvalue $e^{i\boldsymbol{a}\cdot\boldsymbol{k}}$,

$$T_a(e^{i\boldsymbol{k}\cdot\boldsymbol{x}}) = e^{i\boldsymbol{a}\cdot\boldsymbol{p}/\hbar}\,e^{i\boldsymbol{k}\cdot\boldsymbol{x}} = e^{i\boldsymbol{a}\cdot\boldsymbol{k}}\,e^{i\boldsymbol{k}\cdot\boldsymbol{x}}.$$

Similarly a spherical harmonic of order l, m, is an eigenfunction for rotations around the third axis ($T_{n,\phi}$ with $\boldsymbol{n} = (0, 0, 1)$) belonging to the eigenvalue $e^{im\phi}$.

Finally we consider the three quantities

$$(\Psi', x_i\,\Psi')/(\Psi', \Psi'), \quad (i = 1, 2, 3).$$

Here Ψ' denotes the two-component wave function defined in Chap. III, § 2, with $\psi(\boldsymbol{x}, t)$ replaced by $\psi'(\boldsymbol{x}, t) = T\psi(\boldsymbol{x}, t)$ *.

These three quantities are the coordinates, with respect to the dashed system, of the centre of the wave packet described by the wave function $\Psi(\boldsymbol{x}, t)$ in the original reference system.

In fact, using eq. (7) one has

$$(\Psi', \Psi') = (\Psi, \Psi)$$

and

$$(\Psi', \boldsymbol{x}\Psi') = (\Psi, T^{-1}\,\boldsymbol{x}T\Psi).$$

Now, for a translation, **

$$T_a^{-1}\,x_i\,T_a = e^{-i\boldsymbol{a}\cdot\boldsymbol{p}/\hbar}\,x_i\,e^{i\boldsymbol{a}\cdot\boldsymbol{p}/\hbar} = \left(x_i\,e^{-i\boldsymbol{a}\cdot\boldsymbol{p}/\hbar} + [e^{-i\boldsymbol{a}\cdot\boldsymbol{p}/\hbar}, x_i]\right)e^{i\boldsymbol{a}\cdot\boldsymbol{p}/\hbar} = x_i - a_i,$$

* The following remarks, of course, apply also to non-relativistic wave mechanics, in which case the three quantities above are

$$(\psi', x_i\psi')/(\psi', \psi') = \int \psi'^*(\boldsymbol{x}, t)\,x_i\psi'(\boldsymbol{x}, t)\,d^3x/\int \psi'^*(\boldsymbol{x}, t)\,\psi'(\boldsymbol{x}, t)\,d^3x.$$

** Use the relation

$$[G(p), q] = \left[G(p), i\hbar\frac{\partial}{\partial p}\right] = -\,i\hbar\,\partial G/\partial p.$$

and so

$$(\Psi', \boldsymbol{x}\,\Psi') = (\Psi, \boldsymbol{x}\,\Psi) - \boldsymbol{a}(\Psi, \Psi)\,.$$

Similarly, for a rotation,

$$(\Psi', x_i\,\Psi') = a_{ik}(\Psi, x_k\,\Psi)\,,$$

where a_{ik} are the coefficients of the transformation for the coordinates of a fixed point ($x_i' = a_{ik}\,x_k$).

Therefore, the coordinates $\langle x_i \rangle'$ of the centre of the wave packet with respect to the primed coordinate axes are obtained from those with respect to the original axes by the same linear transformation which must be used for the coordinates of a fixed point.

The "coordinates" of the spectral centre do not change under a translation, since

$$(\Psi', \boldsymbol{p}\,\Psi') = (\Psi, \mathrm{e}^{-\mathrm{i}\boldsymbol{a}\cdot\boldsymbol{p}/\hbar}\,\boldsymbol{p}\,\mathrm{e}^{\mathrm{i}\boldsymbol{a}\cdot\boldsymbol{p}/\hbar}\,\Psi) = (\Psi, \boldsymbol{p}\,\Psi)\,,$$

whereas, under a rotation,

$$(\Psi', p_i\,\Psi') = (\Psi, \mathrm{e}^{-\mathrm{i}\phi\boldsymbol{n}\cdot\boldsymbol{L}/\hbar}\,p_i\,\mathrm{e}^{\mathrm{i}\phi\boldsymbol{n}\cdot\boldsymbol{L}/\hbar}\,\Psi) = a_{ik}(\Psi, p_k\,\Psi)\,,$$

where a_{ik} are again the coefficients of the linear transformation for the coordinates of a fixed point.

In general, the matrix element $(\Psi_1', A\Psi_2')$ (A = operator involving $\boldsymbol{p}$ and $\boldsymbol{x}$), giving the numerical value of some quantity as measured with reference to the primed system, is equal to

$$(\Psi_1, T^{-1}\,AT\Psi_2)\,.$$

CHAPTER V

SPIN ZERO PARTICLE IN ELECTROMAGNETIC FIELD

1. Wave equation and charge conservation

In this section we shall consider a particle of spin zero and charge $-e$ ($e =$ absolute value of the electron charge), for example a π^- meson.

The wave equation may be deduced from the classical formula

$$(p_\mu + eA_\mu/c)(p_\mu + eA_\mu/c) + m^2c^2 = 0$$

by the usual replacement

$$p_\mu \to -\mathrm{i}\hbar\, \partial/\partial x_\mu\,.$$

This gives

$$\left[\left(\frac{\partial}{\partial x_\mu} + \frac{\mathrm{i}e}{\hbar c} A_\mu\right)\left(\frac{\partial}{\partial x_\mu} + \frac{\mathrm{i}e}{\hbar c} A_\mu\right) - \kappa^2\right] \psi = 0, \tag{1}$$

or

$$\left[\Box - \kappa^2 + \frac{\mathrm{i}e}{\hbar c}\left(2A_\mu \frac{\partial}{\partial x_\mu} + \frac{\partial A_\mu}{\partial x_\mu}\right) - (e/\hbar c)^2\, \mathbf{A}^2\right] \psi = 0\,. \tag{2}$$

From this one sees that, in general, the wave function must be complex. Its real and imaginary parts are coupled by terms linear in the electromagnetic four-potential:

$$[\Box - \kappa^2 - (e/\hbar c)^2\, \mathbf{A}^2]\, \mathrm{Re}\, \psi = (e/\hbar c)(2A_\mu\, \partial/\partial x_\mu + \partial A_\mu/\partial x_\mu)\, \mathrm{Im}\, \psi\,,$$

$$[\Box - \kappa^2 - (e/\hbar c)^2\, \mathbf{A}^2]\, \mathrm{Im}\, \psi = -(e/\hbar c)(2A_\mu\, \partial/\partial x_\mu + \partial A_\mu/\partial x_\mu)\, \mathrm{Re}\, \psi\,.$$

Hence the necessity of using complex wave functions to describe charged particles of spin zero (as mentioned in Chap. II, § 1).

Multiplying eq. (2) by ψ^*, and the corresponding equation for ψ^* by ψ, and subtracting, we obtain

$$\psi^* \Box \psi - \psi \Box \psi^* + (2\mathrm{i}e/\hbar c)(A_\mu\, \partial\psi^*\psi/\partial x_\mu + \psi^*\psi\, \partial A_\mu/\partial x_\mu) = 0\,,$$

or

$$\frac{\partial}{\partial x_\mu}\left(\psi^* \frac{\partial \psi}{\partial x_\mu} - \frac{\partial \psi^*}{\partial x_\mu} \psi + \frac{2\mathrm{i}e}{\hbar c} A_\mu \psi^* \psi\right) = 0\,.$$

This is the continuity equation

$$\frac{\partial j_\mu}{\partial x_\mu} = 0 \tag{3}$$

for the four-current

$$j_\mu = \frac{\hbar}{2m\mathrm{i}}\left(\psi^* \frac{\partial \psi}{\partial x_\mu} - \frac{\partial \psi^*}{\partial x_\mu}\psi\right) + \frac{e}{mc} A_\mu \psi^* \psi \,. \tag{4}$$

From eq. (3) the conservation of charge follows, namely

$$\int \rho \, \mathrm{d}^3\mathrm{x} = \text{const}\,, \tag{5}$$

where

$$\rho = \frac{j_4}{\mathrm{i}c} = \frac{\hbar \mathrm{i}}{2mc^2}\left(\psi^* \frac{\partial \psi}{\partial t} - \frac{\partial \psi^*}{\partial t}\psi\right) + \frac{e}{mc^2} A_0 \psi^* \psi \,. \tag{6}$$

2. Feshbach-Villars metric

The results (3) and (5) hold, also, for the "four-current" $j_\mu[\psi_1, \psi_2]$ and the "charge density" $\rho[\psi_1, \psi_2]$ obtained by replacing ψ by ψ_2 and ψ^* by ψ_1^* in eqs. (4) and (6) respectively.

Thus we have

$$\frac{\hbar \mathrm{i}}{2mc^2}\int\left(\psi_1^* \frac{\partial \psi_2}{\partial t} - \frac{\partial \psi_1^*}{\partial t}\psi_2 - \frac{2\mathrm{i}e}{\hbar} A_0 \psi_1^* \psi_2\right) \mathrm{d}^3\mathrm{x} = \text{const}\,. \tag{7}$$

For $A_0 = 0$ this reduces to

$$(\hbar \mathrm{i}/2mc^2) \int (\psi_1^* \, \partial\psi_2/\partial t - \partial\psi_1^*/\partial t \; \psi_2) \, \mathrm{d}^3\mathrm{x} = \text{const}\,,$$

in which the expression on the left hand side is the product (Ψ_1, Ψ_2) according to the Feshbach-Villars metric.

It seems natural, therefore, to attempt to define a suitable two-component wave function for the case when an electromagnetic field is present, and to require the identity of the product

$$(\Psi_1, \Psi_2) = \tfrac{1}{2} \int \Psi_1^\dagger \tau_3 \Psi_2 \, \mathrm{d}^3\mathrm{x}$$

with the left-hand side of eq. (7).

To do this we must take

$$\Psi = \begin{pmatrix} u \\ v \end{pmatrix},$$

where

$$u = \frac{1}{\sqrt{2}}\left[\psi - \frac{\hbar}{mc}\left(\frac{\partial}{\partial x_4} + \frac{ie}{\hbar c}A_4\right)\psi\right],$$

$$v = \frac{1}{\sqrt{2}}\left[\psi + \frac{\hbar}{mc}\left(\frac{\partial}{\partial x_4} + \frac{ie}{\hbar c}A_4\right)\psi\right].$$

Note that, whereas two stationary solutions of eq. (2) are not orthogonal in the sense of the Schroedinger metric, they are so if the Feshbach-Villars metric and the above definitions of u and v are adopted. In fact, for two stationary solutions, eq. (7), with $A_0 = A_0(\boldsymbol{x})$ (static), becomes

$$(2mc^2)^{-1}\,\mathrm{e}^{\mathrm{i}(E_1-E_2)t/\hbar}\int (E_1 + E_2 + 2eA_0)\,\psi_1^*(\boldsymbol{x})\,\psi_2(\boldsymbol{x})\,\mathrm{d}^3\mathrm{x} = \text{const}\,.$$

For $E_1 \neq E_2$ this relation can hold only if

$$\int (E_1 + E_2 + 2eA_0)\,\psi_1^*(\boldsymbol{x})\,\psi_2(\boldsymbol{x})\,\mathrm{d}^3\mathrm{x} = 0\,. \tag{8}$$

This condition is clearly different from the orthogonality relation for two functions according to Schroedinger,

$$\int \psi_1^*(\boldsymbol{x})\,\psi_2(\boldsymbol{x})\,\mathrm{d}^3\mathrm{x} = 0\,,$$

although it approximates to the latter for

$$|E_1 + E_2| \gg |eA_0|\,.$$

This is an additional condition, which, apart from $(E_{\mathrm{kin}}/mc^2)^2 \ll 1$, should be satisfied in order that, in the presence of an electromagnetic field, the Schroedinger metric may be considered as the non-relativistic limit of that of Feshbach and Villars.

Eq. (8), on the other hand, coincides with the vanishing of the product (Ψ_1, Ψ_2), when Ψ_1 and Ψ_2 are stationary wave functions $(E_1 \neq E_2)$,

$$(\Psi_1, \Psi_2) = (2mc^2)^{-1}\,\mathrm{e}^{\mathrm{i}(E_1-E_2)t/\hbar}\int (E_1 + E_2 + 2eA_0)\,\psi_1^*(\boldsymbol{x})\,\psi_2(\boldsymbol{x})\,\mathrm{d}^3\mathrm{x} = 0\,.$$

Following Feshbach and Villars, the Klein-Gordon equation for a particle subject to an electromagnetic field can be written in the form

$$\hbar\mathrm{i}\frac{\partial}{\partial t}\Psi = H\Psi\,,$$

where

$$H = -\frac{\hbar^2}{2m}(\tau_3 + \mathrm{i}\tau_2)\left(\nabla_x + \frac{ie}{\hbar c}\boldsymbol{A}\right)^2 + mc^2\tau_3 - eA_0\,.$$

This Hamiltonian obviously reduces to (3) of Chap. III, § 2, when the electromagnetic field is zero. Similarly, the current density can be written in the form

$$j = (\hbar/4m\mathrm{i}) \times$$
$$\times [\Psi^\dagger \tau_3(\tau_3 + \mathrm{i}\tau_2)\nabla_x \Psi - (\nabla_x \Psi^\dagger)\tau_3(\tau_3+\mathrm{i}\tau_2)\Psi] + (e/2mc) A\, \Psi^\dagger \tau_3(\tau_3 + \mathrm{i}\tau_2)\Psi .$$

The charge * is given by the norm of the two-component function Ψ,

$$Q = \int \rho \, \mathrm{d}^3 x = \tfrac{1}{2} \int \Psi^\dagger \, \tau_3 \, \Psi \, \mathrm{d}^3 x .$$

It is very important to note that, whereas in the absence of an electromagnetic field the charge density ρ is $\geqslant 0$ for a stationary wave of positive frequency, this is no longer true if $A_0 \neq 0$. In fact, for a stationary solution one has

$$\rho = [(E + eA_0)/mc^2] \, \psi^* \psi \, ,$$

which is not necessarily positive if $E > 0$ (i.e. for a wave of positive frequency). The charge density is certainly positive, and may be interpreted as a probability density, if apart from $E > 0$ also $eA_0 > 0$. This last condition is realized in the case of a π^- in the electrostatic field of a nucleus, since then

$$eA_0 = Ze^2/4\pi r > 0 \, .$$

On the other hand, when $(eA_0 + E) < 0$, the one-particle model is inadequate, and one enters the domain of validity of field theory.

3. Diagonalization of the Feshbach-Villars Hamiltonian for particles in a static electromagnetic field

The Hamiltonian

$$H = \frac{1}{2m} \left(p + \frac{e}{c} A \right)^2 (\tau_3 + \mathrm{i}\tau_2) + mc^2 \tau_3$$

($A = A(x)$, static vector potential) is transformed into

$$H' = UHU^{-1} = \tau_3 \, c \sqrt{\left(p + \frac{e}{c} A \right)^2 + m^2 c^2}$$

by the transformation

$$U = \mathrm{e}^{\mathrm{i}S} \, ,$$

with

* Henceforth we shall refer to ρ as the charge density, although the true charge density is $- e\rho$.

$$S = \tfrac{1}{2}\,\tau_1\,\mathrm{tg}^{-1}\left[\frac{-\mathrm{i}\left(\boldsymbol{p}+\frac{e}{c}\boldsymbol{A}\right)^2}{\left(\boldsymbol{p}+\frac{e}{c}\boldsymbol{A}\right)^2 + 2m^2c^2}\right].$$

On replacing $\boldsymbol{p}$ by $\boldsymbol{p} + (e/c)\,\boldsymbol{A}$ in the relations given on pages 68 and 69, one obtains the corresponding ones for the case of a particle of spin zero in a static magnetic field.

One may attempt * to diagonalize the Hamiltonian

$$H = \frac{1}{2m}\,\boldsymbol{p}^2(\tau_3 + \mathrm{i}\tau_2) + mc^2\tau_3 - eA_0(\boldsymbol{x})$$

for a particle in a weak electrostatic field with Fourier components smaller than $mc/\hbar$, by a sequence of transformations, each of which removes odd matrices to a higher order in $1/m$.

It is convenient to write the Hamiltonian in the form

$$H = mc^2\tau_3 + \mathfrak{E} + \mathfrak{O}\,,$$

with

$$\mathfrak{E} = -\,eA_0 + \frac{1}{2m}\,\boldsymbol{p}^2\,\tau_3\ \text{(even matrix)},\qquad \mathfrak{O} = \frac{\mathrm{i}}{2m}\,\boldsymbol{p}^2\,\tau_2\ \text{(odd matrix)}\,.$$

The transformation

$$U = \exp\left(\frac{1}{2mc^2}\,\tau_3\,\mathfrak{O}\right)$$

gives

$$H' = UHU^{-1} = mc^2\,\tau_3 + \mathfrak{E} + \frac{\tau_3}{2mc^2}\,(\mathfrak{O}^2 + [\mathfrak{O}, \mathfrak{E}]) + \ldots$$

$$= \tau_3\left[mc^2 + \frac{1}{2m}\,\boldsymbol{p}^2 - \frac{1}{8m^3c^2}\,(\boldsymbol{p}^2)^2 + \ldots\right] - eA_0$$

$$-\,\frac{\mathrm{i}\tau_2}{4m^3c^2}\,(\boldsymbol{p}^2)^2 - \frac{e\tau_1}{4m^2c^2}\,[\boldsymbol{p}^2, A_0] + \ldots,$$

free of odd matrices to the order $1/m^2$.

---

* In analogy with the method followed in Part II, Chap. VII, § 5 for particles of spin $\frac{1}{2}$.

A further transformation

$$U' = \exp\left(\frac{1}{(2mc^2)^2}\,[\mathfrak{O}, \mathfrak{E}]\right)$$

gives

$$H'' = U'H'U'^{-1} = mc^2\,\tau_3 + \mathfrak{E} + \frac{\tau_3}{2mc^2}\,\mathfrak{O}^2 - \frac{1}{4(mc^2)^2}\,[\mathfrak{E}, [\mathfrak{O}, \mathfrak{E}]] + \ldots$$

$$= \tau_3\left[mc^2 + \frac{1}{2m}\,\boldsymbol{p}^2 - \frac{1}{8m^3c^2}\,(\boldsymbol{p}^2)^2 + \ldots\right] - eA_0$$

$$- \frac{\mathrm{i}\,e^2\,\tau_2}{8m^3c^4}\,[A_0, [\boldsymbol{p}^2, A_0]] + \ldots,$$

free of odd matrices to the order $1/m^3$.

By means of another transformation

$$U'' = \exp\left[-\frac{1}{2(mc^2)^3}\,\tau_3\left(\tfrac{1}{3}\,\mathfrak{O}^2 + \tfrac{1}{4}\,[\mathfrak{E}, [\mathfrak{O}, \mathfrak{E}]]\right)\right],$$

the Hamiltonian becomes *

$$H''' = U''H''U''^{-1} = mc^2\tau_3 + \mathfrak{E} + \frac{\tau_3}{2mc^2}\,\mathfrak{O}^2 + \ldots$$

$$= \tau_3\left[mc^2 + \frac{1}{2m}\,\boldsymbol{p}^2 - \frac{1}{8m^3c^2}\,(\boldsymbol{p}^2)^2 + \ldots\right] - eA_0 + \ldots,$$

free of odd matrices to the order $1/m^4$.

In the expression in square brackets one recognizes the first three terms of the expansion of

$$c\sqrt{\boldsymbol{p}^2 + m^2c^2}$$

in power series of $\boldsymbol{p}^2/m^2c^2$. It would, however, be *incorrect* to infer that the diagonal Hamiltonian, obtained by pursuing the above diagonalization procedure to all orders in $1/m$, would be

$$\tau_3\,c\sqrt{\boldsymbol{p}^2 + m^2c^2} - eA_0.$$

---

* A_0 is a function of $\boldsymbol{x}$, which, in the representation in which the Hamiltonian is H''', represents the *mean position* $(\boldsymbol{x}_{\text{av.}})''' = \boldsymbol{x}$. This, in the old representation, corresponds to

$$\boldsymbol{x}_{\text{av.}} = U^{-1}\,U'^{-1}\,U''^{-1}\,\boldsymbol{x}U''\,U'\,U \cong \boldsymbol{x} + \frac{\mathrm{i}\tau_1\,\boldsymbol{p}}{2\kappa\,mc} - \frac{\mathrm{i}e\tau_2}{4\hbar c\kappa^3}\,(\nabla_{\boldsymbol{x}}\,A_0) + \ldots$$

Following Snyder and Weinberg * it may be shown ** that

$$\mathfrak{H} = \mathfrak{U} H \mathfrak{U}^{-1} = \tfrac{1}{2}(\mathfrak{H}_+ + \mathfrak{H}_-)\,\tau_3 + \tfrac{1}{2}(\mathfrak{H}_+ - \mathfrak{H}_-)\,,$$

where ***

$$\mathfrak{U} = \frac{1}{2\sqrt{2(\odot + \odot^+)\,mc^2}}\big[(\odot + \odot^+ + 2mc^2) + (\odot + \odot^+ - 2mc^2)\,\tau_1 + \mathrm{i}\,(\odot^+ - \odot)\,\tau_2 + (\odot^+ - \odot)\,\tau_3\big] =$$

$$= \frac{1}{\sqrt{2(\odot + \odot^+)\,mc^2}}\begin{pmatrix} \odot^+ + mc^2 & \odot^+ - mc^2 \\ \odot - mc^2 & \odot + mc^2 \end{pmatrix},$$

and the operator $\odot$ and its adjoint $\odot^+$ (Schroedinger metric) are solutions of the operator equations

$$\begin{aligned} \odot^2 - e\,[\odot, A_0] &= c^2(p^2 + m^2c^2)\,, \\ (\odot^+)^2 + e\,[\odot^+, A_0] &= c^2(p^2 + m^2c^2)\,, \end{aligned} \tag{9}$$

and

$$\begin{aligned} \mathfrak{H}_+ &= (\odot + \odot^+)^{\frac{1}{2}}\,(\odot - eA_0)\,(\odot + \odot^+)^{-\frac{1}{2}}\,, \\ \mathfrak{H}_- &= (\odot + \odot^+)^{\frac{1}{2}}\,(\odot^+ + eA_0)\,(\odot + \odot^+)^{-\frac{1}{2}}\,. \end{aligned}$$

Note that, since A_0 is real, $\odot$ and $\odot^+$ are real operators ($\odot = \odot^*$, $\odot^+ = (\odot^+)^*$). Using eqs. (9) it may be shown that, besides being real, $\mathfrak{H}_\pm$ are Hermitian.

Proceeding as in the case of free particles (see pages 68 and 69), we separate the K.-G. wave function $\psi(x)$ for a spin zero particle in the electrostatic field $A_0(\boldsymbol{x})$, into its positive and negative frequency components †

$$\psi(x) = \psi_+(x) + \psi_-(x) = \sum_i \big[A_i\, \psi_+^i(\boldsymbol{x})\, \mathrm{e}^{-\mathrm{i}\omega_+^i t} + B_i\, \psi_-^i(\boldsymbol{x})\, \mathrm{e}^{-\mathrm{i}\omega_-^i t}\big]\,,$$

$$(\omega_+^i > 0\,,\quad \omega_-^i < 0)\,.$$

---

* H. Snyder and J. Weinberg, Phys. Rev. **57** (1940), 307.

** Using eqs. (9) and remembering that, for any two operators x and y,

$$\left[\frac{1}{x}, y\right] = -\frac{1}{x}\,[x, y]\,\frac{1}{x}\,,$$

the identity

$$\mathfrak{U} H = \tfrac{1}{2}\,[(\mathfrak{H}_+ + \mathfrak{H}_-)\,\tau_3 + (\mathfrak{H}_+ - \mathfrak{H}_-)]\,\mathfrak{U}$$

is easily verified.

*** Letting $e \to 0$, one has $\odot = \odot^+ = c\sqrt{p^2 + m^2c^2}$, and $\mathfrak{U}$ coincides with the transformation given on p. 67 for free particles.

† For some electrostatic potentials complex frequencies are possible. In such cases the present method fails, and the Hamiltonian cannot be diagonalized. See L. I. Schiff, H. Snyder and J. Weinberg, Phys. Rev. **57** (1940), 315.

The transformed Feshbach-Villars wave function $\Psi' = \mathfrak{U}\Psi$ is of the form

$$\Psi' = \begin{pmatrix} \varphi_+(x) \\ \varphi_-^*(x) \end{pmatrix},$$

with

$$\varphi_+(x) = \sqrt{\frac{\odot + \odot^+}{mc^2}}\, \psi_+(x)\,, \qquad \varphi_-^*(x) = \sqrt{\frac{\odot + \odot^+}{mc^2}}\, \psi_-(x)\,.$$

These latter obey the equations

$$\mathfrak{H}_\pm\, \varphi_\pm(x) = i\hbar \frac{\partial}{\partial t}\, \varphi_\pm(x)\,. \tag{10}$$

For steady wave functions

$$\mathfrak{H}_\pm\, \varphi_\pm^i(\boldsymbol{x}) = \pm\, \omega_\pm^i\, \varphi_\pm^i(\boldsymbol{x}), \quad (\pm\, \omega_\pm^i > 0)\,.$$

In the case of an electrostatic potential with small Fourier components $(|[\boldsymbol{p}^2, eA_0]| \ll m^3c^4)$ and for small momenta $(\boldsymbol{p}^2 \ll m^2c^2)$, one has

$$\odot^2 \cong (\odot^+)^2 \cong c^2(\boldsymbol{p}^2 + m^2c^2)\,, \quad \odot \cong \odot^+ \cong mc^2 + \frac{1}{2m}\boldsymbol{p}^2\,,$$

$$\mathfrak{H}_\pm \cong mc^2 + \frac{1}{2m}\boldsymbol{p}^2 \mp eA_0\,,$$

and eqs. (10) reduce to the Schroedinger equations for positive energy particles of charge $-\,e$ and $+\,e$, respectively. This is not surprising, since, in contrast to

$$\varphi_+^i(\boldsymbol{x}) = \sqrt{\frac{\odot + \odot^+}{mc^2}}\, \psi_+^i(\boldsymbol{x})\,,$$

the transformation

$$\varphi_-^i(\boldsymbol{x}) = \sqrt{\frac{\odot + \odot^+}{mc^2}}\, (\psi_-^i(\boldsymbol{x}))^*$$

is *not linear*. On the other hand, we know that the complex conjugate of the K.-G. wave function for a particle of energy E and charge $-\,e$ is a wave function for a particle of energy $-\,E$ and charge e.

Note that, in contrast to the corresponding K.-G. wave functions, $\varphi_\pm$, as eigenfunctions of the Hermitian operators $\mathfrak{H}_\pm$, obey Schroedinger-type orthonormality relations

$$\int (\varphi_\pm^j)^*\, \varphi_\pm^i\, d^3x = \delta_{ji}\,.$$

Following the procedure used for free particles, the charge and the energy may be expressed in the form

$$Q = \tfrac{1}{2} \int (\varphi_+^* \varphi_+ - \varphi_-^* \varphi_-) \, d^3x = \tfrac{1}{2} \sum_i (|A_i|^2 - |B_i|^2) ,$$

$$E = \tfrac{1}{2} \int (\varphi_+^* \mathfrak{H}_+ \varphi_+ + \varphi_-^* \mathfrak{H}_- \varphi_-) \, d^3x = \tfrac{1}{2} \sum_i (|\omega_+^i| \, |A_i|^2 + |\omega_-^i| \, |B_i|^2) .$$

Finally we notice that relations for the case when both a static magnetic field and an electrostatic field are present, may be obtained by replacing $\boldsymbol{p}$ by

$$\boldsymbol{p} + (e/c) \boldsymbol{A}$$

in the relations for the case of a particle in an electrostatic field. The relative $\odot$ and $\odot^+$ are not real ($\odot \neq \odot^*$, $\odot^+ \neq (\odot^+)^*$), nor are $\mathfrak{H}_\pm$ real, although they are still Hermitian. Therefore the equation for $\varphi_-(x)$ is

$$\mathfrak{H}_-^* \varphi_-(x) = i\hbar \frac{\partial}{\partial t} \varphi_-(x) ,$$

which, for weak slowly varying potentials and small momentum, reduces to the Schroedinger equation for a particle with charge $+ e$, since

$$\left[\left(\boldsymbol{p} + \frac{e}{c} \boldsymbol{A} \right)^2 \right]^* = \left(\boldsymbol{p} - \frac{e}{c} \boldsymbol{A} \right)^2 .$$

4. Gauge invariance

Eq. (3) for the four-current is not invariant under a gauge transformation of the second kind $A'_\mu = A_\mu + \partial \Lambda / \partial x_\mu$. Indeed one has

$$j'_\mu = j_\mu + (e/mc)(\partial \Lambda / \partial x_\mu) \, \psi^* \psi .$$

Gauge invariance of j_μ is assured by subjecting the wave function to the corresponding gauge transformation of the first kind

$$\psi \to e^{-ie\Lambda/\hbar c} \psi , \quad \psi^* \to e^{ie\Lambda/\hbar c} \psi^* .$$

(We now assume Λ to be real.).

Clearly, the Lagrangian density

$$\mathfrak{L} = - \frac{\hbar^2}{2m} \left[\left(\frac{\partial \psi^*}{\partial x_\mu} - \frac{ie}{\hbar c} A_\mu \psi^* \right) \left(\frac{\partial \psi}{\partial x_\mu} + \frac{ie}{\hbar c} A_\mu \psi \right) + \kappa^2 \psi^* \psi \right] , \quad (11)$$

from which eq. (1) follows, is also invariant under the gauge transformations of the first kind combined with that of the second kind, and so is the action integral.

Conversely, let us assume that the action integral is invariant under combined gauge transformations. This gives

$$0 = \int_{V(x)} \delta \mathfrak{L} \, \mathrm{d}^4 x =$$

$$= \int_{V(x)} \left[\frac{\partial \mathfrak{L}}{\partial A_\mu} \delta A_\mu + \frac{\partial \mathfrak{L}}{\partial \psi} \delta \psi + \frac{\partial \mathfrak{L}}{\partial \left(\frac{\partial \psi}{\partial x_\mu} \right)} \delta \frac{\partial \psi}{\partial x_\mu} + \frac{\partial \mathfrak{L}}{\partial \psi^*} \delta \psi^* + \frac{\partial \mathfrak{L}}{\partial \left(\frac{\partial \psi^*}{\partial x_\mu} \right)} \delta \frac{\partial \psi^*}{\partial x_\mu} \right] \mathrm{d}^4 x \,,$$

where

$$\delta A_\mu = \partial \, \delta \Lambda / \partial x_\mu \,, \quad \delta \psi = - (\mathrm{i}e/\hbar c) \, \delta \Lambda \, \psi \,, \quad \delta \psi^* = (\mathrm{i}e/\hbar c) \, \delta \Lambda \, \psi^* \,.$$

For the above equation to hold for an arbitrary $\delta \Lambda(x)$ and an arbitrary region $V(x)$, it is necessary that *

$$\frac{\partial \mathfrak{L}}{\partial A_\mu} = \frac{\mathrm{i}e}{\hbar c} \left[\frac{\partial \mathfrak{L}}{\partial \left(\frac{\partial \psi}{\partial x_\mu} \right)} \psi - \frac{\partial \mathfrak{L}}{\partial \left(\frac{\partial \psi^*}{\partial x_\mu} \right)} \psi^* \right]$$

and **

$$\frac{\partial}{\partial x_\mu} \left(\frac{\partial \mathfrak{L}}{\partial A_\mu} \right) = 0 \,,$$

as is found by partial integration.

The last equation gives the continuity equation

$$\frac{\partial j_\mu}{\partial x_\mu} = 0$$

for the four-current

$$j_\mu = - \frac{c}{e} \frac{\partial \mathfrak{L}}{\partial A_\mu} = \frac{1}{\hbar \mathrm{i}} \left[\frac{\partial \mathfrak{L}}{\partial \left(\frac{\partial \psi}{\partial x_\mu} \right)} \psi - \frac{\partial \mathfrak{L}}{\partial \left(\frac{\partial \psi^*}{\partial x_\mu} \right)} \psi^* \right] .$$

For the Lagrangian (11), j_μ, as defined by either of these expressions, is found to coincide with (4).

---

* So that the boundary terms arising from partial integration vanish.

** So that the integral obtained by partial integration and using the Euler eqs. (1), Chap. II, be zero.

CHAPTER VI

π-MESIC ATOMS

1. Particle of spin zero in electrostatic field

The Klein-Gordon equation describes correctly only particles of spin zero. The necessity of adopting a different equation for electrons (spin $\frac{1}{2}$) was made apparent, among other things, by the fact that the fine structure of the hydrogen atom, as calculated from the Klein-Gordon equation, was in disagreement with the experimental data.

Here we shall undertake the calculation of the fine structure of π-mesic atoms. Roughly speaking, a π-mesic atom consists of a nucleus of charge Ze and a pion (π^-) of charge $-e$ captured in a Bohr orbit *.

The radius of the first Bohr orbit is

$$a_\pi = 4\pi\hbar^2/Ze^2m_\pi = (1/273\,Z)\,0.529 \times 10^{-8}\text{ cm},$$

where m_π is the pion mass ($m_\pi = 273\,m_e$), and rationalized electrical units are used. Thus the first Bohr orbit of a π-mesic atom is 273 times smaller than the first electronic Bohr orbit. A π^- captured by a nucleus does not "feel" the electron screening. Moreover, if $R \simeq (1.5 \times 10^{-13})\,A^{\frac{1}{3}}$ is the radius of the nucleus and A is the atomic number, we see that

$$a_\pi < R,$$

if

$$ZA^{\frac{1}{3}} > 10^5/774\,.$$

This condition is satisfied by $A \simeq 80$ and $Z \simeq 40$. Then the π^- in the ground state of the π-mesic atom travels inside nuclear matter, and it is not really correct to consider the nucleus as a point charge, as we shall do in the following.

Let us now consider a stationary solution of the Klein-Gordon equation**

---

* See, for instance, the review by D. West, *Mesonic Atoms*, Reports on Progress in Physics **21** (1958), 271.

** In the sequel the motion of the nucleus is not taken into account.

with $A_0 = A_0(\boldsymbol{x})$, $\boldsymbol{A} = 0$,

$$\{\Delta + [(E + eA_0)^2 - E_0^2]/\hbar^2c^2\}\, \psi = 0 \tag{1}$$

(E = total energy, E_0 = rest energy). This equation may be written in a form more convenient for the discussion of the non-relativistic approximation:

$$\left(-\frac{\hbar^2}{2m}\Delta - \frac{eA_0\,E}{E_0}\right)\psi = \frac{(E-E_0)(E+E_0) + e^2A_0^2}{2E_0}\,\psi\,, \tag{2}$$

which, for

$$\begin{aligned} |E - E_0| &\ll 2E_0\,, \\ e^2A_0^2 &\ll 2E_0\,|E - E_0|\,, \end{aligned} \tag{3}$$

reduces to the Schroedinger equation for a particle in an electrostatic field *.

Let us now see under what circumstances the conditions (3) are satisfied and, therefore, the non-relativistic approximation is allowable.

For a π^- in the ground state one has

$$E - E_0 = -\,Ze^2/8\pi a_\pi = -\,\tfrac{1}{2}\,Z^2 m_\pi c^2\,(e^2/4\pi\hbar c)^2\,.$$

Then, if $Z = 40$,

$$|E - E_0|/E_0 = \tfrac{1}{2}\,Z^2\alpha^2 \simeq 1/23$$

(α = fine structure constant $= e^2/4\pi\hbar c = 1/137$) and the first of the inequalities (3) is satisfied.

On the other hand, for a pion in the first Bohr orbit we have

$$eA_0 = Z^2\alpha^2 m_\pi c^2 < m_\pi c^2\,.$$

Thus

$$e^2A_0^2/2E_0\,|E - E_0| = Z^2\alpha^2\,.$$

Therefore the second inequality (3) is satisfied for light nuclei only. Moreover, for such nuclei one has $eA_0 \ll m_\pi c^2$, and the term eA_0 in the expression for the charge density is really a small perturbation.

2. Eigenfunctions of π-mesic atoms

If A_0 is spherically symmetric, eq. (1) has solutions of the type

$$\psi_{nlm} = N_{nlm}\,R_{nl}(r)\,Y_l^{(m)}(\theta, \phi)\,.$$

---

* Using (3),

$$eA_0E \simeq eA_0E_0,\quad E + E_0 \simeq 2E_0\,,$$

and eq. (2) becomes the Schroedinger equation

$$-\,(\hbar^2/2m)\,\Delta\psi = (E - E_0 + eA_0)\,\psi\,.$$

The normalization constants N_{nlm} must be determined in such a way that

$$(\Psi, \Psi) = 1\,,$$

which, in the non-relativistic approximation, reduces to

$$\int |\psi_{nlm}|^2\,\mathrm{d}^3x = 1\,.$$

In order to determine the functions $R_{nl}(r)$ and $Y_l^{(m)}(\theta, \phi)$, we insert ψ_{nlm} for ψ in eq. (1) and separate the variables in the usual way *. Eq. (1) thus divides into two equations. The one for $Y_l^{(m)}$ is the same equation as encountered in looking for the eigenfunctions of $\boldsymbol{L}^2$. Therefore, $Y_l^{(m)}(\theta, \phi)$ may be taken to be a spherical harmonic (see Chap. IV, § 2).

The radial equation for $R_{nl}(r)$ is

$$\frac{\mathrm{d}^2R}{\mathrm{d}r^2} + \frac{2}{r}\frac{\mathrm{d}R}{\mathrm{d}r} + \left(A + 2\frac{B}{r} + \frac{C}{r^2}\right) R = 0\,, \tag{4}$$

where

$$A + 2B/r + C/r^2 = \left[(E + Ze^2/4\pi r)^2 - E_0^2\right]/\hbar^2c^2 - l(l+1)/r^2\,,$$

i.e.

$$A = (E^2 - E_0^2)/\hbar^2c^2\,, \quad B = Z\alpha\, E/\hbar c\,, \quad C = -\,l(l+1) + Z^2\alpha^2\,,$$

(α = fine structure constant).

In the case of the Schroedinger equation, one would have

$$A = 2m_\pi(E - E_0)/\hbar^2\,, \quad B = Z\alpha\, m_\pi c/\hbar\,, \quad C = -\,l(l+1)\,.$$

For bound states of π^- around the nucleus $E < E_0$, and therefore $A < 0$.

Introducing the dimensionless variable

$$\rho = 2r\sqrt{-A}\,,$$

eq. (4) takes the form

$$\frac{\mathrm{d}^2R}{\mathrm{d}\rho^2} + \frac{2}{\rho}\frac{\mathrm{d}R}{\mathrm{d}\rho} + (-\tfrac{1}{4} + B/\rho\sqrt{-A} + C/\rho^2)\,R = 0\,. \tag{5}$$

Then, putting

$$R = \mathrm{e}^{-\frac{1}{2}\rho}\,v\,,$$

after substitution in eq. (5) we have

$$v'' + (-1 + 2/\rho)v' + \left[(-1 + B/\sqrt{-A})/\rho + C/\rho^2\right] v = 0\,. \tag{6}$$

Expansion of v in a power series,

$$v = \rho^\lambda \sum a_\nu \rho^\nu\,,$$

* See H. Margenau and G. M. Murphy, *loc. cit.*, pp. 218–223.

and substitution in eq. (6) gives the recurrence formula *

$$a_\nu = (\lambda + \nu - B/\sqrt{-A})\, a_{\nu-1}/[(\lambda + \nu)^2 + (\lambda + \nu) + C]$$

for the coefficients, and the indicial equation

$$\lambda(\lambda + 1) = -C\,.$$

In order that v be finite for an infinite ρ, the coefficients of the series must all be zero from a certain $\nu = n_r$, i.e. the series must reduce to a polynomial of degree n_r. Therefore one must have

$$\lambda + n_r + 1 - B/\sqrt{-A} = 0\,, \tag{7}$$

which ensures that a_{n_r+1} and all following coefficients vanish.

Taking into account the indicial equation, the recurrence formula becomes

$$a_\nu = [(\nu - n_r - 1)/\nu(\nu + 2\lambda + 1)]\, a_{\nu-1}\,.$$

Iterating this formula, we have

$$a_\nu = \frac{(-1)^\nu\, n_r!\,(2\lambda + 1)!}{(n_r - \nu)!\,\nu!\,(\nu + 1 + 2\lambda)!}\, a_0\,.$$

(Here we use the convention that the factorial of a negative number is infinite, so that, if $\nu > n_r$, it is $1/(n_r - \nu)! = 0$. Thus $a_{n_r+1} = 0$.)

Choosing for the arbitrary constant

$$a_0 = 1/n_r!(2\lambda + 1)!\,,$$

the expression for $R_{nl}(r)$ becomes

---

* Indeed eq. (6) gives

$$0 = \Sigma\,[a_\nu(\lambda + \nu)(\lambda + \nu - 1)\,\rho^{\lambda+\nu-2} + 2a_\nu(\lambda + \nu)\,\rho^{\lambda+\nu-2} - a_\nu(\lambda + \nu)\,\rho^{\lambda+\nu-1} + \\ + (-1 + B/\sqrt{-A})\,a_\nu\rho^{\lambda+\nu-1} + Ca_\nu\rho^{\lambda+\nu-2}]\,.$$

Now, collecting equal powers of ρ, we have

$$\sum_{\nu=0} [(\lambda + \nu)(\lambda + \nu - 1) + 2(\lambda + \nu) + C]\, a_\nu\, \rho^{\lambda+\nu-2} + \\ + \sum_{\nu=0} [-1 - (\lambda + \nu) + B/\sqrt{-A}]\, a_\nu\, \rho^{\lambda+\nu-1} = 0\,,$$

and, changing the index in the second sum,

$$\sum_{\nu=1} \{ [(\lambda + \nu)(\lambda + \nu - 1) + 2(\lambda + \nu) + C]\, a_\nu + \\ + [-1 - (\lambda + \nu - 1) + B/\sqrt{-A}]\, a_{\nu-1}\}\, \rho^{\lambda+\nu-2} + a_0\,[\lambda(\lambda + 1) + C]\,\rho^{\lambda-2} = 0\,.$$

This equation must hold for any value of ρ, and therefore the coefficients of different powers of ρ must vanish. This gives the recurrence formula and the indicial equation.

$$R_{nl}(r) = e^{-\frac{1}{2}\varrho} \rho^{\lambda} \sum_{\nu=0}^{n_r} (-1)^{\nu} \rho^{\nu}/\nu!\,(n_r - \nu)!\,(\nu + 1 + 2\lambda)!\,. \tag{8}$$

The constant λ is determined from the indicial equation

$$\lambda(\lambda + 1) = l(l + 1) - \alpha^2 Z^2\,, \tag{9}$$

whose solutions are

$$\lambda = \pm\left[(l + \tfrac{1}{2})^2 - Z^2\alpha^2\right]^{\frac{1}{2}} - \tfrac{1}{2}\,. \tag{10}$$

If the relativistic correction $\alpha^2 Z^2$ is neglected, these reduce to the solutions $\lambda = l$, $\lambda = -(l + 1)$ of the indicial equation for the Schroedinger case.

Of the two values for λ, only the positive is physically acceptable. In fact, if we take the negative root and, for simplicity, consider the non-relativistic case, $\psi_{nlm} \to \infty$ for $\rho = 0$ in such a way that $\int |\psi_{nlm}|^2\, d^3x$ diverges *. This integral, on the other hand, converges if we take the positive root.

For the evaluation of $(\nu + 1 + 2\lambda)!$ when 2λ is not an integer, one must have recourse to a table of the gamma function. Using the property $\Gamma(z) = (z - 1)!$, one may write

$$(\nu + 1 + 2\lambda)! = \Gamma(\nu + 2 + 2\lambda)\,.$$

Finally, we notice that, for a given a_0, the solution of the Klein-Gordon (or of the Schroedinger) equation is fully determined when the normalization constant is such that $(\Psi, \Psi) = 1$. In the pure Kepler case the normalization constant does not depend on m.

NOTE 1. We may verify that, in the non-relativistic case ($\lambda = l$), the polynomials

$$\sum_{\nu=0}^{n-l-1} \left[(-1)^{\nu} \rho^{\nu}/\nu!\,(n - l - 1 - \nu)!\,(2l + 1 + \nu)!\right]$$

in eq. (8), ($n = n_r + l + 1$), are, but for a constant, the associated Laguerre polynomials $L_{n+l}^{(2l+1)}(\rho)$, as given, for instance, in the book of Margenau and Murphy (*loc. cit.*, pp. 77 and 129),

$$L_n^{(m)}(\rho) = (-1)^m (n!)^2\, {}_1F_1(-n + m;\, m + 1;\, \rho)/m!\,(n - m)!\,.$$

Here n and $m \leqslant n$ are integers, and ${}_1F_1$ is a confluent hypergeometric function defined by the series

$${}_1F_1(\alpha; \gamma; x) = \sum_{\nu=0}^{\infty} \Gamma(\alpha + \nu)\, \Gamma(\gamma) x^{\nu}/\Gamma(\alpha)\, \Gamma(\gamma + \nu)\nu!\,,$$

* For a more complete discussion see W. Pauli, Enc. of Phys. Vol. 5/1 (Springer, Berlin, 1958), and L. D. Landau and E. M. Lifshitz, *Quantum Mechanics* (London, 1958), p. 107.

so that

$$ {}_1F_1(-n+m;m+1;\rho) = $$
$$ = \sum_{\nu=0}^{\infty} (-1)^\nu (n-m)!\, m!\, \rho^\nu/\nu!\,(m+\nu)!\,(n-m-\nu)!. $$

Therefore

$$ L_{n+l}^{(2l+1)}(\rho) = $$
$$ = (-1)^{2l+1}\left[(n+l)!\right]^2 \sum_{\nu=0}^{\infty} (-1)^\nu \rho^\nu/(n-l-1-\nu)!\,(2l+1+\nu)!\,\nu!. $$

Then, obviously,

$$ R_{nl}(r) = e^{-\varrho/2}\rho^\lambda (-1)^{2l+1}\left[(n+l)!\right]^{-2} L_{n+l}^{(2l+1)}(\rho). $$

3. Energy levels

From eq. (7), remembering the definition of A and B, one has at once

$$ E/m_\pi c^2 = \left[1+\alpha^2Z^2/(n_r+\sqrt{(l+\tfrac{1}{2})^2-\alpha^2Z^2}+\tfrac{1}{2})^2\right]^{-\frac{1}{2}}. \tag{11} $$

Expanding the right-hand side in a power series of α^2Z^2, we have

$$ \begin{aligned} E/m_\pi c^2 &= \left[1+Z^2\alpha^2/(n_r+l+1)^2+Z^4\alpha^4/(n_r+l+1)^3(l+\tfrac{1}{2})+\ldots\right]^{-\frac{1}{2}} = \\ &= 1-Z^2\alpha^2/2(n_r+l+1)^2+\left[Z^4\alpha^4/(n_r+l+1)^3\right]\left[3/8(n_r+l+1)\right. \\ &\quad \left. -1/(2l+1)\right]+\ldots \end{aligned} $$

Thus, to a first approximation, we get

$$ E-m_\pi c^2 = -\frac{Z^2\alpha^2 m_\pi c^2}{2n^2}\left[1+\frac{Z^2\alpha^2}{n^2}\left(\frac{n}{l+\frac{1}{2}}-\frac{3}{4}\right)\right], \tag{12} $$

where $n = n_r + l + 1$.

The Bohr-Sommerfeld relativistic theory * yields, on the other hand,

$$ E-m_\pi c^2 = -\frac{Z^2\alpha^2 m_\pi c^2}{2n^2}\left[1+\frac{Z^2\alpha^2}{n^2}\left(\frac{n}{l+1}-\frac{3}{4}\right)\right], \tag{13} $$

which differs from eq. (12) by the occurrence of the integer $l+1$ instead of the half-integer $l+\frac{1}{2}$.

Within the approximation ($Z^2\alpha^2 \ll 1$) considered here, eq. (13), with m_π replaced by m_e, and $Z = 1$, agrees with the experimental results for the

* See, for instance, A. Sommerfeld, *Atombau und Spektrallinien*, Band I (Braunschweig, 1951), p. 272.

hydrogen fine structure, and also coincides with those predicted by the Dirac electron theory, whereas eq. (12) does not (see Part II, Chap. IX, § 3).

The first term in (12) gives Rydberg's formula

$$E - m_\pi c^2 = -Z^2\alpha^2 m_\pi c^2/2n^2 = -Z^2 R_\pi ch/n^2 ,$$

where

$$R_\pi = m_\pi c\alpha^2/2h \simeq 273 \times 110\,000 \text{ cm}^{-1} .$$

For the Balmer spectrum, neglecting fine structure, one has

$$1/\lambda = Z^2 R_\pi(\tfrac{1}{4} - 1/n^2) .$$

Thus the maximum wavelength

$$\lambda_{\text{max}} = \left[Z^2 R_\pi(\tfrac{1}{4} - \tfrac{1}{9})\right]^{-1} \simeq 2 \times 10^{-7}\, Z^{-2} \text{ cm}$$

is in the X-ray region.

The relativistic correction partially eliminates the degeneracy of the energy levels. Thus a degenerate level characterized by a certain n, splits into n levels with energies $E_{nl}(l = 0, 1 \ldots n-1)$. There is no splitting for the state $n = 1$, which is a non-degenerate s-state, but its energy is lowered by the amount $-\frac{5}{4} Z^4\alpha^2 R_\pi ch$. The level $n = 2$ splits, since the sublevel $l = 0$ is lowered by $-13Z^4\alpha^2 R_\pi ch/64$, and the sublevel $l = 1$ by $-7Z^4\alpha^2 R_\pi ch/192$ (see Fig. 3).

Similarly $n = 3$ splits into three levels.

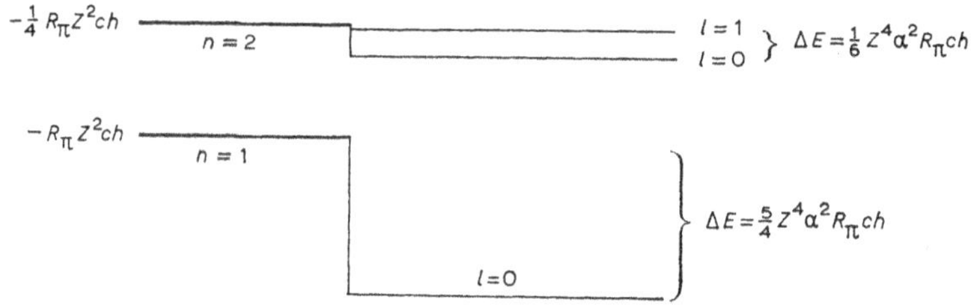

Fig. 3

4. π-Mesic atom in a static magnetic field

For a stationary solution representing a particle of given energy E in a static electromagnetic field

$$\boldsymbol{A} = \boldsymbol{A}(\boldsymbol{x}) , \quad A_0 = A_0(\boldsymbol{x}) ,$$

the Klein-Gordon equation (2) of Chap. V, with the Lorentz gauge $\partial A_\mu/\partial x_\mu = 0$, takes the form

$$\left[\Delta + \frac{2ie}{\hbar c} \boldsymbol{A} \cdot \boldsymbol{\nabla}_x - (e/\hbar c)^2 \boldsymbol{A}^2 + \frac{(E + eA_0)^2 - E_0^2}{\hbar^2 c^2}\right] \psi = 0 . \qquad (14)$$

This differs from eq. (1) by the two terms $(2ie/\hbar c)\, \boldsymbol{A} \cdot \boldsymbol{\nabla}_x\, \psi$ and $-\, (e/\hbar c)^2\, \boldsymbol{A}^2 \psi$. In the non-relativistic approximation, eq. (14) reduces to the Schroedinger equation

$$\left(-\frac{\hbar^2}{2m}\Delta - \frac{ie\hbar}{mc} \boldsymbol{A} \cdot \boldsymbol{\nabla}_x + \frac{e^2}{2mc^2} \boldsymbol{A}^2 - eA_0 + E_0\right) \psi = E\psi .$$

Let us now consider the case of a uniform magnetic field of intensity H in the direction of the z-axis. For this field one has

$$\boldsymbol{A} = \tfrac{1}{2}\, \boldsymbol{H} \times \boldsymbol{x} \equiv (-\tfrac{1}{2}\, Hy,\ \tfrac{1}{2}Hx, 0) ,$$

$$\boldsymbol{A} \cdot \boldsymbol{\nabla}_x = \tfrac{1}{2}\, H \left(x \frac{\partial}{\partial y} - y \frac{\partial}{\partial x}\right) = \tfrac{1}{2}\, H \frac{\partial}{\partial \phi} = \frac{iH}{2\hbar} L_3 .$$

Moreover the perturbation term

$$e^2 A^2/\hbar^2 c^2 = e^2 H^2 r^2 \sin^2 \theta / 4\hbar^2 c^2$$

is important only for very strong magnetic fields and for orbits of very large radius, that is, large total quantum number n. For the moment, therefore, we shall neglect it.

NOTE 2. If one writes eq. (14) in the form

$$\{(\hbar^2/2m)\, \Delta + (ie\hbar/mc)\, \boldsymbol{A} \cdot \boldsymbol{\nabla}_x - (e^2/2mc^2)\, A^2 + \\ + [(E + eA_0)^2 - E_0^2]/2mc^2\}\, \psi = 0 ,$$

one sees that $e^2 r^2 H^2 \sin^2 \theta / 8m_\pi c^2$ is negligible compared with the characteristic energy $\hbar\omega_{\mathrm{L}} = eH\hbar/2m_\pi c$ of the Larmor precession. Indeed

$$e^2 r^2 H^2 / 8m_\pi c^2\, \hbar\omega_{\mathrm{L}} = \alpha\pi r^2 H/e \simeq 10^{-14}\, H_{\mathrm{gauss}} ,$$

having chosen for r the radius a_π of the first Bohr orbit. Thus we see that, even for very strong magnetic fields, e.g. $H \simeq 2 \times 10^5$ gauss (the maximum reached so far), the above term remains negligible in comparison with the total energy.

Spectroscopically, for $H = 2 \times 10^5$ gauss this term would give a contribution of the order of 10^{-10} cm^{-1}, which is definitely outside the possibility of experimental detection. Actually, even though this term is spectroscopically negligible, this is not the case in the theory of atomic diamagnetism (see *Note 3*, and R. E. Peierls, *Quantum Theory of Solids* (Oxford, 1955), p. 144). The modification of the Zeeman effect due to this term is called the quadratic Zeeman effect.

Within the above approximation it is not difficult to show that the eigenfunctions for a π-mesic atom in a uniform magnetic field are of the same type as those for the pure Kepler case. This is due to the fact that *

$$(2\mathrm{i}e/\hbar c)\ \boldsymbol{A}\cdot\boldsymbol{\nabla}_x\ \psi_{nlm} = (\mathrm{i}e/\hbar c)\ N_{nlm}\ R_{nl}(r)\ P_l^{(m)}(\cos\,\theta)\ H\ \partial/\partial\phi\ \mathrm{e}^{\mathrm{i}m\phi} = \\ = -\ (em/\hbar c)\ H\ \psi_{nlm}\,,$$

namely the functions ψ_{nlm} are eigenfunctions of the perturbation $(\mathrm{i}e/\hbar c)\ H\ \partial/\partial\phi$.

Thus for the radial part R we have the equation

$$\frac{\mathrm{d}^2R}{\mathrm{d}r^2} + \frac{2}{r}\frac{\mathrm{d}R}{\mathrm{d}r} + (A + 2B/r + C/r^2)\ R = 0\,,$$

where now

$$A = \frac{E^2 - E_0^2}{\hbar^2c^2} - \frac{em}{\hbar c}\,H\,,\quad B = Z\alpha E/\hbar c\,,\quad C = -\ l(l+1) + Z^2\alpha^2\,.$$

We may now introduce the variable $\rho = 2r\sqrt{-A}$, put $R = \mathrm{e}^{-\frac{1}{2}\rho}\,v$ and proceed just in the same way as in § 2.

NOTE 3. An interesting calculation is the solution of the Schroedinger equation for a particle of charge $e > 0$ and mass M in a uniform magnetic field H in the direction of the z axis,

$$-\frac{\hbar^2}{2M}\left\{\frac{1}{r}\frac{\partial}{\partial r}\left(r\frac{\partial\psi}{\partial r}\right) + \frac{1}{r^2}\frac{\partial^2\psi}{\partial\phi^2} + \frac{\partial^2\psi}{\partial z^2}\right\} + \frac{\hbar e\mathrm{i}}{2Mc}\,H\frac{\partial\psi}{\partial\phi} + \frac{H^2e^2}{8Mc^2}\,r^2\psi = \\ = \hbar\mathrm{i}\,\frac{\partial\psi}{\partial t}\,,$$

where $r = \sqrt{x^2+y^2}$, $\phi = \mathrm{tg}^{-1}(y/x)$.

---

* Note that the operator $(r\sin\theta)^{-2}\partial^2/\partial\phi^2$ has as eigenfunctions not only $\mathrm{e}^{\mathrm{i}m\phi}$, but also $\cos(m\phi)$ and $\sin(m\phi)$, so that the eigenfunctions for a π-mesic atom (without magnetic field) can have $\cos(m\phi)$ or $\sin(m\phi)$ instead of $\mathrm{e}^{\mathrm{i}m\phi}$. The introduction of a magnetic field imposes the choice of the exponential form which is an eigenfunction of $(2\mathrm{i}e/\hbar c)\ \boldsymbol{A}\cdot\boldsymbol{\nabla}_x$. In other words, in contrast to the pure Kepler case, the wave function must now be complex, in agreement with Chap. V, § 1.

The motion in a plane perpendicular to the z axis is quantized. It is easy to verify that the eigenfunctions are

$$\psi(r, \phi, t) = e^{-iE_s t/\hbar} e^{im\phi} \rho^m e^{-\varrho^2/4} \sum_{\nu=0}^{s} (-1)^\nu \rho^{2\nu}/2^\nu \nu!(s-\nu)!(m+\nu)!,$$

with $\rho^2 = eHr^2/\hbar c$ and

$$E_s = (s + \tfrac{1}{2})\, e\hbar H/Mc\,.$$

This energy quantization is responsible for the diamagnetism of a gas of spin zero particles, which does not exist if the motion of the particles is treated according to classical mechanics.

5. Zeeman effect for π-mesic atoms

In calculating the energy levels, one finds that one must have

$$B/\sqrt{-A} = Z\alpha E/(E_0^2 - E^2 + emH\hbar c)^{\frac{1}{2}} = n_r + \left[(l+\tfrac{1}{2})^2 - Z^2\alpha^2\right]^{\frac{1}{2}} + \tfrac{1}{2}.$$

From this it is easy to derive a relation between the energies of the levels in the presence of a magnetic field, which depend on all quantum numbers n, l, m and will be denoted by $E_{nlm}(\boldsymbol{H})$, and those of the degenerate levels in the absence of a magnetic field.

One has

$$E_{nlm}(\boldsymbol{H}) = \left[1 + (2m\hbar\omega_L/m_\pi c^2)\right]^{\frac{1}{2}} E_{nl}(\boldsymbol{H} = 0)\,,$$

where $\omega_L = eH/2m_\pi c$ is the Larmor frequency.

The correction, therefore amounts to multiplication by a factor dependent on the magnetic field and on the magnetic quantum number m.

In all cases of practical interest one may use the non-relativistic approximation, according to which

$$\left[1 + (2m\hbar\omega_L/m_\pi c^2)\right]^{\frac{1}{2}} \simeq 1 + (m\hbar\omega_L/m_\pi c^2) + \ldots$$

Then one has

$$E_{nlm}(\boldsymbol{H}) \simeq E_{nl} + m\hbar\omega_L\,, \tag{15}$$

the correction being additive. From (15) it follows that a degenerate level, characterized by the quantum numbers n and l, is split into $2l+1$ levels. For example, a p-state of energy E_{n1} gives rise, in a magnetic field, to three energy levels

$$E_{n11} = E_{n1} + \hbar\omega_L\,, \quad E_{n10} = E_{n1}\,, \quad E_{n1,-1} = E_{n1} - \hbar\omega_L\,.$$

A d-state gives rise to five energy levels etc.

This is the Zeeman effect for π-mesic atoms *.

Notice that, in the non-relativistic approximation, the introduction of a static magnetic field does not change the radial structure of the eigenfunctions..These obey the equation

$$\frac{d^2R}{dr^2}+\frac{2}{r}\frac{dR}{dr}+\left[A+\frac{2}{r}(n_r+l+1)\sqrt{-A}-\frac{l(l+1)}{r^2}\right]R=0\,,$$

where

$$A=\frac{2m_\pi}{\hbar^2}(E-E_0-m\hbar\omega_{\mathrm{L}})=\frac{2m_\pi}{\hbar^2}(E'-E_0)=A'\,.$$

A' and E' refer to the case without magnetic field; the total energy in the presence of a magnetic field is obtained from E' (for $\boldsymbol{H}=0$) by adding the energy $\hbar\omega_{\mathrm{L}}$ due to the field.

Interpreting the results obtained, one can say that a magnetic moment

$$\boldsymbol{\mu}=-\frac{e}{2m_\pi c}\boldsymbol{L}\,,$$

results from the motion of the particle, its interaction energy with the magnetic field being

$$E_{\mathrm{int}}=-\boldsymbol{\mu}\cdot\boldsymbol{H}=(e/2m_\pi c)\,\boldsymbol{H}\cdot\boldsymbol{L}\,.$$

For the eigenfunction $\psi\sim R_{nl}\,Y_l^{(m)}(\theta,\phi)$, since the quantization axis is in the direction of the magnetic field, this interaction has the value

$$E_{\mathrm{int}}=m\hbar\omega_{\mathrm{L}}\,.$$

Thus the magnetic field changes the frequency $\nu=(E_{nlm}-E_{n'l'm'})/h$, emitted in the transition between two energy levels for $\boldsymbol{H}=0$, by the amount

$$\Delta\nu=(m-m')\,\omega_{\mathrm{L}}/2\pi\,.$$

For dipole transitions (which are the only important ones in atomic and π-mesic-atomic physics) one has the selection rules $\Delta m=0,\pm 1$ (see Part III, Chap. V, § 6). Correspondingly one has $\Delta\nu=0$ (unshifted line) and $\Delta\nu=\pm\,\omega_{\mathrm{L}}/2\pi$ (shifted lines).

As is well known, the unshifted line is linearly polarized in the direction of the magnetic field, and is not observed in this direction, whereas the

* Since π^- has spin zero, π-mesic atoms exhibit only the normal Zeeman effect, whereas for electrons the Zeeman effect may be normal and anomalous.

shifted lines, if observed in the direction of the magnetic field, appear circularly polarized. For the line with greater frequency, the electric field of the emitted light rotates in the same direction as the current generating the magnetic field, while the opposite is true for the one with the smaller frequency. Perpendicularly to the magnetic field one observes, instead of the single line for $\boldsymbol{H} = 0$, a triplet of equidistant lines. The intermediate one is polarized parallel to the magnetic field, the other two perpendicular to it. This result was already explained by the old Lorentz theory, but for the calculation of the intensities of the different components quantum theory must be used (see Part III).

In fact, the Zeeman effect for π-mesic atoms is of purely theoretical interest, being outside any possibility of experimental detection, as is seen from the following orders of magnitude:
Since the Bohr magneton for π^- is

$$e\hbar/2m_\pi c = (0.9273/273) \times 10^{-20} \text{ erg gauss}^{-1} = \\ = (0.579/273) \times 10^{-8} \text{ eV gauss}^{-1},$$

the separation of contiguous Zeeman levels ($\hbar\omega_{\mathrm{L}}$) gives rise to a frequency change of the order of

$$\Delta\omega = \omega_{\mathrm{L}} = \left[(0.9273/273 \times 1.05) \times 10^7 \text{ sec}^{-1} \text{ gauss}^{-1}\right] H_{\text{gauss}} = \\ = \left[(0.88/273) \times 10^7 \text{ sec}^{-1} \text{ gauss}^{-1}\right] H_{\text{gauss}}.$$

Hence for the Balmer spectrum of π-mesic atoms one has

$$\Delta\lambda/\lambda = \lambda\, \Delta(\omega/2\pi c) \lesssim (3.4 \times 10^{-14} \text{ gauss}^{-1})\, H_{\text{gauss}}/Z^2 .$$

For $H = 2 \times 10^5$ gauss this gives

$$\Delta\lambda/\lambda \lesssim 10^{-8},$$

which is outside the possibility of experimental verification, since for X-ray detectors $\Delta\lambda/\lambda \simeq (0.1 - 0.2) \times 10^{-3}$ at most.
For an "electronic" atom one has, on the other hand,

$$\Delta\lambda/\lambda \simeq 10^{-4}/Z^2 .$$

This shift can be detected. It is, however, found that for the hydrogen atom experimental results do not agree with the theoretical predictions discussed in this section. As we have already noted, this is a consequence of our having neglected the spin.

CHAPTER VII

DISCONTINUOUS TRANSFORMATIONS

1. Space inversion

Each of the continuous transformations considered in the preceding chapter involved a parameter (e.g. an angle of rotation) which could be chosen arbitrarily small. At its approaching zero, the transformation reduced to the identity. This is not the case for the transformations studied in this chapter, which produce finite, discontinuous changes from the original situation.

The space inversion or reflection, denoted by P, is the transformation

$$\boldsymbol{x}' = -\boldsymbol{x}\,, \quad \text{i.e.} \quad x_i' = -x_i\,, \quad (i = 1, 2, 3)\,,$$

corresponding to the transition from a left-handed to a right-handed system of axes.

Polar vectors are those vectors $\boldsymbol{a}$ whose components a_1', a_2', a_3' are equal and opposite to those in the non-primed system

$$\boldsymbol{a}' = -\boldsymbol{a}\,, \quad \text{i.e.} \quad a_1' = -a_1\,, \quad a_2' = -a_2\,, \quad a_3' = -a_3\,.$$

Examples of polar vectors are position $\boldsymbol{x}$, velocity $\boldsymbol{v} = \mathrm{d}\boldsymbol{x}/\mathrm{d}t$, momentum, acceleration, electric field, vector potential (since $E = -(1/c)\,\partial \boldsymbol{A}/\partial t + \nabla_x A_0$), current density ($\boldsymbol{j} = \sigma \boldsymbol{E}$) and electric dipole moment (which, by definition, has the direction from negative to positive charge). The operator ∇_x is also a polar vector.

On the other hand, one calls axial vectors those vectors whose components in the dashed system are equal to those in the undashed system. Therefore the direction of such vectors changes when going from a left-handed to a right-handed system.

A typical axial vector is the vector product $\boldsymbol{a} \times \boldsymbol{b}$ of two vectors of the same type (polar or axial), which is defined differently with respect to a left-handed or a right-handed system of axes. More precisely, its direction must be such that the vectors $\boldsymbol{a}$, $\boldsymbol{b}$, $\boldsymbol{a} \times \boldsymbol{b}$ form a triad of the same type (left-handed or right-handed) as the reference system of axes adopted.

One has therefore Fig. 4 if the reference system is left-handed, and Fig. 5 for a right-handed system.

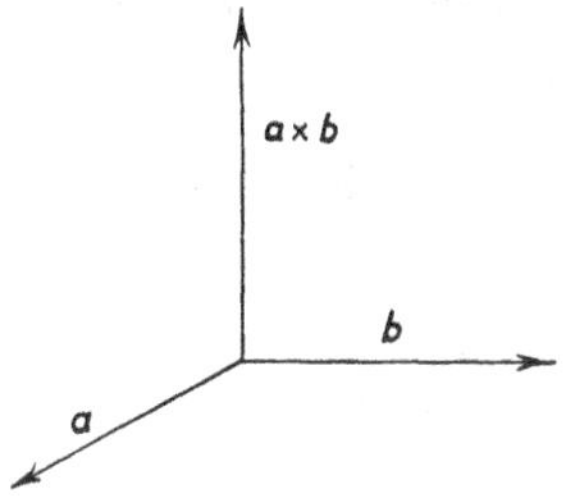

Fig. 4

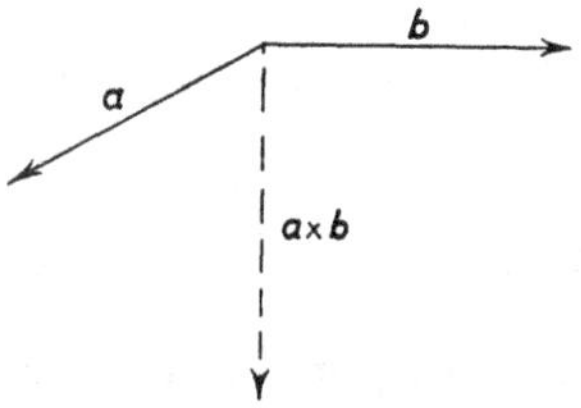

Fig. 5

Thus

$$(\boldsymbol{a} \times \boldsymbol{b})_1 = (\boldsymbol{a}' \times \boldsymbol{b}')'_1, \quad (\boldsymbol{a} \times \boldsymbol{b})_2 = (\boldsymbol{a}' \times \boldsymbol{b}')'_2, \quad (\boldsymbol{a} \times \boldsymbol{b})_3 = (\boldsymbol{a}' \times \boldsymbol{b}')'_3 .$$

This can also be seen formally from the relations

$$a'_2 b'_3 - a'_3 b'_2 = a_2 b_3 - a_3 b_2 , \quad \text{etc.}$$

assuming that $\boldsymbol{a}$ and $\boldsymbol{b}$ are vectors of the same type.

Typical axial vectors are the angular momentum $\boldsymbol{L} = \boldsymbol{x} \times \boldsymbol{p}$, the magnetic field $\boldsymbol{H} = \nabla_x \times \boldsymbol{A}$, the magnetic dipole moment etc.

It is easy to show that Maxwell's equations are invariant under P.

A scalar is a one-component quantity whose value does not depend on the nature (left-handed or right-handed) of the reference system. For example, the inner product of two vectors of the same type and the triple scalar product of three axial vectors, are scalars.

On the other hand, a pseudoscalar is a one-component quantity whose value depends on the nature of the reference system, and changes sign when one goes over from one type to another. For example, the inner product of a polar vector and an axial vector, and the potential of a magnetic dipole $V = \boldsymbol{\mu} \cdot \boldsymbol{x}/r^3$, are both pseudoscalars.

In general, a skew-symmetric tensor of third rank $a_{ikl} = - a_{kil} = a_{kli} \dots$ (i, k, l, being permutations of 1, 2, 3) may be regarded as a pseudoscalar. In fact this tensor has essentially one component, say a_{123}, and one easily sees that for the transformation

$$x'_i = \alpha_{ik} x_k , \quad \alpha_{ik} = - \delta_{ik} , \quad (i = 1, 2, 3) ,$$

one has

$$a'_{123} = \alpha_{1i} \alpha_{2k} \alpha_{3l} a_{ikl} = - a_{123} .$$

The tensor ε_{ikl}, appearing in the Introduction, § 3, is of such a type. Therefore, assuming $\varepsilon_{123} = 1$ in all left-handed systems, one has $\varepsilon'_{123} = -1$ in all right-handed systems.

Now, if we go back to the inner product of the polar vector $\boldsymbol{a}$ and the axial vector $\boldsymbol{b} \times \boldsymbol{c}$ ($\boldsymbol{b}$ and $\boldsymbol{c}$ of the same type), we obtain

$$\boldsymbol{a} \cdot (\boldsymbol{b} \times \boldsymbol{c}) = a_1(b_2c_3 - c_2b_3) + a_2(b_3c_1 - c_3b_1) + a_3(b_1c_2 - c_1b_2) = \\ = - \boldsymbol{a}' \cdot (\boldsymbol{b}' \times \boldsymbol{c}')' ,$$

(having put $a_i = -a'_i, \quad b_i = \pm b'_i, \quad c_i = \pm c'_i$).

If, however, one starts from the relation

$$\boldsymbol{a} \cdot (\boldsymbol{b} \times \boldsymbol{c}) = \varepsilon_{ikl}\, a_i b_k c_l$$

and appends primes to all quantities, one obtains

$$[\boldsymbol{a} \cdot (\boldsymbol{b} \times \boldsymbol{c})]' = \varepsilon'_{ikl}\, a'_i b'_k c'_l = \varepsilon_{ikl}\, a_i b_k c_l ,$$

(as $\varepsilon'_{ikl} = -\varepsilon_{ikl}$), which seems to contradict what has been obtained previously.

Really the components of the vector product must be written

$$d_i = \varepsilon_{123}\, \varepsilon_{ikl}\, b_k c_l$$

in order that

$$d'_i = \varepsilon'_{123}\, \varepsilon'_{ikl}\, b'_k\, c'_l = d_i .$$

Therefore

$$\boldsymbol{a} \cdot (\boldsymbol{b} \times \boldsymbol{c}) = \varepsilon_{123}\, \varepsilon_{ikl}\, a_i b_k c_l .$$

Notice that the skew-symmetric tensor $\varepsilon_{\mu\nu\varrho\sigma}$ (which is the extension of ε_{ikl} to four-dimensions) is the typical pseudoscalar of relativistic theories.

2. Transformation of wave functions

Let us now ask whether the wave function of a spinless particle is a scalar, i.e. obeys the transformation law

$$\psi'(\boldsymbol{x}', t) = \psi(\boldsymbol{x}, t)$$

not only under rotations, translations etc. but also under P, or is a pseudoscalar with the transformation

$$\psi'(\boldsymbol{x}', t) = -\psi(\boldsymbol{x}, t)$$

under P, namely a skew-symmetric tensor of the fourth rank $\phi_{\mu\nu\varrho\sigma}$. ($\phi_{\mu\nu\varrho\sigma}$ is zero if any two indices are equal, and is equal to $+\phi_{1234}$ or $-\phi_{1234}$ according as the indices $\mu\nu\rho\sigma$ are an even or an odd permutation of 1, 2, 3, 4).

It is easy to see that the transformation law of a one-component wave function under space inversion can only be that of a scalar or a pseudoscalar. In fact, the transformation law must be

$$\psi'(\boldsymbol{x}', t) = \eta_{\mathrm{P}}\, \psi(\boldsymbol{x}, t)\,, \tag{1}$$

for $\boldsymbol{x}' = -\boldsymbol{x}$, with

$$\eta_{\mathrm{P}}^{*}\, \eta_{\mathrm{P}} = 1\,,$$

if one requires the charge density and the energy to be scalars, the current density and the momentum polar vectors and the angular momentum an axial vector (see Chap. II, § 3, eqs. (15), (16), § 4, eq. (19), and Chap. V, § 1, eqs. (4), (6).

On performing another inversion $\boldsymbol{x}'' = -\boldsymbol{x}'$, we have

$$\psi''(\boldsymbol{x}'', t) = \eta_{\mathrm{P}}\, \psi'(\boldsymbol{x}', t) = \eta_{\mathrm{P}}^{2}\, \psi(\boldsymbol{x}, t)\,.$$

However, two successive inversions are equivalent to the identity, i.e. they lead back to the original system. Therefore $\psi''(\boldsymbol{x}'', t)$ must be equal to $\psi(\boldsymbol{x}, t)$, and $\eta_{\mathrm{P}}^{2} = 1$.

Thus

$$\eta_{\mathrm{P}} = \pm 1\,,$$

corresponding to the case of a wave function which is a scalar or a skew-symmetric tensor of fourth rank, respectively.

It is readily seen that the Klein-Gordon and Schroedinger Lagrangian densities are invariant under space inversions. This property, however, does not lead to a continuity equation and to a conservation law, such as those for energy and momentum. In some cases, however, the parity of the state of a physical system is conserved in transitions between states if the interactions causing them are invariant under space inversion. That is, the probability that one of such transitions may occur is zero if the parity of the final state is different from that of the initial state *.

In general wave functions are not invariant in form under P. For example

$$\mathrm{P}(\mathrm{e}^{\mathrm{i}\boldsymbol{k}\cdot\boldsymbol{x}}) = \mathrm{e}^{-\mathrm{i}\boldsymbol{k}\cdot\boldsymbol{x}'} = \mathrm{e}^{\mathrm{i}\boldsymbol{k}'\cdot\boldsymbol{x}'} = \mathrm{e}^{\mathrm{i}\boldsymbol{k}\cdot\boldsymbol{x}}\,,$$

from which one sees that the original function is $\psi(\xi) = \mathrm{e}^{\mathrm{i}\boldsymbol{k}\cdot\xi}$ with $\xi = \boldsymbol{x}$, while the transformed function is $\psi'(\xi) = \mathrm{e}^{-\mathrm{i}\boldsymbol{k}\cdot\xi}$ with $\xi = \boldsymbol{x}'$.

The spherical harmonics $Y_l^{(m)}$ are form-invariant if l is even and, except for a sign, if l is odd. In fact

$$\mathrm{P}[Y_l^{(m)}(\theta, \phi)] = (-1)^l\, Y_l^{(m)}(\theta', \phi')\,,$$

* See, for instance, the review: E. Corinaldesi, *Particles and Symmetries*, Nucl. Phys. **7** (1958), 305.

where $Y_l^{(m)}(\theta, \phi)$ and $Y_l^{(m)}(\theta', \phi')$ are the same function $Y_l^{(m)}(\Theta, \Phi)$, taken respectively for $\Theta = \theta$, $\Phi = \phi$ and for $\Theta = \theta' = \pi - \theta$, $\Phi = \phi' = \phi + \pi$.

Wave functions which P leaves invariant in form but for a factor P' may be regarded as eigenfunctions of P belonging to the eigenvalue P'. For instance, $Y_l^{(m)}$ is an eigenfunction of P with eigenvalue $(-1)^l$. A plane wave $e^{ik\cdot x}$ is an eigenfunction of P (belonging to the eigenvalue one) only for $k = 0$, namely for a particle at rest.

Since P applied twice is equivalent to the identity, one has

$$P'^2 = 1 \quad \text{or} \quad P' = \pm 1 .$$

This means that $+1$ and -1 are the only possible eigenvalues of P.

The eigenvalue P' above is called the *parity* of the eigenfunction. It must not be confused with the constant η_P, whose value is $+1$ if the wave function is a scalar and -1 if it is a pseudoscalar (namely a skew-symmetric tensor of fourth rank, which can be expressed as the product of the Ricci-Levi Civita symbol $\varepsilon_{\mu\nu\varrho\sigma}$ and a scalar function). Indeed, the wave functions which are eigenfunctions of P in the first case are $\propto Y_l^{(m)}$ with eigenvalue $(-1)^l$, and $(e^{ik\cdot x})_{k=0}$ with eigenvalue 1. In the second case, on the other hand, they are $\propto \varepsilon_{\mu\nu\varrho\sigma} Y_l^{(m)}$ with eigenvalue $-(-1)^l$, and $\varepsilon_{\mu\nu\varrho\sigma}(e^{ik\cdot x})_{k=0}$ with eigenvalue -1. For this reason η_P, the eigenvalue of P for a wave function describing a particle at rest, is called the *intrinsic parity* of the particle. It is not necessary to take this into account unless one is dealing with processes involving creation and destruction of particles.

NOTE 1. Example of experimental verification of parity conservation.

Consider the following ideal experiment. A beam of linearly po-

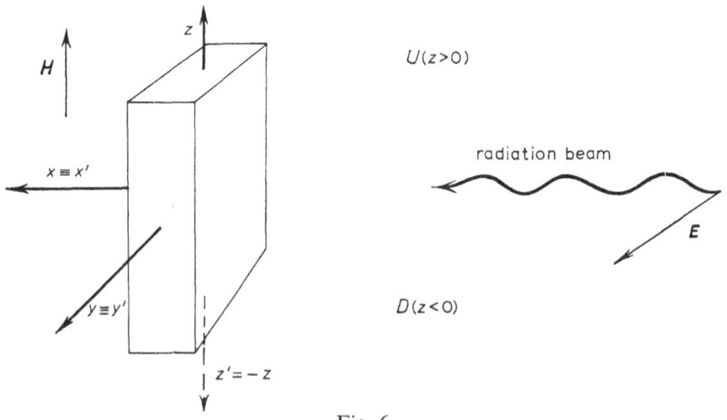

Fig. 6

larized radiation hits a photocathode consisting of π-mesic atoms. The photocathode lies within a constant magnetic field $\boldsymbol{H}$ (Fig. 6).

The electric field of the radiation is polarized perpendicular to $\boldsymbol{H}$, in the direction of the y axis. One can show that a count of the photomesons emitted in the half-spaces $D(z < 0)$ and $U(z > 0)$ would provide an indication of a possible non-conservation of parity in the interaction of π-mesons with the electromagnetic field, that is, the incompleteness of the scheme presented in the previous chapter.

In fact, the description of the static magnetic field and of the incident light wave is the same with respect to both the systems of axes x, y, z and x', y', z'. For example the components of the static magnetic field are in both systems $(0, 0, H)$, those of the electric field of the incident wave are $(0, E, 0)$ and those of the propagation vector are $(1, 0, 0)$.

If the interaction causing the production of photomesons is invariant under space inversion, the intensity of the photomesons must be a function invariant under the transformation $x' = x$, $y' = y$, $z' = -z$. Therefore no asymmetry between the two half-spaces D and U may exist.

3. Time reversal and the operation PT

Time reversal consists in the transformation

$$t' = -t\,,$$

namely in the change of the direction of the time axis. Under this transformation, denoted by T, velocity, momentum, angular momentum, current density (which is proportional to a velocity), magnetic field ($\boldsymbol{H} = e\boldsymbol{x} \times \boldsymbol{v}/cr^3$) and vector potential ($A = \frac{1}{2}\boldsymbol{H} \times \boldsymbol{x}$ for a uniform field) change sign. T leaves unchanged acceleration and electric field, scalar potential (A_0) and charge density.

It is easily seen that Maxwell's equations are invariant under T. The trajectory of a point mass remains unchanged, but is described backwards.

While the Klein-Gordon equation is invariant under time reversal if the wave function is transformed as a scalar, this is not the case for its non-relativistic limit, the Schroedinger equation. In fact, this latter involves the first time derivative multiplied by the imaginary unit, and so, when the direction of the time axis is reversed, the wave function must be replaced by its complex conjugate, which may be multiplied by a numerical factor η_{T}.

Also the Klein-Gordon equation is invariant under this operation,

$$\psi'(\boldsymbol{x}, t') = \eta_{\mathrm{T}}\, \psi^*(\boldsymbol{x}, t)\,. \tag{2}$$

Since two successive time reversals must lead back to the original wave function, η_{T} must be of modulus one,

$$\eta_{\mathrm{T}}\eta_{\mathrm{T}}^* = 1\,.$$

Furthermore, both the relativistic and the non-relativistic expression for the energy and the charge density are left unchanged by (2), while those for the momentum, the angular momentum and the current density change sign, as one must require for physical reasons *. Thus (2) is the correct transformation law for the wave function.

$\eta_{\mathrm{T}} = +1$ if ψ is a scalar,

$= -1$ if it is a skew-symmetric tensor $\phi_{\mu\nu\varrho\sigma} = \varepsilon_{\mu\nu\varrho\sigma}\,\phi$.

Note that $\varepsilon^*_{\mu\nu\varrho\sigma} = \varepsilon_{\mu\nu\varrho\sigma}$ according to the definition adopted here ($\varepsilon_{1234} = \pm 1$).

Notice that steady wave functions

$$\psi(x) = \psi(\boldsymbol{x})\, \mathrm{e}^{-\mathrm{i}Et/\hbar}\,,$$

with $\psi(\boldsymbol{x}) = \psi^*(\boldsymbol{x})$, are invariant in form under T.

In fact

$$\psi'(\boldsymbol{x}, t') = \psi^*(\boldsymbol{x})\, \mathrm{e}^{\mathrm{i}Et/\hbar} = \psi(\boldsymbol{x})\, \mathrm{e}^{-\mathrm{i}Et'/\hbar}\,,$$

that is

$$\psi'(\boldsymbol{x}, \tau) = \psi(\boldsymbol{x}, \tau)\,.$$

Such stationary wave functions may be regarded as eigenfunctions of T with eigenvalue one. If $\psi(\boldsymbol{x}, t)$ is a spherical harmonic, it can be real only for $m = 0$ (remember that $Y_l^{(m)*} = (-1)^m\, Y_l^{(-m)}$). Hence stationary eigenfunctions of the angular momentum are eigenfunctions of T only if they belong to the eigenvalue $l_3 = 0$. In fact, the operation of time reversal changes the sign of the components of the angular momentum and, therefore, transforms a wave function with the quantum numbers l, m into one with the quantum numbers l and $-m$.

The product PT of space inversion and time reversal, i.e.

$$x'_\mu = -x_\mu\,, \quad (\mu = 1, 2, 3, 4)\,,$$

leaves velocity, charge density, vector and scalar potentials unchanged, and changes the sign of angular momentum and electric and magnetic fields.

* If the wave function were transformed as a scalar, the charge density would change sign, while the current density would not.

For the convenience of the reader we summarize in the table below the transformation properties of various physical quantities.

For wave functions of particles of spin zero one has

$$\psi'(x') = \psi'(\boldsymbol{x}', t') = \eta_{\mathrm{P}}\, \psi^*(\boldsymbol{x}, t) = \eta_{\mathrm{P}}\, \psi^*(x)$$

under PT.

TABLE I

	P	T	PT
$\boldsymbol{p}$	−	−	+
$\boldsymbol{x} \times \boldsymbol{p}$	+	−	−
ρ	+	+	+
$\boldsymbol{j}$	−	−	+
$\boldsymbol{E}$	−	+	−
$\boldsymbol{H}$	+	−	−
A_0	+	+	+
$\boldsymbol{A}$	−	−	+

NOTE 2. Example of experimental detection of non-invariance under time reversal.

A π-mesic atom effects transitions $n = 4 \to n = 3$, $n = 3 \to n = 2$ and is left in a p-state. One can show that a measurement of the quantity $(\boldsymbol{k}_1 \times \boldsymbol{k}_2) \cdot \boldsymbol{L}$, where $\boldsymbol{k}_1$ and $\boldsymbol{k}_2$ are the momenta of the photons emitted and $\boldsymbol{L}$ is the angular momentum – in the centre of mass system – of the recoiling atom, could provide evidence for a possible non-invariance of the interaction under time reversal.

In fact, if the mean value of $(\boldsymbol{k}_1 \times \boldsymbol{k}_2) \cdot \boldsymbol{L}$ were different from zero, since this quantity must change sign under time reversal, it would mean that the development of the system differs from that of the time reversed system and therefore the interaction is not invariant under time reversal.

4. Charge conjugation

The Klein-Gordon equation for a spin zero particle in an electromagnetic field, and the corresponding Lagrangian density, are invariant under the operation of charge conjugation, denoted by C:

$$\psi'(x) = \eta_C \, \psi^*(x) , \tag{3}$$

where η_C is arbitrary but $|\eta_C| = 1$, accompanied by a change of sign *either* of the electromagnetic field

$$F_{\mu\nu} \to - F_{\mu\nu} , \quad \text{i.e.} \quad A_\mu \to - A_\mu , \tag{4}$$

or of the charge

$$e \to - e . \tag{5}$$

In both cases (4) and (5) the four-current

$$j_\mu = (\hbar/2mi) \left(\psi^* \frac{\partial \psi}{\partial x_\mu} - \frac{\partial \psi^*}{\partial x_\mu} \psi \right) + \frac{e}{mc} A_\mu \psi^* \psi$$

changes sign, whereas $- ej_\mu$ changes sign according to (4) and remains unchanged according to (5).

Finally the interaction Lagrangian $- ej_\mu A_\mu / c$ remains unchanged in either case.

NOTE 3. Example of experimental detection of non-invariance under charge conjugation.

One can show that the existence of quadrupole transitions $\Delta l = 2$ with emission of a photon for a bound system consisting of a meson of spin zero and charge e and its antiparticle (same mass and spin, charge $- e$) would indicate that the electromagnetic interaction is not invariant under charge conjugation.

In fact, to perform the operation of charge conjugation amounts to exchanging the two particles, i.e. the operation $\boldsymbol{p} \to - \boldsymbol{p}$, $\boldsymbol{x} \to - \boldsymbol{x}$ (where $\boldsymbol{p}$ and $\boldsymbol{x}$ are the relative momentum and the relative position in the centre of mass system), which is a space inversion. Therefore the system considered is an eigenstate of C belonging to the eigenvalue $(- 1)^l$, since by assumption it is in an eigenstate of P with the same eigenvalue.

A photon is a system in an eigenstate of C with eigenvalue $- 1$ (since, under charge conjugation, $A_\mu \to - A_\mu$). A transition $\Delta l = 2$ means therefore that the system (particle + electromagnetic radiation) goes from an eigenstate of C with the eigenvalue $(- 1)^l$ to one with the eigenvalue $(- 1)^{l+2+1} = - (- 1)^l$, i.e. which is opposite. The existence of such a transition would indicate that the interaction causing it is not invariant under C.

PART II

PARTICLES OF SPIN ONE-HALF

CHAPTER I

SPIN OPERATORS

1. Commutation and anticommutation relations

It has been stressed several times that the Klein-Gordon equation does not provide a correct description of the behaviour of electrons. These particles, together with protons, neutrons, μ-mesons and hyperons, possess an internal degree of freedom, the *spin*.

The spin is a vector $\boldsymbol{s} \equiv (s_1, s_2, s_3)$, whose component $\boldsymbol{s} \cdot \boldsymbol{n}$ along an arbitrary axis, defined by the unit vector $\boldsymbol{n}$, is a dichotomic variable. A measurement of $\boldsymbol{s} \cdot \boldsymbol{n}$ yields one of only two possible results, $+\hbar/2$ and $-\hbar/2$.

Pauli was the first to recognize in these properties those of an angular momentum, and developed a non-relativistic theory of the electron spin. In this theory $\boldsymbol{s}$ is regarded as an Hermitian * operator, whose components obey the commutation relations

$$[s_i, s_k] = i\hbar\, \varepsilon_{ikl}\, s_l\,, \tag{1}$$

identical with those written in Part I, Chap. IV, § 1, for the components of the orbital angular momentum. Because of this, simultaneous measurements of different components of $\boldsymbol{s}$ are incompatible. The exceptional case mentioned in Part I, Chap. IV, § 1, does not arise, as the operators s_1, s_2 and s_3 do not have the eigenvalue zero.

Since the eigenvalues of any component s_i are $s_i' = \pm\hbar/2$, one must have

$$s_i^2 = \tfrac{1}{4}\hbar^2 I\,, \tag{2}$$

where I denotes the identity operator; henceforth, for the sake of brevity,

* The distinction made in Part I, Chap. III, § 2, between self-adjoint and Hermitian operators was necessary because the product of two Feshbach-Villars wave functions was defined as $\frac{1}{2}\int \Psi_2^\dagger \tau_3 \Psi_1\, d^3x$ rather than $\frac{1}{2}\int \Psi_2^\dagger \Psi_1\, d^3x$. For particles of spin $\frac{1}{2}$ the product of two wave functions is defined as $\int \Psi_2^\dagger \Psi_1\, d^3x$ (see Chap. II and Chap. IV, § 3). Thus an Hermitian matrix (involving neither $\boldsymbol{x}$ nor its derivatives) is a self-adjoint operator. The term "self-adjoint" is not used in this part.

I will be often replaced by one. From eqs. (1) and (2) it follows that

$$\sum_l i\hbar\, \varepsilon_{ikl}(s_i s_l + s_l s_i) = s_i [s_i, s_k] + [s_i, s_k] s_i = [s_i{}^2, s_k] = 0$$

(no sum convention of equal indices is implied in this formula!).

Thus *

$$s_i s_l + s_l s_i = 0\,, \quad i \neq l\,. \tag{3}$$

Denoting by

$$\{a, b\} = ab + ba$$

the *anticommutator* of two operators a and b, eqs. (2) and (3) may be condensed into one formula

$$\{s_i, s_k\} = \tfrac{1}{2}\,\hbar^2\,\delta_{ik}\,.$$

It is convenient to introduce the adimensional Hermitian operator $\boldsymbol{\sigma}$ defined by

$$\boldsymbol{s} = \tfrac{1}{2}\,\hbar\boldsymbol{\sigma}\,.$$

Its components satisfy the relations

$$[\sigma_i, \sigma_k] = 2\mathrm{i}\,\varepsilon_{ikl}\,\sigma_l\,, \tag{4}$$

$$\{\sigma_i, \sigma_k\} = 2\delta_{ik}\,. \tag{5}$$

Adding eqs. (4) and (5) one obtains

$$\sigma_i \sigma_k = \mathrm{i}\varepsilon_{ikl}\,\sigma_l + \delta_{ik}\,, \tag{6}$$

and, in particular,

$$\sigma_1\sigma_2 = \mathrm{i}\sigma_3\,, \quad \sigma_2\sigma_3 = \mathrm{i}\sigma_1\,, \quad \sigma_3\sigma_1 = \mathrm{i}\sigma_2\,.$$

Let us now introduce the eigenstates $|+\rangle$ and $|-\rangle$ of the operator σ_3, belonging to the eigenvalues $\sigma_3' = \pm 1$:

$$\sigma_3\,|+\rangle = |+\rangle\,, \quad \sigma_3\,|-\rangle = -\,|-\rangle\,.$$

Such eigenstates may be used as a basic set to obtain a *representation* of the operators σ_1, σ_2 and σ_3.

---

* It should be stressed that the anticommutation relation (3) follows from the dichotomic character of the projection of $\boldsymbol{s}$ along an arbitrary direction, irrespective of the validity of eq. (1). In fact

$$\tfrac{1}{4}\,\hbar^2 = (\boldsymbol{s}\cdot\boldsymbol{n})^2 = n_i n_k s_i s_k = \tfrac{1}{2}\,n_i n_k (s_i s_k + s_k s_i)$$
$$= \sum_i (n_i s_i)^2 + \tfrac{1}{2}\sum_{i\neq k} n_i n_k (s_i s_k + s_k s_i) = \tfrac{1}{4}\,\hbar^2 + \tfrac{1}{2}\sum_{i\neq k} n_i n_k (s_i s_k + s_k s_i)\,,$$

from which, as $\boldsymbol{n}$ is arbitrary, one has

$$s_i s_k + s_k s_i = 0\,, \qquad i \neq k\,.$$

Notice first that the operators

$$\sigma_\pm = \tfrac{1}{2}(\sigma_1 \pm \mathrm{i}\sigma_2)$$

transform the state $|-\rangle$ into the state $|+\rangle$ and $|+\rangle$ into $|-\rangle$, but for numerical factors, which may be chosen in such a way as to make the representation of σ_1 and σ_2 as simple as possible. In fact, using the relations (6) we get

$$\sigma_3(\sigma_1 + \mathrm{i}\sigma_2)\,|-\rangle = (\sigma_1 + \mathrm{i}\sigma_2)\,|-\rangle = 2\alpha\,|+\rangle\,,$$
$$\sigma_3(\sigma_1 - \mathrm{i}\sigma_2)\,|+\rangle = -\,(\sigma_1 - \mathrm{i}\sigma_2)\,|+\rangle = -\,2\beta\,|-\rangle.$$

However, since

$$\sigma_-\sigma_+\,|-\rangle = \alpha\beta\,|-\rangle \quad \text{and} \quad \sigma_-\sigma_+ = \tfrac{1}{2}(I - \sigma_3)\,,$$

α and β are not completely arbitrary, but must satisfy

$$\alpha\beta = 1\,.$$

Moreover, if the normalization of the eigenstates $|\pm\rangle$ is

$$\langle+|+\rangle = \langle-|-\rangle = 1\,,$$

one has

$$\beta^* = (\langle-|\,\sigma_-\,|+\rangle)^* = \langle+|\,\sigma_+\,|-\rangle = \alpha\,.$$

Choosing $\alpha = \beta = 1$, thereby fully defining the states $|+\rangle$ and $|-\rangle$, one has

$$\langle+|\,\sigma_+\,|-\rangle = \langle-|\,\sigma_-\,|+\rangle = 1\,,$$
$$\langle+|\,\sigma_+\,|+\rangle = \langle-|\,\sigma_+\,|-\rangle = \langle+|\,\sigma_-\,|+\rangle = \langle-|\,\sigma_-\,|-\rangle =$$
$$= \langle-|\,\sigma_+\,|+\rangle = \langle+|\,\sigma_-\,|-\rangle = 0\,,$$
$$\langle+|\,\sigma_3\,|+\rangle = 1\,, \quad \langle-|\,\sigma_3\,|-\rangle = -\,1\,.$$

Therefore the matrices representing σ_+ and σ_- with respect to the basic set $|+\rangle$ and $|-\rangle$ are

$$\sigma_+ = \begin{pmatrix} 0 & 1 \\ 0 & 0 \end{pmatrix}, \quad \sigma_- = \begin{pmatrix} 0 & 0 \\ 1 & 0 \end{pmatrix}.$$

Hence one has the Hermitian matrices

$$\sigma_1 = \sigma_+ + \sigma_- = \begin{pmatrix} 0 & 1 \\ 1 & 0 \end{pmatrix}, \quad \sigma_2 = -\,\mathrm{i}(\sigma_+ - \sigma_-) = \begin{pmatrix} 0 & -\,\mathrm{i} \\ \mathrm{i} & 0 \end{pmatrix},$$
$$\sigma_3 = \begin{pmatrix} 1 & 0 \\ 0 & -\,1 \end{pmatrix}, \tag{7}$$

(*Pauli matrices*), which satisfy the same commutation and anticommutation relations as the operators they represent, and are identical in form with the τ's introduced in Part I, Chap. III, § 2.

2. Two-component wave functions

As far as the spin variables are concerned, a generic state $|\Psi\rangle$ may be expressed as a linear combination of the states of the basic set,

$$|\Psi\rangle = |+\rangle\langle+|\Psi\rangle + |-\rangle\langle-|\Psi\rangle .$$

Here $\langle+|\Psi\rangle$ and $\langle-|\Psi\rangle$ are the components along the "directions" of the states $|+\rangle$ and $|-\rangle$ respectively. They are themselves states as regards the degrees of freedom other than the spin. On the other hand, if A is a spin operator, the components $\langle+|\, A\, |\Psi\rangle$ and $\langle-|\, A\, |\Psi\rangle$ of the state $A\,|\Psi\rangle$ may be expressed in the form

$$\langle+|\, A\, |\Psi\rangle = \langle+|\, A\, |+\rangle\langle+|\Psi\rangle + \langle+|\, A\, |-\rangle\langle-|\Psi\rangle ,$$

$$\langle-|\, A\, |\Psi\rangle = \langle-|\, A\, |+\rangle\langle+|\Psi\rangle + \langle-|\, A\, |-\rangle\langle-|\Psi\rangle .$$

These last equations may be condensed into the matrix equation

$$\begin{pmatrix}\langle+|\, A\, |\Psi\rangle\\ \langle-|\, A\, |\Psi\rangle\end{pmatrix} = \begin{pmatrix}\langle+|\, A\, |+\rangle & \langle+|\, A\, |-\rangle\\ \langle-|\, A\, |+\rangle & \langle-|\, A\, |-\rangle\end{pmatrix}\begin{pmatrix}\langle+|\Psi\rangle\\ \langle-|\Psi\rangle\end{pmatrix} .$$

All this suggests the construction of a wave mechanics employing two-component wave functions

$$\Psi(x) = \begin{pmatrix}\psi_+(x)\\ \psi_-(x)\end{pmatrix} . \tag{8}$$

Here ψ_+ and ψ_-, the components * along the eigenstates $|+\rangle$ and $|-\rangle$ of σ_3, are, roughly speaking, probability amplitudes for the particle having a certain position and a certain value of s_3.

The spin components will be represented by the Pauli matrices (7).

---

* So far we have established that the wave mechanics of particles of spin $\frac{1}{2}$ must employ wave functions with *at least* two components. We shall see later that the relativistic theory of particles of spin $\frac{1}{2}$ and mass different from zero (electrons, protons etc., but not neutrinos) involves a four-component wave function.

We may introduce the eigenvectors * χ_+ and χ_- belonging to the eigenvalues + 1 and − 1 as

$$\chi_+ = \begin{pmatrix} 1 \\ 0 \end{pmatrix}, \quad \chi_- = \begin{pmatrix} 0 \\ 1 \end{pmatrix},$$

and write

$$\Psi = \begin{pmatrix} \psi_+ \\ \psi_- \end{pmatrix} = \chi_+\psi_+ + \chi_-\psi_- .$$

Note that χ_+ and χ_- have unit normalisation, i.e.

$$\chi_+^\dagger \chi_+ = \chi_-^\dagger \chi_- = 1 ,$$

and are orthogonal

$$\chi_+^\dagger \chi_- = \chi_-^\dagger \chi_+ = 0 .$$

It is easy to determine the eigenvectors of $\boldsymbol{\sigma} \cdot \boldsymbol{n}$, i.e. the projection of $\boldsymbol{\sigma}$ in a direction other than that of the third axis. The eigenvector

$$\begin{pmatrix} a \\ b \end{pmatrix}$$

belonging to the eigenvalue one must satisfy the equation

$$\boldsymbol{\sigma} \cdot \boldsymbol{n} \begin{pmatrix} a \\ b \end{pmatrix} = \begin{pmatrix} n_3 & n_1 - in_2 \\ n_1 + in_2 & -n_3 \end{pmatrix} \begin{pmatrix} a \\ b \end{pmatrix} = \begin{pmatrix} an_3 + b(n_1 - in_2) \\ a(n_1 + in_2) - bn_3 \end{pmatrix} = \begin{pmatrix} a \\ b \end{pmatrix},$$

having solutions

$$a = n_1 - in_2 , \quad b = 1 - n_3 .$$

Therefore

$$\chi_{\boldsymbol{\sigma} \cdot \boldsymbol{n} = 1} = [2(1 - n_3)]^{-\frac{1}{2}} \begin{pmatrix} n_1 - in_2 \\ 1 - n_3 \end{pmatrix} .$$

Similarly

$$\chi_{\boldsymbol{\sigma} \cdot \boldsymbol{n} = -1} = [2(1 - n_3)]^{-\frac{1}{2}} \begin{pmatrix} n_3 - 1 \\ n_1 + in_2 \end{pmatrix} .$$

---

* Some authors prefer to use a different notation,

$$\psi(\boldsymbol{x}, \sigma_3) = \begin{cases} \psi_+(\boldsymbol{x}) & \sigma_3 = +1 \\ \psi_-(\boldsymbol{x}) & \sigma_3 = -1 , \end{cases}$$

$$\psi(\boldsymbol{x}, \sigma_3) = \chi_+(\sigma_3)\, \psi_+(\boldsymbol{x}) + \chi_-(\sigma_3)\, \psi_-(\boldsymbol{x})$$

with

$$\chi_+(\sigma_3) = \begin{cases} 1 & \text{if} \quad \sigma_3 = 1 \\ 0 & \text{if} \quad \sigma_3 = -1 , \end{cases}$$

$$\chi_-(\sigma_3) = \begin{cases} 0 & \text{if} \quad \sigma_3 = 1 \\ 1 & \text{if} \quad \sigma_3 = -1 . \end{cases}$$

In these formulae the factor $[2(1-n_3)]^{-\frac{1}{2}}$ has been introduced so that the eigenvectors have norm one.

3. Spin, rotation operator, parity

An important property of the spin, following from its interpretation as an angular momentum, is that it may be considered as the generator of infinitesimal rotations. We shall show that the operator

$$R_\phi = \mathrm{e}^{\frac{1}{2}\mathrm{i}\boldsymbol{\sigma}\cdot\boldsymbol{n}\phi}$$

induces a finite counter-clockwise rotation of the coordinate axes through an angle ϕ around the direction of the unit vector $\boldsymbol{n}$.

To this end we shall study how this operator acts on $\boldsymbol{\sigma}$. For an infinitesimal transformation

$$R_{\delta\phi} = 1 + \tfrac{1}{2}\,\mathrm{i}\,\boldsymbol{\sigma}\cdot\boldsymbol{n}\,\delta\phi + \dots, \quad R_{\delta\phi}^{-1} = 1 - \tfrac{1}{2}\,\mathrm{i}\boldsymbol{\sigma}\cdot\boldsymbol{n}\,\delta\phi + \dots.$$

Neglecting quantities of the second order in $\delta\phi$, we have

$$R_{\delta\phi}^{-1}\,\sigma_i\,R_{\delta\phi} = \sigma_i + \tfrac{1}{2}\,\mathrm{i}n_k\,\delta\phi\,[\sigma_i, \sigma_k] = \sigma_i + \delta\phi(\boldsymbol{\sigma}\times\boldsymbol{n})_i\,,$$

namely

$$R_{\delta\phi}^{-1}\,\boldsymbol{\sigma}\,R_{\delta\phi} = \boldsymbol{\sigma} + \boldsymbol{\sigma}\times\boldsymbol{n}\,\delta\phi\,.$$

The right-hand side is nothing but the transformation of the fixed vector $\boldsymbol{\sigma}$ under an infinitesimal counter-clockwise rotation $\delta\phi$ around $\boldsymbol{n}$ of the coordinate axes.

We now verify that the unitary operator $R_\phi = \mathrm{e}^{\mathrm{i}\boldsymbol{\sigma}\cdot\boldsymbol{n}\phi/2}$ induces a finite rotation of angle ϕ around $\boldsymbol{n}$.

It is useful to rewrite R_ϕ in the form *

$$R_\phi = \cos(\tfrac{1}{2}\,\phi) + \mathrm{i}\,\boldsymbol{\sigma}\cdot\boldsymbol{n}\sin(\tfrac{1}{2}\,\phi)\,. \tag{9}$$

Then one has

$$R_\phi^{-1}\,\boldsymbol{\sigma}\,R_\phi = \boldsymbol{\sigma}\cos^2(\tfrac{1}{2}\,\phi) + \mathrm{i}\left[\boldsymbol{\sigma}(\boldsymbol{\sigma}\cdot\boldsymbol{n}) - (\boldsymbol{\sigma}\cdot\boldsymbol{n})\,\boldsymbol{\sigma}\right]\sin(\tfrac{1}{2}\,\phi)\cos(\tfrac{1}{2}\,\phi) + \\ + (\boldsymbol{\sigma}\cdot\boldsymbol{n})\,\boldsymbol{\sigma}(\boldsymbol{\sigma}\cdot\boldsymbol{n})\sin^2(\tfrac{1}{2}\,\phi)\,,$$

---

* The exponential $\mathrm{e}^{\mathrm{i}\boldsymbol{\sigma}\cdot\boldsymbol{n}\phi/2}$ can be expressed as a power series. Separating even and odd powers,

$$\mathrm{e}^{\mathrm{i}\boldsymbol{\sigma}\cdot\boldsymbol{n}\phi/2} = \Sigma(\mathrm{i}\boldsymbol{\sigma}\cdot\boldsymbol{n}\phi/2)^{2\nu}/(2\nu)! + \Sigma(\mathrm{i}\boldsymbol{\sigma}\cdot\boldsymbol{n}\phi/2)^{2\nu+1}/(2\nu+1)!\,.$$

As $(\boldsymbol{\sigma}\cdot\boldsymbol{n})^2 = 1$, this becomes

$$\mathrm{e}^{\mathrm{i}\boldsymbol{\sigma}\cdot\boldsymbol{n}\phi/2} = \Sigma(-1)^\nu(\phi/2)^{2\nu}/(2\nu)! + \mathrm{i}(\boldsymbol{\sigma}\cdot\boldsymbol{n})\,\Sigma(-1)^\nu(\phi/2)^{2\nu+1}/(2\nu+1)! \\ = \cos(\tfrac{1}{2}\,\phi) + \mathrm{i}\boldsymbol{\sigma}\cdot\boldsymbol{n}\sin(\tfrac{1}{2}\,\phi)\,.$$

which, since *

$$\boldsymbol{\sigma}(\boldsymbol{\sigma}\cdot\boldsymbol{n}) - (\boldsymbol{\sigma}\cdot\boldsymbol{n})\,\boldsymbol{\sigma} = 2\mathrm{i}\,\boldsymbol{n}\times\boldsymbol{\sigma}\,,$$
$$(\boldsymbol{\sigma}\cdot\boldsymbol{n})\,\boldsymbol{\sigma}(\boldsymbol{\sigma}\cdot\boldsymbol{n}) = 2\boldsymbol{n}(\boldsymbol{n}\cdot\boldsymbol{\sigma}) - \boldsymbol{\sigma}\,,$$

gives

$$R_\phi^{-1}\,\boldsymbol{\sigma}\,R_\phi = \boldsymbol{\sigma}\cos\phi + \boldsymbol{n}(\boldsymbol{n}\cdot\boldsymbol{\sigma})(1-\cos\phi) + \boldsymbol{\sigma}\times\boldsymbol{n}\sin\phi = \boldsymbol{\sigma}'\,. \tag{10}$$

This is in accordance with the formula ** for the transformation of a vector under a rotation of the system of coordinate axes through an angle ϕ around the direction $\boldsymbol{n}$.

As a special case of eq. (10), the formulae for a rotation around the third axis are

$$\mathrm{e}^{-\mathrm{i}\sigma_3\phi/2}\,\sigma_1\,\mathrm{e}^{\mathrm{i}\sigma_3\phi/2} = \sigma_1\cos\phi + \sigma_2\sin\phi\,,$$
$$\mathrm{e}^{-\mathrm{i}\sigma_3\phi/2}\,\sigma_2\,\mathrm{e}^{\mathrm{i}\sigma_3\phi/2} = \sigma_2\cos\phi - \sigma_1\sin\phi\,,$$
$$\mathrm{e}^{-\mathrm{i}\sigma_3\phi/2}\,\sigma_3\,\mathrm{e}^{\mathrm{i}\sigma_3\phi/2} = \sigma_3\,.$$

Now, one has the eigenvalue equations

$$\sigma_3\,|+\rangle = |+\rangle\,,\quad \sigma_3\,|-\rangle = -\,|-\rangle$$

with respect to the original system of axes, and, with respect to a rotated system,

$$\sigma_3'\,|+\rangle' = |+\rangle'\,,\quad \sigma_3'\,|-\rangle' = -\,|-\rangle'\,,$$

where $|+\rangle'$ and $|-\rangle'$ are the eigenstates of the component of $\boldsymbol{\sigma}$ in the direction of the z' axis.

But

$$\sigma_3' = R_\phi^{-1}\,\sigma_3\,R_\phi\,,$$

from which it is seen that one must have

$$|\pm\rangle = R_\phi\,|\pm\rangle'\,.$$

We must, however, find the transformation properties of the wave function, whose elements are the components of the spin state $|\Psi\rangle$ with respect to the basic states $|\pm\rangle$.

---

* These equations are easily verified:

$$\sigma_i\sigma_k n_k - \sigma_k n_k\sigma_i = n_k(\sigma_i\sigma_k - \sigma_k\sigma_i) = 2\mathrm{i}\varepsilon_{ikl}n_k\sigma_l = 2\mathrm{i}(\boldsymbol{n}\times\boldsymbol{\sigma})_i\,;$$
$$\sigma_i\sigma_k n_k + \sigma_k n_k\sigma_i = n_k(\sigma_i\sigma_k + \sigma_k\sigma_i) = 2n_i$$

and

$$(\boldsymbol{\sigma}\cdot\boldsymbol{n})\,\sigma_i(\boldsymbol{\sigma}\cdot\boldsymbol{n}) = \boldsymbol{\sigma}\cdot\boldsymbol{n}(2n_i - \sigma_k n_k\sigma_i) = 2n_i(\boldsymbol{\sigma}\cdot\boldsymbol{n}) - \sigma_i\,.$$

** See, for instance, E. Madelung, *Die Mathematischen Hilfsmittel des Physikers* (Springer, Berlin, 1950), p. 155. There, however, a formula is given for the rotation of a vector around $\boldsymbol{n}$ through an angle ϕ. Here the system of coordinate axes rotates, while the vector $\boldsymbol{\sigma}$ is fixed. Therefore $\sin\phi$ occurs with a different sign in eq. (10) and in the formula given in the reference.

Denoting by $\langle\pm|\Psi\rangle'$ the components of $|\Psi\rangle$ along the new basic states $|\pm\rangle'$, we have

$$\langle\pm|\Psi\rangle' = \langle\pm|\, e^{i\boldsymbol{\sigma}\cdot\boldsymbol{n}\phi/2}\,|+\rangle\langle+|\Psi\rangle + \langle\pm|\, e^{i\boldsymbol{\sigma}\cdot\boldsymbol{n}\phi/2}\,|-\rangle\langle-|\Psi\rangle\,,$$

from which

$$\psi'_{\pm}(\boldsymbol{x}') = \langle\pm|\, e^{i\boldsymbol{\sigma}\cdot\boldsymbol{n}\phi/2}\,|+\rangle\, \psi_+(\boldsymbol{x}) + \langle\pm|\, e^{i\boldsymbol{\sigma}\cdot\boldsymbol{n}\phi/2}\,|-\rangle\, \psi_-(\boldsymbol{x})\,.$$

This can be expressed in matrix form as

$$\Psi'(\boldsymbol{x}') = \begin{pmatrix}\psi'_+(\boldsymbol{x}')\\ \psi'_-(\boldsymbol{x}')\end{pmatrix} = \begin{pmatrix}\langle+|\, e^{i\boldsymbol{\sigma}\cdot\boldsymbol{n}\phi/2}\,|+\rangle & \langle+|\, e^{i\boldsymbol{\sigma}\cdot\boldsymbol{n}\phi/2}\,|-\rangle\\ \langle-|\, e^{i\boldsymbol{\sigma}\cdot\boldsymbol{n}\phi/2}\,|+\rangle & \langle-|\, e^{i\boldsymbol{\sigma}\cdot\boldsymbol{n}\phi/2}\,|-\rangle\end{pmatrix}\begin{pmatrix}\psi_+(\boldsymbol{x})\\ \psi_-(\boldsymbol{x})\end{pmatrix} =$$
$$= e^{i\boldsymbol{\sigma}\cdot\boldsymbol{n}\phi/2}\,\Psi(\boldsymbol{x})\,,$$

where, in the last expression on the right, the components σ_1, σ_2 and σ_3 of $\boldsymbol{\sigma}$ are the Pauli matrices.

An eigenvector of $\boldsymbol{\sigma}\cdot\boldsymbol{n}$ belonging to the eigenvalue ± 1 is multiplied by $e^{\pm i\phi/2}$ as the result of a rotation through an angle ϕ around $\boldsymbol{n}$. In particular, χ_+ and χ_- are eigenvectors of the operator for a rotation around the third axis,

$$e^{i\sigma_3\phi/2}\begin{pmatrix}1\\0\end{pmatrix} = e^{i\phi/2}\begin{pmatrix}1\\0\end{pmatrix}, \quad e^{i\sigma_3\phi/2}\begin{pmatrix}0\\1\end{pmatrix} = e^{-i\phi/2}\begin{pmatrix}0\\1\end{pmatrix}.$$

It is important to note that, under a rotation of 360^0 ($\phi = 2\pi$), the spin function is multiplied by -1, owing to the fact that a half-integral multiple of ϕ occurs in the exponent of the rotation operator. This is not the case for the rotation operator acting on the "external" degrees of freedom:

$$\psi_{\pm}(\boldsymbol{x}') = e^{-i\boldsymbol{L}\cdot\boldsymbol{n}\phi/\hbar}\,\psi_{\pm}(\boldsymbol{x})\,.$$

For instance, the spherical harmonics, which are eigenfunctions of the third component of the orbital angular momentum, are multiplied by $e^{im\phi}$ (with m integer) under a rotation around the third axis, and are therefore left unaltered by a rotation of 2π.

The spin functions of particles of spin $\frac{1}{2}$ are not single valued representations of the rotation group *.

The coordinate-and-spin functions

$$\mathfrak{Y}_l^{(m+\frac{1}{2})} = Y_l^{(m)}(\theta,\phi)\begin{pmatrix}1\\0\end{pmatrix}, \quad \mathfrak{Y}_l^{(m-\frac{1}{2})} = Y_l^{(m)}(\theta,\phi)\begin{pmatrix}0\\1\end{pmatrix}$$

---

* This is true for the spin functions of all particles of half-integral spin ($\frac{1}{2}$, $\frac{3}{2}$ . . .), see Van der Waerden, *loc. cit.*

are eigenfunctions of $l_3 + \frac{1}{2}\sigma_3 = (L_3 + s_3)/\hbar$ belonging to the eigenvalues $m + \frac{1}{2}$ and $m - \frac{1}{2}$ respectively. Under a rotation α around the z axis they change (in form) by the factor $\mathrm{e}^{\mathrm{i}\,(m\pm\frac{1}{2})\alpha}$

$$\mathfrak{Y}_l^{(m\pm\frac{1}{2})'}(\theta, \phi) = \mathrm{e}^{\mathrm{i}\,(m\pm\frac{1}{2})\alpha}\, \mathfrak{Y}_l^{(m\pm\frac{1}{2})}(\theta, \phi)\,.$$

The components of $\boldsymbol{\sigma}$, and therefore also the eigenvectors of σ_3, are left unchanged by space inversion. Therefore, as for particles of spin zero, the wave function $\Psi(\boldsymbol{x})$ is transformed under space inversion according to

$$\Psi'(\boldsymbol{x}', t) = \eta_{\mathrm{P}}\, \Psi(\boldsymbol{x}, t)$$

with $|\eta_{\mathrm{P}}| = 1$. Here, however, it cannot be stated that the square of the inversion operator P is one, but, owing to the two-valuedness of the spin functions, only $\mathrm{P}^4 = 1$. Hence, instead of $\eta_{\mathrm{P}}^2 = 1$, which we had for spinless particles, for particles of spin $\frac{1}{2}$ we have $\eta_{\mathrm{P}}^4 = 1$, which restricts the values of the intrinsic parity to the four possibilities

$$\eta_{\mathrm{P}} = \pm 1\,, \quad \pm \mathrm{i}\,.$$

Particles of spin $\frac{1}{2}$ may, therefore, have an imaginary intrinsic parity.

NOTE 1. A wave function of the type (8) has intrinsic parity $+ 1$ or $- 1$ according as $\psi_\pm(x)$ are scalars or skew-symmetric tensors like ε_{ikl}, ($i, k, l = 1, 2, 3$, non-relativistic theory!).

From such a wave function, one with intrinsic parity $\pm$ i is easily constructed, e.g. in the form

$$\tfrac{1}{2}\left[1 - (\varepsilon_{ikl}\, \sigma_i \sigma_k \sigma_l/3!)\right] \Psi(x)\,.$$

In fact

$$\frac{1}{2}\left(1 - \frac{\varepsilon_{ikl}\, \sigma_i \sigma_k \sigma_l}{3!}\right) = \frac{1 - \mathrm{i}}{2}$$

in a left-handed system. But, on transforming to a right-handed system

$$\frac{1}{2}\left(1 - \frac{\varepsilon_{ikl}\, \sigma_i \sigma_k \sigma_l}{3!}\right)' = \frac{1}{2}\left(1 - \frac{\varepsilon'_{ikl}}{3!}\, \sigma_i \sigma_k \sigma_l\right) = \tfrac{1}{2}(1 + \mathrm{i}) = \tfrac{1}{2}\mathrm{i}(1 - \mathrm{i})\,.$$

CHAPTER II

THE WEYL EQUATION

1. Wave equation

The non-relativistic wave mechanics of particles of spin $\frac{1}{2}$ rests on the assumption that the two-component wave function obeys the Schroedinger equation

$$[(-\hbar^2/2m)\,\Delta + V]\,\Psi = \hbar \mathrm{i}\,\partial\Psi/\partial t\,. \tag{1}$$

This equation splits into two equations for the components ψ_+ and ψ_-, which are independent if the potential does not involve the spin *.

The construction of a relativistic equation for the two-component wave function is not so simple. We impose the condition that such an equation must be linear in order that the superposition principle remain valid, and of the first order with respect to both space and time derivatives. It should be required also that the components ψ_+ and ψ_- are coupled even in the absence of external spin-dependent forces, in such a way that, even for a free particle, an interdependence of the translational and spin degrees of freedom will be established **.

The equation sought must be invariant under space-time translations. Therefore it may not contain $\boldsymbol{x}$ and t (except as arguments of the wave function), but only their derivatives. Finally it must be invariant under spatial rotations. These conditions are required in order that, for a free particle, the four-momentum and the angular momentum be constants of

* This is not the case for a particle in an electromagnetic field. Eq. (1) then becomes

$$[(-\hbar^2/2m)(\nabla_x + \mathrm{i}e\boldsymbol{A}/\hbar c)^2 - eA_0 + (e\hbar/2mc)\,\boldsymbol{\sigma}\cdot\boldsymbol{H}]\,\Psi = \hbar\mathrm{i}\,\partial\Psi/\partial t$$

($-e$ = charge of the particle, $\boldsymbol{\mu} = -e\hbar\boldsymbol{\sigma}/2mc$ = magnetic moment associated with the spin angular momentum, see eq. (5), Chap. VIII).

** The Klein-Gordon equation

$$(\Box - \kappa^2)\begin{pmatrix}\psi_+\\ \psi_-\end{pmatrix} = 0$$

would not fulfil these requirements, since it would split into independent equations for ψ_+ and ψ_-.

motion. The rotation invariance can be satisfied only if the operator $\boldsymbol{\sigma}$ occurs in a scalar product $\boldsymbol{\sigma} \cdot \boldsymbol{a}$, $\boldsymbol{a}$ being a vector. This cannot be $\boldsymbol{x}$ (translation invariance!), nor can it be a fixed vector. Therefore the dependence of $\boldsymbol{\sigma}$ must occur through the intermediacy of $\boldsymbol{\sigma} \cdot \nabla_x$.

The above conditions restrict the form of the relativistic equation to

$$\left(\frac{1}{c}\frac{\partial}{\partial t} + \alpha\, \boldsymbol{\sigma} \cdot \nabla_x + \beta\right) \Psi = 0\,, \tag{2}$$

α and β being constants.

If this equation is multiplied by

$$\left(\alpha\, \boldsymbol{\sigma} \cdot \nabla_x - \beta - \frac{1}{c}\frac{\partial}{\partial t}\right)$$

from the left, one has *

$$\alpha^2 \Delta\, \Psi - \frac{1}{c^2}\frac{\partial^2 \Psi}{\partial t^2} - 2\frac{\beta}{c}\frac{\partial \Psi}{\partial t} - \beta^2\, \Psi = 0\,, \tag{3}$$

from which the spin has disappeared. A further requirement, that the linear motion of a particle be described by a plane wave, so that eq. (3) must have solutions of the type $\sim e^{i(k,x)}$, gives

$$k_0^2 + 2i\,\beta\, k_0 - \alpha^2 \boldsymbol{k}^2 - \beta^2 = 0\,.$$

The solutions of this equation are

$$k_0 = -\, i\beta \pm \alpha\, |\boldsymbol{k}|\,,$$

which are in general incompatible ** with the assumption that $k_4 = ik_0$ and $\boldsymbol{k}$ are the components of a four-vector, except in the case $\alpha = 1, \beta = 0$, for which

$$k_0 = \pm\, |\boldsymbol{k}|\,.$$

This is the energy-momentum relation for a particle of zero rest mass.

---

* Use the identity

$$(\boldsymbol{\sigma} \cdot \nabla_x)(\boldsymbol{\sigma} \cdot \nabla_x) = \sigma_i \sigma_k \nabla_i \nabla_k = \Delta + i\varepsilon_{ikl}\sigma_l \nabla_i \nabla_k = \Delta\,,$$

which is due to the antisymmetry of ε_{ikl} and the symmetry of $\nabla_i \nabla_k$ in the two indices i and k.

** For $\alpha = 1$ and $\beta = imc/\hbar$, eq. (2) would take the attractive form

$$\left[\frac{1}{c}\frac{\partial}{\partial t} + \boldsymbol{\sigma} \cdot \nabla_x + \frac{imc}{\hbar}\right] \Psi = 0,$$

but the relation between k_0 and $\boldsymbol{k}$ would then be $k_0 = \kappa \pm |\, \boldsymbol{k}\, |$.

For $\alpha = 1$, $\beta = 0$, eq. (3) becomes the Klein-Gordon equation for massless particles:

$$\Box \Psi = 0 . \tag{4}$$

Thus one is forced to the conclusion that a relativistic two-component theory of particles of spin $\frac{1}{2}$ is feasible only if the rest mass is zero, in which case the wave function obeys the equation

$$\left(\frac{1}{c}\frac{\partial}{\partial t} + \boldsymbol{\sigma} \cdot \nabla_x\right) \Psi = 0 . \tag{5}$$

It seems that this equation, established long ago by Weyl *, is appropriate for the description of neutrinos **.

For a solution of the type

$$\Psi(x) \propto e^{i(\boldsymbol{k} \cdot \boldsymbol{x} - k_0 x_0)} ,$$

one has

$$\boldsymbol{\sigma} \cdot \boldsymbol{k} = k_0 .$$

On the other hand, since $k_0 = \pm |\boldsymbol{k}|$, this gives

$$\boldsymbol{\sigma} \cdot \boldsymbol{k}/|\boldsymbol{k}| = \pm 1 . \tag{6}$$

Therefore the spin is parallel to $\boldsymbol{k}$ for a positive frequency solution (neutrino) and is antiparallel for a solution of negative frequency (antineutrino) ***.

From eq. (6) it follows that a typical solution of eq. (5) is

$$\Psi(x) = [2(\boldsymbol{k}^2 - k_3 |\boldsymbol{k}|)]^{-\frac{1}{2}} \times$$
$$\times \left[a\, e^{i(\boldsymbol{k} \cdot \boldsymbol{x} - \omega t)} \begin{pmatrix} k_1 - i k_2 \\ |\boldsymbol{k}| - k_3 \end{pmatrix} + b\, e^{i(\boldsymbol{k} \cdot \boldsymbol{x} + \omega t)} \begin{pmatrix} k_3 - |\boldsymbol{k}| \\ k_1 + i k_2 \end{pmatrix} \right] .$$

The general solution of eq. (5) is then a superposition of solutions of this type.

On multiplying eq. (5) by $\Psi^\dagger \equiv (\psi_+^* \, \psi_-^*)$ from the left, and the Hermitian conjugate of (5),

$$\frac{1}{c}\frac{\partial \Psi^\dagger}{\partial t} + \nabla_x \Psi^\dagger \cdot \boldsymbol{\sigma} = 0 ,$$

* H. Weyl, Zs. für Physik **56** (1929), 330.

** See the article by T. D. Lee and C. N. Yang, Parity Nonconservation and a Two-component Theory of the Neutrino, Phys. Rev. **105** (1957), 1671.

*** Actually it is customary to call an antineutrino the particle whose spin is parallel to $\boldsymbol{k}$ and is emitted in the β^--decay $n \to p + e^- + \textit{antineutrino}$.

by Ψ from the right, and adding, one has

$$\frac{1}{c}\frac{\partial}{\partial t}(\Psi^\dagger \Psi) + \nabla_x(\Psi^\dagger \boldsymbol{\sigma} \Psi) = 0 .$$

This is a continuity equation which may be written in the form

$$\partial j_\mu / \partial x_\mu = 0 , \tag{7}$$

where the four-current $j_\mu = (\boldsymbol{j}, j_4)$ is

$$\boldsymbol{j} = c\, \Psi^\dagger \boldsymbol{\sigma} \Psi , \quad j_4 = \mathrm{i}\, c\, \Psi^\dagger \Psi .$$

It will be shown that j_μ, ($\mu = 1, 2, 3, 4$), behave as the components of a four-vector under a Lorentz transformation. On the other hand, the quantities

$$\Psi^\dagger \frac{\partial \Psi}{\partial x_\mu} - \frac{\partial \Psi^\dagger}{\partial x_\mu} \Psi , \quad (\mu = 1, 2, 3, 4) ,$$

which, as is easy to verify using eq. (4), satisfy the continuity equation

$$\frac{\partial}{\partial x_\mu}\left[\Psi^\dagger \frac{\partial \Psi}{\partial x_\mu} - \frac{\partial \Psi^\dagger}{\partial x_\mu} \Psi\right] = 0 ,$$

are not the components of a four-vector (as may be verified after reading § 2). They are proportional to the densities of the momentum, T_{4i}, ($i = 1, 2, 3$), and of the energy, T_{44}.

Note that the energy density $(\mathrm{i}\hbar/2)(\Psi^\dagger \partial\Psi/\partial t - \partial\Psi^\dagger/\partial t\, \Psi)$ is not positive definite.

The energy-momentum tensor $T_{\mu\nu}$ may, of course, be deduced by Noether's theorem from the Lagrangian of the Weyl equation, which the reader may construct as an exercise (after studying the following chapter on the Dirac equation).

2. Transformation properties

2.1. Space inversion

Eq. (5) cannot be invariant under space inversion, owing to the occurrence of the pseudoscalar $\boldsymbol{\sigma} \cdot \nabla_x$. This is in fact the inner product of the polar vector ∇_x with $\boldsymbol{\sigma}$, which, as an angular momentum, is an axial vector. For a long time this appeared a sufficient reason for rejecting the Weyl equation, which was resumed following the discovery of parity violation in processes of emission and absorption of neutrinos.

The property of parallelism or antiparallelism of $\boldsymbol{k}$ with $\boldsymbol{\sigma}$, which distinguishes between neutrinos and antineutrinos, is also non-invariant under space inversion, nor is the continuity equation (7) since $\boldsymbol{j} = c\,\Psi^\dagger\,\boldsymbol{\sigma}\,\Psi$ is an axial vector and $\Psi^\dagger\Psi$ a scalar.

2.2. *Time reversal*

It is easy to see that eq. (5) is invariant under time reversal, if the transformation

$$\Psi'(\boldsymbol{x}, t') = \eta_{\mathrm{T}}\,\sigma_1\,\sigma_3\,\Psi^*\,, \tag{8}$$

with

$$|\eta_{\mathrm{T}}|^2 = 1\,,$$

is assumed for the wave function.

Indeed one has

$$\boldsymbol{\sigma}\,\sigma_1\,\sigma_3 = -\,(\sigma_1\,\sigma_3\,\boldsymbol{\sigma})^*$$

and therefore

$$\left(\frac{1}{c}\frac{\partial}{\partial t'} + \boldsymbol{\sigma}\cdot\boldsymbol{\nabla}_x\right)\Psi' = -\,\eta_{\mathrm{T}}\,\sigma_1\,\sigma_3\left(\frac{1}{c}\frac{\partial}{\partial t} + \boldsymbol{\sigma}\cdot\boldsymbol{\nabla}_x\right)^*\Psi^* = 0\,.$$

One easily sees that eq. (8) gives the correct transformation for the current density

$$\boldsymbol{j}' = (\boldsymbol{j}')^* = c\Psi^\dagger\,\sigma_3\sigma_1(\boldsymbol{\sigma}\,\sigma_1\,\sigma_3)^*\,\Psi = -\,c\,\Psi^\dagger\,\boldsymbol{\sigma}\,\Psi = -\boldsymbol{j}\,.$$

On the other hand,

$$\rho' = \rho'^* = \Psi^\dagger\,\sigma_3\,\sigma_1\,\sigma_1\,\sigma_3\,\Psi = \rho\,.$$

2.3. *Spatial rotations*

Invariance under space rotations may be established by showing that the equation

$$\left(\frac{1}{c}\frac{\partial}{\partial t} + \boldsymbol{\sigma}\cdot\boldsymbol{\nabla}_{x'}\right)\Psi'(\boldsymbol{x}', t) = 0 \tag{9}$$

in a reference system rotated counterclockwise through an angle ϕ around the direction of the unit vector $\boldsymbol{n}$, follows from the equation in the original reference system if

$$\Psi'(\boldsymbol{x}', t) = \mathrm{e}^{\mathrm{i}\boldsymbol{\sigma}\cdot\boldsymbol{n}\phi/2}\,\Psi(\boldsymbol{x}, t)\,.$$

In fact, multiplying eq. (5) by $\exp(\mathrm{i}\,\boldsymbol{\sigma}\cdot\boldsymbol{n}\phi/2)$ from the left, one has

$$0 = \left(\frac{1}{c}\frac{\partial}{\partial t} + \mathrm{e}^{\mathrm{i}\boldsymbol{\sigma}\cdot\boldsymbol{n}\phi/2}\,\boldsymbol{\sigma}\,\mathrm{e}^{-\mathrm{i}\boldsymbol{\sigma}\cdot\boldsymbol{n}\phi/2}\,\boldsymbol{\nabla}_x\right)\mathrm{e}^{\mathrm{i}\boldsymbol{\sigma}\cdot\boldsymbol{n}\phi/2}\,\Psi(x) =$$

$$= \left(\frac{1}{c}\frac{\partial}{\partial t} + \boldsymbol{\sigma}\cdot\boldsymbol{\nabla}_{x'}\right)\Psi'(\boldsymbol{x}', t)\,.$$

Use has been made of the fact that

$$e^{i\boldsymbol{\sigma}\cdot\boldsymbol{n}\phi/2}\,\sigma_i\,e^{-i\boldsymbol{\sigma}\cdot\boldsymbol{n}\phi/2}\,\nabla_i = (a^{-1})_{ik}\,\sigma_k\,\nabla_i = \sigma_k\,a_{ki}\,\nabla_i = \sigma_k\,\nabla'_k$$

(here a_{ik} are the coefficients of the transformation $x'_i = a_{ik}\,x_k$ for the space rotation considered, and $(a^{-1})_{ik}$ are the elements of the inverse matrix of $\| a_{ik} \|$).

Notice that the matrices σ_1, σ_2 and σ_3 in eq. (9) are identical with those in eq. (5).

2.4. *Lorentz transformations*

For simplicity we consider a transformation with relative velocity along the first axis *, for which

$$x'_1 = \gamma(x_1 - vt)\,, \quad t' = \gamma\left(t - \frac{v}{c^2}\,x_1\right), \quad x'_2 = x_2\,, \quad x'_3 = x_3\,.$$

Expressing the operators $\partial/\partial t'$ and $\boldsymbol{\nabla}_{x'}$ in terms of $\partial/\partial t$ and $\boldsymbol{\nabla}_x$, and assuming the transformation formula for the wave function

$$\Psi'(x') = \left(\gamma + 1 - \gamma\frac{v}{c}\,\sigma_1\right)\Psi(x)\,, \tag{10}$$

one has **

$$\left(\frac{1}{c}\frac{\partial}{\partial t'} + \boldsymbol{\sigma}\cdot\boldsymbol{\nabla}_{x'}\right)\Psi'(x') = \left(\gamma + 1 + \gamma\frac{v}{c}\,\sigma_1\right)\left(\frac{1}{c}\frac{\partial}{\partial t} + \boldsymbol{\sigma}\cdot\boldsymbol{\nabla}_x\right)\Psi(x)\,.$$

---

* The extension of the proof to the case of an arbitrary Lorentz transformation may be found in the article by R. H. Good, Jr., *Theory of Particles with Zero Rest Mass*, in Lectures on Theoretical Physics, Vol. I, Boulder (1958) (Interscience, New York, 1959), p. 30.

** In fact

$$\left(\frac{1}{c}\frac{\partial}{\partial t'} + \boldsymbol{\sigma}\cdot\boldsymbol{\nabla}_{x'}\right)\Psi'(x') = \frac{\gamma}{c}\left(\frac{\partial}{\partial t} + v\frac{\partial}{\partial x_1}\right)\Psi' + \sigma_1\gamma\left(\frac{\partial}{\partial x_1} + \frac{v}{c^2}\frac{\partial}{\partial t}\right)\Psi' + \sigma_2\frac{\partial\Psi'}{\partial x_2} +$$

$$+ \sigma_3\frac{\partial\Psi'}{\partial x_3} = \gamma\left(1 + \sigma_1\frac{v}{c}\right)\left(\frac{1}{c}\frac{\partial}{\partial t} + \sigma_1\frac{\partial}{\partial x_1}\right)\Psi' + \left(\sigma_2\frac{\partial}{\partial x_2} + \sigma_3\frac{\partial}{\partial x_3}\right)\Psi'\,.$$

Expressing $\Psi'(x')$ in terms of $\Psi(x)$ and remembering that $\sigma_i\sigma_k = i\varepsilon_{ikl}\sigma_l + \delta_{ik}$, one finds

$$\left(\frac{1}{c}\frac{\partial}{\partial t'} + \boldsymbol{\sigma}\cdot\boldsymbol{\nabla}_{x'}\right)\Psi'(x') = \left(\gamma^2 + \gamma + \gamma\sigma_1\frac{v}{c} - \gamma^2\frac{v^2}{c^2}\right)\left(\frac{1}{c}\frac{\partial\Psi}{\partial t} + \sigma_1\frac{\partial\Psi}{\partial x_1}\right) +$$

$$+ \left(\gamma + 1 + \gamma\frac{v}{c}\,\sigma_1\right)\left(\sigma_2\frac{\partial\Psi}{\partial x_2} + \sigma_3\frac{\partial\Psi}{\partial x_3}\right) = \left(\gamma + 1 + \gamma\frac{v}{c}\,\sigma_1\right)\left(\frac{1}{c}\frac{\partial\Psi}{\partial t} + \boldsymbol{\sigma}\cdot\boldsymbol{\nabla}_x\Psi\right).$$

Therefore the equation

$$\left(\frac{1}{c}\frac{\partial}{\partial t'} + \boldsymbol{\sigma}\cdot\nabla_{x'}\right)\Psi'(x') = 0$$

leads to eq. (5) of § 1, and vice versa. The σ's are the same in the two equations.

It should be stressed, however, that the transformation formula has so far been determined only up to a constant. As we show below, a numerical factor must be introduced in eq. (10) ,which is replaced by

$$\Psi'(x') = [2(1+\gamma)]^{-\frac{1}{2}}\left(\gamma + 1 - \gamma\frac{v}{c}\sigma_1\right)\Psi(x)$$

and

$$\Psi'(x')^{\dagger} = \Psi^{\dagger}(x)\,[2(1+\gamma)]^{-\frac{1}{2}}\left(\gamma + 1 - \gamma\frac{v}{c}\sigma_1\right).$$

We can now write

$$[2(1+\gamma)]^{-\frac{1}{2}}\left(\gamma + 1 - \gamma\frac{v}{c}\sigma_1\right) = \cos\chi + \mathrm{i}\,\sigma_1\sin\chi = \mathrm{e}^{\mathrm{i}\sigma_1\chi},$$

with

$$\cos\chi = \cosh(\mathrm{i}\chi) = [\tfrac{1}{2}(\gamma+1)]^{\frac{1}{2}}, \quad \mathrm{i}\sin\chi = \sinh(\mathrm{i}\chi) = -\,[\tfrac{1}{2}(\gamma-1)]^{\frac{1}{2}}.$$

Therefore a Lorentz transformation along the x_1 axis, namely a rotation in the plane x_1, t, is described by the operator $\mathrm{e}^{\mathrm{i}\sigma_1\chi}$. Similarly $\mathrm{e}^{\mathrm{i}\sigma_2\chi}$ and $\mathrm{e}^{\mathrm{i}\sigma_3\chi}$ are the operators for Lorentz transformations along the x_2 and x_3 axes, respectively. These operators are similar to $\mathrm{e}^{\mathrm{i}\sigma_1\phi/2}$, $\mathrm{e}^{\mathrm{i}\sigma_2\phi/2}$ and $\mathrm{e}^{\mathrm{i}\sigma_3\phi/2}$ for space rotations around the first, second and third axes. The angle χ, however, is not real, so that $\mathrm{e}^{\mathrm{i}\sigma_i\chi}$, ($i = 1, 2, 3$), are not unitary. They are, however, unitary in the subspace of the solutions of the Weyl equation. In fact, for any two solutions Ψ_1 and Ψ_2, the quantities $c\Psi_2^{\dagger}\boldsymbol{\sigma}\Psi_1$ and $\mathrm{i}c\Psi_2^{\dagger}\Psi_1$ transform as the components of a four-vector (see below), and obey the continuity eq. (7). The product $(\Psi_2, \Psi_1) = \int \Psi_2^{\dagger}\Psi_1\,\mathrm{d}^3\mathrm{x}$ is therefore equal to the product $(\Psi_2', \Psi_1') = \int \Psi_2'^{\dagger}\Psi_1'\,\mathrm{d}^3\mathrm{x}'$ of the transformed wave functions.

It is easy to show that the continuity equation for the four-current j_μ, which is not invariant under space inversion, is so with respect to Lorentz transformations. To show this, it is sufficient to prove that j_1, j_2, j_3 and j_4 are transformed as the components of a four-vector.

For a rotation in the plane x_1, t,

$$j_1' = c\,\Psi^{\dagger\prime}\sigma_1\Psi' = [c/2(\gamma+1)]\,\Psi^\dagger\sigma_1(\gamma+1)(\gamma+1-2\gamma\beta\sigma_1+\gamma^2\beta^2/(\gamma+1))\Psi$$
$$= c\,\Psi^\dagger\sigma_1(\gamma-\gamma\beta\sigma_1)\Psi = \gamma(j_1+\mathrm{i}\beta j_4)\,,$$
$$j_2' = c\,\Psi^{\dagger\prime}\sigma_2\Psi' = [c/2(\gamma+1)]\,\Psi^\dagger\sigma_2 2(\gamma+1)\Psi = j_2\,.$$

Similarly one finds

$$j_3' = j_3\,,\quad j_4' = \gamma(j_4 - \mathrm{i}\beta j_1)\,.$$

In order to obtain this result, the factor $[2(\gamma+1)]^{-\frac{1}{2}}$ has been introduced in eq. (10) *. Note also that

$$\rho = j_4/\mathrm{i}c = \Psi^\dagger\Psi = \overset{*}{\psi}_+\psi_+ + \overset{*}{\psi}_-\psi_-$$

is positive definite, and may be interpreted as a probability density.

* A similar situation presents itself in relativity, when the invariance of the surface of a spherical electromagnetic wave is imposed (i.e. that $\boldsymbol{x}^2 - c^2t^2 = 0$ be transformed into $\boldsymbol{x}'^2 - ct'^2 = 0$). This determines the Lorentz transformation up to a factor, which is fixed by the more restrictive condition $\boldsymbol{x}^2 - c^2t^2 = \boldsymbol{x}'^2 - c^2t'^2$.

Chapter III

THE DIRAC EQUATION

1. Wave equation and four-current

The characteristic property of the two-component wave functions satisfying the Weyl equation is the parallelism of $\boldsymbol{k}$ with $\boldsymbol{\sigma}$ for positive frequency solutions and the antiparallelism for negative frequency solutions.

It is known from experiment, and it is implied in the non-relativistic theory, that the spin of the electron * can be parallel or antiparallel to any arbitrary direction, and in particular to $\boldsymbol{k}$. Therefore, the electron has one degree of freedom more than a particle described by the Weyl equation. It is therefore natural to describe it by a four-component wave function instead of a two-component one.

The electron wave function may thus be represented by a one-column matrix using either the two-index (spinor) notation ψ_{ik}, $(i, k = 1, 2)$,

$$\psi = \begin{pmatrix} \psi_{11} \\ \psi_{21} \\ \psi_{12} \\ \psi_{22} \end{pmatrix}, \tag{1}$$

or the one-index notation ψ_i, $(i = 1, 2, 3, 4)$,

$$\psi = \begin{pmatrix} \psi_1 \\ \psi_2 \\ \psi_3 \\ \psi_4 \end{pmatrix}.$$

In Dirac's theory ** the four-component wave function obeys the equation ***

* The same is true for nucleons (protons and neutrons), μ-mesons etc.

** P. A. M. Dirac, Proc. Roy. Soc. (London) **117** (1928), 610; **118** (1928), 351.

*** The form of this equation is determined by linearity and invariance requirements, just as the Weyl equation (2), Chap. II, of which it is the direct extension in the case of a four-component wave function.

$$\left(\frac{1}{c}\frac{\partial}{\partial t} + \boldsymbol{\alpha}\cdot\nabla_x + \mathrm{i}\,\kappa\,\beta\right)\psi = 0\,, \tag{2}$$

very similar to eq. (2), Chap. II, § 1. Now, however, α_1, α_2, α_3 and β are 4×4 matrices

The requirement that each component ψ_i, $(i = 1, 2, 3, 4)$, must obey the Klein-Gordon equation, leads to certain conditions for $\boldsymbol{\alpha}$ and β. On multiplying eq. (2) by $(-\,\partial/c\,\partial t + \boldsymbol{\alpha}\cdot\nabla_x + \mathrm{i}\,\kappa\,\beta)$ from the left, we obtain

$$\left[\tfrac{1}{2}(\alpha_k\alpha_l + \alpha_l\alpha_k)\,\nabla_k\nabla_l + \mathrm{i}\,\kappa(\beta\alpha_k + \alpha_k\beta)\,\nabla_k - \kappa^2\beta^2 - c^{-2}\,\partial^2/\partial t^2\right]\psi = 0\,.$$

In order that this reduce to the Klein-Gordon equation, the relations

$$\begin{gathered}\alpha_k\,\alpha_l + \alpha_l\,\alpha_k = 2\delta_{kl}\,I\,,\quad \beta^2 = I\,,\\ \beta\,\alpha_k + \alpha_k\,\beta = 0\end{gathered} \tag{3}$$

must be fulfilled, I being the 4×4 unit matrix.

The conditions (3) and (3′) do not determine uniquely the matrices $\boldsymbol{\alpha}$ and β. In fact, if S is a 4×4 matrix having an inverse S^{-1}, the matrices

$$\alpha'_k = S\,\alpha_k\,S^{-1}\,,\quad \beta' = S\,\beta\,S^{-1}$$

also satisfy eqs. (3).

Thus an infinite number of representations of the anticommutation relations (3) exist. As we shall see below, the matrices α_k and β must be Hermitian, since only then a continuity equation for the four-current can be constructed. Now, in order that α'_k and β' be Hermitian, the matrix S above must be unitary, i.e.

$$S^{-1} = S^\dagger\,.$$

To deduce the continuity equation, we first write eq. (2) in a more compact form, which will be particularly suitable for the discussion of relativistic invariance. By multiplication with $-\mathrm{i}\,\beta$ from the left and introduction of the new matrices γ_μ, $(\mu = 1, 2, 3, 4)$,

$$\gamma_k = -\,\mathrm{i}\,\beta\alpha_k\,,\quad (k = 1, 2, 3)\,,\quad \gamma_4 = \beta\,, \tag{4}$$

eq. (2) takes the form

$$\left(\gamma_\mu\frac{\partial}{\partial x_\mu} + \kappa\right)\psi = 0\,. \tag{5}$$

Using eqs. (3) and (4), it is easy to verify that the γ's satisfy the anticommutation relations

$$\gamma_\mu\,\gamma_\nu + \gamma_\nu\,\gamma_\mu = 2\delta_{\mu\nu}\,I\,. \tag{6}$$

The matrix

$$\gamma_5 = \gamma_1\, \gamma_2\, \gamma_3\, \gamma_4 = \mathrm{i}\; \alpha_1\, \alpha_2\, \alpha_3$$

is very important in the theory of weak interactions and in that of the pion-nucleon interaction. As γ_5 anticommutes with any γ_μ, and $\gamma_5^2 = I$, we may extend (6) writing

$$\gamma_\mu\, \gamma_\nu + \gamma_\nu\, \gamma_\mu = 2\, \delta_{\mu\nu}\, I\,, \quad (\mu, \nu = 1, 2, 3, 4, 5)\,. \tag{6'}$$

Like $\boldsymbol{\alpha}$ and β, the γ's are determined but for a unitary transformation

$$\gamma'_\mu = S\, \gamma_\mu\, S^{-1}\,,$$

which leaves (6′) unchanged. Conversely, any set of four matrices γ_μ, satisfying eq. (6), may be obtained from another set with the same property by a unitary transformation.

Clearly, since the α_k's and β are Hermitian, the γ's are also Hermitian,

$$\gamma_k^\dagger = \mathrm{i}\; \alpha_k\, \beta = -\,\mathrm{i}\, \beta\, \alpha_k = \gamma_k\,, \quad \gamma_4^\dagger = \beta^\dagger = \gamma_4\,.$$

Now the Hermitian conjugate of eq. (5) is

$$\frac{\partial \psi^\dagger}{\partial x_k}\, \gamma_k - \frac{\partial \psi^\dagger}{\partial x_4}\, \gamma_4 + \kappa\, \psi^\dagger = 0\,.$$

Multiplying this equation by γ_4 on the right and using eq. (6), we have

$$\frac{\partial \bar{\psi}}{\partial x_\mu}\, \gamma_\mu - \kappa \bar{\psi} = 0\,, \tag{7}$$

where $\bar{\psi}$ is the *adjoint* wave function * defined by

$$\bar{\psi} = \psi^\dagger\, \gamma_4\,.$$

On multiplying eq. (5) by $\bar{\psi}$ on the left and eq. (7) by ψ on the right, and summing, we have

$$\frac{\partial \bar{\psi}}{\partial x_\mu}\, \gamma_\mu\, \psi + \bar{\psi}\, \gamma_\mu\, \frac{\partial \psi}{\partial x_\mu} = 0\,, \tag{8}$$

* The components of $\bar{\psi}$ are clearly

$$\bar{\psi}_\beta = \psi_\alpha{}^* \gamma_{4,\, \alpha\beta} = (\gamma_4{}^T)_{\beta\alpha} \psi_\alpha{}^*\,.$$

By A^T we denote the transpose of A. It is obtained by exchanging rows and columns. Thus $A^\dagger = (A^*)^T$.

or *

$$\frac{\partial}{\partial x_\mu} (\bar{\psi}\, \gamma_\mu\, \psi) = 0\,. \tag{9}$$

This is the continuity equation for the four-current

$$j_\mu = \mathrm{i}\, c\, \bar{\psi}\, \gamma_\mu\, \psi\,.$$

From the fourth component,

$$j_4 = \mathrm{i}\, c\, \bar{\psi}\, \gamma_4\, \psi = \mathrm{i}\, c\, \psi^\dagger \psi\,,$$

one has the charge density

$$\rho = j_4/\mathrm{i}c = \psi_1^* \psi_1 + \psi_2^* \psi_2 + \psi_3^* \psi_3 + \psi_4^* \psi_4\,,$$

which is positive definite and can be interpreted as a probability density for a single particle. As we shall see, this interpretation is only approximately correct, as was the case for particles of spin zero.

From the Hermiticity of the α's it follows that the spatial components of j_μ are real,

$$\boldsymbol{j}^* = (c\, \psi^\dagger\, \boldsymbol{\alpha}\, \psi)^* = c\, \psi^T\, \boldsymbol{\alpha}^*\, \psi^* = (c\, \psi^\dagger\, \boldsymbol{\alpha}\, \psi)^T = \boldsymbol{j}\,,$$

as it must be.

2. Pauli's, Kramers' and Majorana's representations

Here three different representations of the γ's are given, each of which will prove suitable for treating a particular aspect of the theory.

In the *Pauli representation* ** the α's and β are

$$\alpha_1 = \begin{pmatrix} \cdot & \cdot & \cdot & 1 \\ \cdot & \cdot & 1 & \cdot \\ \cdot & 1 & \cdot & \cdot \\ 1 & \cdot & \cdot & \cdot \end{pmatrix}, \quad \alpha_2 = \begin{pmatrix} \cdot & \cdot & \cdot & -\mathrm{i} \\ \cdot & \cdot & \mathrm{i} & \cdot \\ \cdot & -\mathrm{i} & \cdot & \cdot \\ \mathrm{i} & \cdot & \cdot & \cdot \end{pmatrix},$$

$$\alpha_3 = \begin{pmatrix} \cdot & \cdot & 1 & \cdot \\ \cdot & \cdot & \cdot & -1 \\ 1 & \cdot & \cdot & \cdot \\ \cdot & -1 & \cdot & \cdot \end{pmatrix}, \quad \beta = \begin{pmatrix} 1 & \cdot & \cdot & \cdot \\ \cdot & 1 & \cdot & \cdot \\ \cdot & \cdot & -1 & \cdot \\ \cdot & \cdot & \cdot & -1 \end{pmatrix}.$$

(The dots in the above denote zeros.)

---

* Obviously, if the γ's were not Hermitian, the first term on the left hand side of eq. (8) would have $\gamma_\mu{}^\dagger$ instead of γ_μ, and eq. (9) could not be written.

** P. A. M. Dirac, Proc. Roy. Soc. **117** (1928), 610, and W. Pauli, *loc. cit.*, p. 142.

These may be written in a more compact form by introducing the matrices

$$\rho_1 = \begin{pmatrix} 0 & 1 \\ 1 & 0 \end{pmatrix}, \quad \rho_2 = \begin{pmatrix} 0 & -\mathrm{i} \\ \mathrm{i} & 0 \end{pmatrix}, \quad \rho_3 = \begin{pmatrix} 1 & 0 \\ 0 & -1 \end{pmatrix}, \quad I_\varrho = \begin{pmatrix} 1 & 0 \\ 0 & 1 \end{pmatrix},$$

identical with the σ's, and defining the product between ρ's and σ's as, e.g.,

$$\rho_1 \times \sigma_1 = \begin{pmatrix} 0 & \sigma_1 \\ \sigma_1 & 0 \end{pmatrix} = \begin{pmatrix} . & . & . & 1 \\ . & . & 1 & . \\ . & 1 & . & . \\ 1 & . & . & . \end{pmatrix}.$$

Then

$$\boldsymbol{\alpha} = \rho_1 \times \boldsymbol{\sigma}, \quad \text{i.e.} \quad \alpha_i = \rho_1 \times \sigma_i = \begin{pmatrix} 0 & \sigma_i \\ \sigma_i & 0 \end{pmatrix}, \quad \beta = \rho_3 \times I_\sigma = \begin{pmatrix} I_\sigma & 0 \\ 0 & -I_\sigma \end{pmatrix}.$$

The Dirac equation takes the form

$$\left(\frac{1}{c} \frac{\partial}{\partial t} + \rho_1 \times \boldsymbol{\sigma} \cdot \boldsymbol{\nabla}_x + \mathrm{i}\,\kappa\, \rho_3 \times I_\sigma \right) \psi = 0 .$$

If the components of the wave function are labelled by two indices as in eq. (1), the σ's act on the first index. E.g. σ_1 exchanges the first two components of the wave function as if they were those of the Pauli non-relativistic theory, and acts in a similar way on the third and fourth component,

$$I_\varrho \times \sigma_1 \begin{pmatrix} \psi_{11} \\ \psi_{21} \\ \psi_{12} \\ \psi_{22} \end{pmatrix} = \begin{pmatrix} \begin{pmatrix} 0 & 1 \\ 1 & 0 \end{pmatrix} & 0 \\ 0 & \begin{pmatrix} 0 & 1 \\ 1 & 0 \end{pmatrix} \end{pmatrix} \begin{pmatrix} \psi_{11} \\ \psi_{21} \\ \psi_{12} \\ \psi_{22} \end{pmatrix} = \begin{pmatrix} \psi_{21} \\ \psi_{11} \\ \psi_{22} \\ \psi_{12} \end{pmatrix}.$$

The ρ's, on the other hand, act on the second index. E.g. ρ_1 exchanges the first with the third component and the second with the fourth, i.e.

$$\rho_1 \times I_\sigma \begin{pmatrix} \psi_{11} \\ \psi_{21} \\ \psi_{12} \\ \psi_{22} \end{pmatrix} = \begin{pmatrix} 0 & 1 \\ 1 & 0 \end{pmatrix} \begin{pmatrix} \begin{pmatrix} \psi_{11} \\ \psi_{21} \end{pmatrix} \\ \begin{pmatrix} \psi_{12} \\ \psi_{22} \end{pmatrix} \end{pmatrix} = \begin{pmatrix} \psi_{12} \\ \psi_{22} \\ \psi_{11} \\ \psi_{21} \end{pmatrix}.$$

Therefore, the ρ's commute with the σ's since they act on different indices.

The γ-matrices are

$$\boldsymbol{\gamma} = \rho_2 \times \boldsymbol{\sigma}, \quad \gamma_4 = \rho_3 \times I_\sigma, \quad \gamma_5 = -\rho_1 \times I_\sigma,$$

i.e.

$$\gamma_1 = \begin{pmatrix} \cdot & \cdot & \cdot & -\mathrm{i} \\ \cdot & \cdot & -\mathrm{i} & \cdot \\ \cdot & \mathrm{i} & \cdot & \cdot \\ \mathrm{i} & \cdot & \cdot & \cdot \end{pmatrix}, \quad \gamma_2 = \begin{pmatrix} \cdot & \cdot & \cdot & -1 \\ \cdot & \cdot & 1 & \cdot \\ \cdot & 1 & \cdot & \cdot \\ -1 & \cdot & \cdot & \cdot \end{pmatrix},$$

$$\gamma_3 = \begin{pmatrix} \cdot & \cdot & -\mathrm{i} & \cdot \\ \cdot & \cdot & \cdot & \mathrm{i} \\ \mathrm{i} & \cdot & \cdot & \cdot \\ \cdot & -\mathrm{i} & \cdot & \cdot \end{pmatrix}, \quad \gamma_4 = \begin{pmatrix} 1 & \cdot & \cdot & \cdot \\ \cdot & 1 & \cdot & \cdot \\ \cdot & \cdot & -1 & \cdot \\ \cdot & \cdot & \cdot & -1 \end{pmatrix},$$

$$\gamma_5 = \begin{pmatrix} \cdot & \cdot & -1 & \cdot \\ \cdot & \cdot & \cdot & -1 \\ -1 & \cdot & \cdot & \cdot \\ \cdot & -1 & \cdot & \cdot \end{pmatrix}.$$

Moreover

$$\bar{\psi} = \psi^\dagger \gamma_4 = (\psi_1^* \; \psi_2^* \; \psi_3^* \; \psi_4^*) \begin{pmatrix} 1 & \cdot & \cdot & \cdot \\ \cdot & 1 & \cdot & \cdot \\ \cdot & \cdot & -1 & \cdot \\ \cdot & \cdot & \cdot & -1 \end{pmatrix} = (\psi_1^* \; \psi_2^* \; -\psi_3^* \; -\psi_4^*) \,.$$

The *Kramers representation* * is

$$\alpha_i' = \rho_3 \times \sigma_i = \begin{pmatrix} \sigma_i & 0 \\ 0 & -\sigma_i \end{pmatrix}, \quad \beta' = \rho_1 \times I_\sigma \,.$$

Therefore for the γ's one has

$$\boldsymbol{\gamma}' = -\rho_2 \times \boldsymbol{\sigma} \,, \quad \gamma_4' = \rho_1 \times I_\sigma \,, \quad \gamma_5' = -\rho_3 \times I_\sigma \,, \tag{10}$$

and the Dirac equation becomes

$$\nabla_x \cdot \begin{pmatrix} 0 & \mathrm{i}\boldsymbol{\sigma} \\ -\mathrm{i}\boldsymbol{\sigma} & 0 \end{pmatrix} \begin{pmatrix} \Psi_1 \\ \Psi_2 \end{pmatrix} + \frac{\partial}{\partial x_4} \begin{pmatrix} 0 & I_\sigma \\ I_\sigma & 0 \end{pmatrix} \begin{pmatrix} \Psi_1 \\ \Psi_2 \end{pmatrix} + \frac{mc}{\hbar} \begin{pmatrix} \Psi_1 \\ \Psi_2 \end{pmatrix} = 0 \,, \tag{11}$$

where

$$\Psi_1 = \begin{pmatrix} \psi_{11} \\ \psi_{21} \end{pmatrix}, \quad \Psi_2 = \begin{pmatrix} \psi_{12} \\ \psi_{22} \end{pmatrix}.$$

Finally the *Majorana representation* **, which is particularly suitable for the study of charge conjugation, uses the matrices

---

* H. A. Kramers, Proc. of Amsterdam Academy **40** (1937), 814 and Hand. u. Jahrb. d. Chem. Physik I (1934), 63, 64.

** E. Majorana, Nuovo Cimento **14** (1937), 171.

$$\alpha_1'' = \rho_3 \times \sigma_1\,, \quad \alpha_2'' = -\rho_1 \times I_\sigma\,, \quad \alpha_3'' = \rho_3 \times \sigma_3\,, \quad \beta'' = \rho_3 \times \sigma_2$$

and

$$\gamma_1'' = -I_\varrho \times \sigma_3 = \begin{pmatrix} -1 & . & . & . \\ . & 1 & . & . \\ . & . & -1 & . \\ . & . & . & 1 \end{pmatrix},$$

$$\gamma_2'' = -\rho_2 \times \sigma_2 = \begin{pmatrix} . & . & . & 1 \\ . & . & -1 & . \\ . & -1 & . & . \\ 1 & . & . & . \end{pmatrix}, \quad \gamma_3'' = I_\varrho \times \sigma_1 = \begin{pmatrix} . & 1 & . & . \\ 1 & . & . & . \\ . & . & . & 1 \\ . & . & 1 & . \end{pmatrix},$$

$$\gamma_4'' = \beta'' = \rho_3 \times \sigma_2 = \begin{pmatrix} . & -\mathrm{i} & . & . \\ \mathrm{i} & . & . & . \\ . & . & . & \mathrm{i} \\ . & . & -\mathrm{i} & . \end{pmatrix}, \quad \gamma_5'' = \rho_1 \times \sigma_2 = \begin{pmatrix} . & . & . & -\mathrm{i} \\ . & . & \mathrm{i} & . \\ . & -\mathrm{i} & . & . \\ \mathrm{i} & . & . & . \end{pmatrix}.$$

Notice that all the elements of γ_1'', γ_2'' and γ_3'' are real, while those of γ_4'' are all purely imaginary, i.e. *

$$(\gamma_k'')^* = \gamma_k''\,, \quad (\gamma_4'')^* = -\gamma_4''\,,$$

just like $x_k^* = x_k\,, \quad x_4^* = -x_4$.

This has the consequence that in the Majorana representation ψ and ψ^* obey the same equation. In fact

$$0 = \left[\left(\gamma_\mu \frac{\partial}{\partial x_\mu} + \kappa\right)\psi\right]^* = \left(\gamma_\mu \frac{\partial}{\partial x_\mu} + \kappa\right)\psi^*\,.$$

One can verify that

$$S_{\mathrm{K,P}} = (1/\sqrt{2})(I_\varrho + \mathrm{i}\,\rho_2)\,\rho_1 \tag{12}$$

is the unitary matrix transforming Pauli's γ's into Kramers', while

$$S_{\mathrm{M,K}} = (1/\sqrt{2})\,\mathrm{e}^{\mathrm{i}\pi/4}(I + \mathrm{i}\,\rho_2 \times \sigma_2) \tag{13}$$

transforms Kramers' γ's into Majorana's.

---

* A^* denotes a matrix whose elements are the complex conjugates of the elements of A. Do not confuse A^* with the Hermitian conjugate $A^\dagger = (A^*)^T$.

Therefore

$$S_{M,P} = \tfrac{1}{2}\, e^{i\pi/4}(I + i\,\rho_2 \times \sigma_2)(I_\varrho + i\,\rho_2)\,\rho_1 \tag{14}$$

is the transformation from Pauli's to Majorana's representations.

These same matrices also transform the wave functions. E.g., if ψ_P is a solution of the Dirac equation in Pauli's representation, the corresponding solution ψ_K in Kramers' representation is $\psi_K = S_{K,P}\,\psi_P$.

CHAPTER IV

THE CONSTANTS OF MOTION

1. The energy-momentum tensor

The Dirac equations for ψ and $\bar{\psi}$ may be deduced from a variational principle. We adopt as a Lagrangian density the function

$$\mathfrak{L} = -\frac{\hbar c}{2}\left[\bar{\psi}\left(\gamma_\mu \frac{\partial}{\partial x_\mu} + \kappa\right)\psi - \left(\frac{\partial \bar{\psi}}{\partial x_\mu}\gamma_\mu - \kappa\,\bar{\psi}\right)\psi\right], \tag{1}$$

where ψ and $\bar{\psi}$ are regarded as independent functions.

The stationary principle

$$\delta A = \int_V \delta\mathfrak{L}\, \mathrm{d}^4 x = 0,$$

for variations *

$$\bar{\psi} \to \bar{\psi} + \delta\bar{\psi}$$

such that $\delta\bar{\psi} = 0$ on the boundary of V, leads to the Dirac equation for ψ. Similarly the adjoint equation follows from the assumption that the action integral is stationary for the variation $\psi \to \psi + \delta\psi$ with $\delta\psi = 0$ on the boundary.

It is worth noting that, if ψ is a solution of the Dirac equation, $\mathfrak{L} = 0$. (The Klein-Gordon Lagrangian does *not* have this property.)

Moreover, using the continuity equation (9), Chap. III, the action integral is invariant under gauge transformations of the first kind

$$\psi \to \mathrm{e}^{\mathrm{i}\alpha}\,\psi\,, \quad \bar{\psi} \to \mathrm{e}^{-\mathrm{i}\alpha}\,\bar{\psi}\,,$$

where $\alpha = \alpha(\boldsymbol{x}, t)$ is an arbitrary function which vanishes on the boundary of V. Conversely, if this invariance property of the action is postulated, it is

* While in the case of spin zero particles there were two independent functions ψ and ψ^*, here, in fact, we have eight functions ψ_α and $\bar{\psi}_\alpha$, ($\alpha = 1, 2, 3, 4$). Less concisely, we might write $\bar{\psi}_\alpha \to \bar{\psi}_\alpha + \delta\bar{\psi}_\alpha$ with $\delta\bar{\psi}_\alpha = 0$ on the boundary of V. The variations of the components of the wave function are independent of one another.

readily shown * that the four-current $j_\mu = \mathrm{i}\, c\, \bar{\psi}\, \gamma_\mu\, \psi$ obeys the continuity equation

$$\frac{\partial j_\mu}{\partial x_\mu} = 0\,.$$

In analogy with what has been done for particles of spin zero, the energy-momentum tensor may be derived from the invariance of the Lagrangian density under space-time translations. In fact, by Noether's theorem, one has

$$\frac{\partial T_{\mu\nu}}{\partial x_\mu} = 0 \tag{2}$$

with

$$T_{\mu\nu} = -\sum_{\alpha=1}^{4} \left[\left(\partial \mathfrak{L} / \partial \frac{\partial \psi_\alpha}{\partial x_\mu} \right) \frac{\partial \psi_\alpha}{\partial x_\nu} + \frac{\partial \bar{\psi}_\alpha}{\partial x_\nu} \left(\partial \mathfrak{L} / \partial \frac{\partial \bar{\psi}_\alpha}{\partial x_\mu} \right) \right] + \mathfrak{L}\, \delta_{\mu\nu}\,.$$

Since $\mathfrak{L} = 0$ for wave functions which are actually solutions of the Dirac equation, the energy-momentum tensor reduces to

$$T_{\mu\nu} = \frac{\hbar c}{2} \left(\bar{\psi}\, \gamma_\mu \frac{\partial \psi}{\partial x_\nu} - \frac{\partial \bar{\psi}}{\partial x_\nu}\, \gamma_\mu\, \psi \right). \tag{3}$$

Note that T_{ik}, $(i, k = 1, 2, 3)$, and T_{44} are real, while T_{4i} and $T_{i4}, (i = 1, 2, 3)$, are purely imaginary, as is appropriate for an energy-momentum tensor **. This would not have been the case if the Lagrangian density

* Again this property is *not* shared by the action integral for free spin zero particles. In fact

$$\int\limits_V \left(\frac{\partial \psi'^*}{\partial x_\mu} \frac{\partial \psi'}{\partial x_\mu} + \kappa^2 \psi'^* \psi' \right) \mathrm{d}^4x - \int\limits_V \left(\frac{\partial \psi^*}{\partial x_\mu} \frac{\partial \psi}{\partial x_\mu} + \kappa^2 \psi^* \psi \right) \mathrm{d}^4x = 0$$

$$(\psi' = \psi \mathrm{e}^{\mathrm{i}\alpha(x)}\,, \quad \alpha(x) = 0 \quad \text{on the boundary of } V)$$

leads to.

$$\int \left[\mathrm{i}\alpha \frac{\partial}{\partial x_\mu} \left(\frac{\partial \psi^*}{\partial x_\mu} \psi - \psi^* \frac{\partial \psi}{\partial x_\mu} \right) - \frac{\partial \alpha}{\partial x^\mu} \frac{\partial \alpha}{\partial x^\mu} \psi^* \psi \right] \mathrm{d}^4x = 0\,,$$

from which, for arbitrary α, it does *not* follow that

$$\frac{\partial}{\partial x_\mu} \left(\frac{\partial \psi^*}{\partial x_\mu} \psi - \psi^* \frac{\partial \psi}{\partial x_\mu} \right) = 0\,.$$

** To show that $T_{ik} = T_{ik}{}^*$, $(i, k = 1, 2, 3)$, one has at once

$$T_{ik}{}^* = \frac{\hbar c}{2} \left(\psi^\dagger \gamma_4 \gamma_i \frac{\partial \psi}{\partial x_k} - \frac{\partial \psi^\dagger}{\partial x_k} \gamma_4 \gamma_i \psi \right)^{*T} = \frac{\hbar c}{2} \left(\frac{\partial \psi^\dagger}{\partial x_k} \gamma_i \gamma_4 \psi - \psi^\dagger \gamma_i \gamma_4 \frac{\partial \psi}{\partial x_k} \right) = T_{ik}\,.$$

The proof of $T_{44} = T_{44}{}^*$ and $T_{4i} = -\,T_{4i}{}^*$ is analogous.

$$\mathfrak{L} = -\hbar c\,\bar{\psi}\left(\gamma_\mu \frac{\partial}{\partial x_\mu} + \kappa\right)\psi$$

had been adopted. Although this yields the Dirac equation, the corresponding energy-momentum tensor is unacceptable.

From eq. (2) it follows that the energy

$$E = -\int T_{44}\,\mathrm{d}^3\mathrm{x} = \tfrac{1}{2}\hbar\mathrm{i}\int\left(\psi^\dagger \frac{\partial\psi}{\partial t} - \frac{\partial\psi^\dagger}{\partial t}\psi\right)\mathrm{d}^3\mathrm{x} \tag{4}$$

and the momentum

$$P_k = \frac{1}{\mathrm{i}c}\int T_{4k}\,\mathrm{d}^3\mathrm{x} = -\frac{1}{2}\hbar\mathrm{i}\int\left(\psi^\dagger \frac{\partial\psi}{\partial x_k} - \frac{\partial\psi^\dagger}{\partial x_k}\psi\right)\mathrm{d}^3\mathrm{x} \tag{5}$$

are constants of motion.

It is convenient to condense (4) and (5) in the four-momentum

$$P_\mu = \frac{\hbar}{2\mathrm{i}}\sum_\alpha\int\left(\psi_\alpha^* \frac{\partial\psi_\alpha}{\partial x_\mu} - \frac{\partial\psi_\alpha^*}{\partial x_\mu}\psi_\alpha\right)\mathrm{d}^3\mathrm{x}\,. \tag{6}$$

It may be interesting to notice that, but for the different number of components of the wave functions, there is a formal analogy between the expression (6) for the four-momentum and that for the four-current of free particles of spin zero,

$$j_\mu = \frac{\hbar}{2m\mathrm{i}}\left(\varphi^* \frac{\partial\varphi}{\partial x_\mu} - \frac{\partial\varphi^*}{\partial x_\mu}\varphi\right).$$

From eq. (4) it is apparent that the energy is positive-definite only if the wave function involves solely positive frequencies. On the other hand, the charge density $\rho = \psi^\dagger\psi$ is always positive-definite. The case is just the opposite for spin zero particles, for which the energy is positive-definite while the charge density is not.

Wave functions involving only positive frequencies, for which $\rho > 0$ and $E > 0$, may be interpreted as probability amplitudes for a single particle. In general, however, only field theory, where the number of particles is not a constant of the motion, may give a correct description of particles of spin $\frac{1}{2}$ subject to forces.

2. Angular momentum

The energy-momentum tensor (3), which is not symmetric, must be replaced by the symmetrized tensor

$$T'_{\mu\nu} = \tfrac{1}{2}(T_{\mu\nu} + T_{\nu\mu})$$

before the angular momentum may be defined. This replacement does not change the expression (6) for the four-momentum *, and the new tensor, like $T_{\mu\nu}$, obeys the equation **

$$\frac{\partial T'_{\mu\nu}}{\partial x_\mu} = 0\,.$$

General methods for the construction of symmetric energy-momentum tensors were developed by Belinfante *** and Rosenfeld †.

Now let us consider the tensor

$$\mathfrak{H}_{\mu\nu\lambda} = x_\nu T'_{\mu\lambda} - x_\lambda T'_{\mu\nu}$$

of the third rank, skew-symmetric with respect to the last two indices (generalized angular momentum density).

One has

$$\frac{\partial}{\partial x_\mu} \mathfrak{H}_{\mu\nu\lambda} = T'_{\nu\lambda} - T'_{\lambda\nu}$$

---

* In fact, the expression for the energy is clearly the same if $T_{\mu\nu}$ is replaced by $T_{\mu\nu}'$. For the momentum one has

$$P_k' = \tfrac{1}{2}\left[P_k + \frac{\hbar}{2\mathrm{i}}\int\left(\bar{\psi}\gamma_k \frac{\partial\psi}{\partial x_4} - \frac{\partial\bar{\psi}}{\partial x_4}\gamma_k\psi\right)\mathrm{d}^3\mathrm{x}\right] =$$

$$= \tfrac{1}{2}P_k - \frac{\hbar}{4\mathrm{i}}\int\left(\psi^\dagger\gamma_k\gamma_4\frac{\partial\psi}{\partial x_4} + \frac{\partial\psi^\dagger}{\partial x_4}\gamma_4\gamma_k\psi\right)\mathrm{d}^3\mathrm{x}$$

and, using the Dirac equation,

$$P_k' = \tfrac{1}{2}P_k + \frac{\hbar}{4\mathrm{i}}\int\left(\psi^\dagger\gamma_k\gamma_i\frac{\partial\psi}{\partial x_i} - \frac{\partial\psi^\dagger}{\partial x_i}\gamma_i\gamma_k\psi\right)\mathrm{d}^3\mathrm{x}\,.$$

Integrating by parts, and taking into account the anticommutation relations for the γ's, one has

$$P_k' = \tfrac{1}{2}P_k - \frac{1}{2}\hbar\mathrm{i}\int\psi^\dagger\frac{\partial\psi}{\partial x_k}\mathrm{d}^3\mathrm{x} = P_k\,.$$

** Note that, in general, $\partial T_{\nu\mu}/\partial x_\mu = 0$ does not follow from $\partial T_{\mu\nu}/\partial x_\mu = 0$. In our case, however, both these equations hold.

In fact,

$$\frac{\partial T_{\nu\mu}}{\partial x_\mu} = \frac{\hbar c}{2}\left[\bar{\psi}\gamma_\nu\frac{\partial^2\psi}{\partial x_\mu{}^2} - \frac{\partial^2\bar{\psi}}{\partial x_\mu{}^2}\gamma_\nu\psi\right] = 0\,,$$

as

$$\Box\,\psi = \kappa^2\psi, \qquad \Box\,\bar{\psi} = \kappa^2\bar{\psi}\,.$$

*** F. J. Belinfante, Physica **6** (1939), 887.

† L. Rosenfeld, Mém. de l'Acad. Roy. de Belgique **6** (1940), 30.

and, as $T'_{\mu\nu}$ is symmetric,

$$\frac{\partial}{\partial x_\mu}\,\mathfrak{H}_{\mu\nu\lambda} = 0\,.$$

Thus

$$M_{\nu\lambda} = \frac{1}{\mathrm{i}c}\int \mathfrak{H}_{4\nu\lambda}\,\mathrm{d}^3\mathrm{x}$$

is a constant of motion and may be identified with the angular momentum tensor *. In particular, the angular momentum components

$$M_i = \frac{1}{2\mathrm{i}c}\,\varepsilon_{ikl}\int \mathfrak{H}_{4kl}\,\mathrm{d}^3\mathrm{x} = \frac{1}{\mathrm{i}c}\,\varepsilon_{ikl}\int x_k\,T'_{4l}\,\mathrm{d}^3\mathrm{x}\,, \quad (i = 1, 2, 3)\,,$$

are constants of motion. (Remember that $T'_{4l}/\mathrm{i}c$ is the density of the l-th component of the momentum, so that the physical meaning of M_i is evident.)

By a suitable transformation using the Dirac equation, M_i may be written as

$$\varepsilon_{ikl}\,\frac{\hbar}{2\mathrm{i}}\int x_k\left[\left(\psi^\dagger\frac{\partial\psi}{\partial x_l} - \frac{\partial\psi^\dagger}{\partial x_l}\,\psi\right) + \tfrac{1}{2}\sum_{m(\neq l)}\left(\psi^\dagger\gamma_l\gamma_m\frac{\partial\psi}{\partial x_m} - \frac{\partial\psi^\dagger}{\partial x_m}\,\gamma_m\gamma_l\,\psi\right)\right]\mathrm{d}^3\mathrm{x}\,.$$

Finally, by partial integration, and remembering that $\gamma_l\gamma_m + \gamma_m\gamma_l = 0$ for $m \neq l$, one finds

$$M_i = \varepsilon_{ikl}\int\left(\psi^\dagger\,x_k\,\frac{\hbar}{\mathrm{i}}\frac{\partial}{\partial x_l}\,\psi + \psi^\dagger\,\frac{\hbar}{4\mathrm{i}}\,\gamma_k\gamma_l\,\psi\right)\mathrm{d}^3\mathrm{x}\,.$$

Thus the angular momentum appears as the sum of the orbital angular momentum

$$\boldsymbol{L} = \int \psi^\dagger\left(\boldsymbol{x}\times\frac{\hbar}{\mathrm{i}}\nabla_x\right)\psi\,\mathrm{d}^3\mathrm{x}$$

and of the spin

$$\boldsymbol{S} = \int \psi^\dagger\left(\frac{\hbar}{4\mathrm{i}}\,\boldsymbol{\gamma}\times\boldsymbol{\gamma}\right)\psi\,\mathrm{d}^3\mathrm{x} = \int \psi^\dagger\left(\frac{\hbar}{4\mathrm{i}}\,\boldsymbol{\alpha}\times\boldsymbol{\alpha}\right)\psi\,\mathrm{d}^3\mathrm{x}\,.$$

The components of the spin have the form of mean values of the matrices

$$\boldsymbol{s} \equiv \begin{cases} s_1 = \hbar\,\gamma_2\gamma_3/2\mathrm{i} = \hbar\,\alpha_2\alpha_3/2\mathrm{i}\,, \\ s_2 = \hbar\,\gamma_3\gamma_1/2\mathrm{i} = \hbar\,\alpha_3\alpha_1/2\mathrm{i}\,, \\ s_3 = \hbar\,\gamma_1\gamma_2/2\mathrm{i} = \hbar\,\alpha_1\alpha_2/2\mathrm{i}\,. \end{cases}$$

---

* On the other hand, the generalized angular momentum density $\mathfrak{H}_{\mu\nu\lambda}$ can also be derived from eq. (22), Part I, Chap. II, § 4 ($\mathfrak{L}$ being our original Lagrangian density (1) leading to a non-symmetric energy-momentum tensor), the last term of which gives the spin.

These matrices are Hermitian, for example

$$s_1^\dagger = \tfrac{1}{2}\,\mathrm{i}\,\hbar\,\gamma_3^\dagger\gamma_2^\dagger = -\tfrac{1}{2}\,\mathrm{i}\,\hbar\,\gamma_2\gamma_3 = s_1\,.$$

It is easy to verify that the components of the spin matrices obey the commutation relations

$$[s_i, s_j] = \mathrm{i}\,\hbar\,\varepsilon_{ijl}\,s_l$$

of an angular momentum, and the anticommutation relations

$$\{s_i, s_j\} = \tfrac{1}{2}\,\hbar^2\,\delta_{ij}$$

(spin $\frac{1}{2}$!).

From these it follows that

$$s_i s_j = \tfrac{1}{4}\,\hbar^2\,\delta_{ij} + \tfrac{1}{2}\,\mathrm{i}\,\hbar\,\varepsilon_{ijl}\,s_l\,.$$

In analogy with Chap. I, § 1, the 4 × 4 dimensionless matrices

$$\sigma_i = 2\,s_i/\hbar = \varepsilon_{ikl}\,\gamma_k\gamma_l/2\mathrm{i}$$

may be introduced. The form of these matrices (not to be confused with Pauli's matrices) depends, of course, on the representation used for the γ's. Hereafter σ_i denotes a Pauli matrix or $\varepsilon_{ikl}\gamma_k\gamma_l/2\mathrm{i}$ according as it acts on a two-component or a four-component function.

Note the commutation relations * of the spin components with the γ's,

$$\gamma_4\,s_i - s_i\,\gamma_4 = 0\,,$$
$$\gamma_i\,s_k - s_k\,\gamma_i = \mathrm{i}\,\hbar\,\varepsilon_{ikm}\,\gamma_m\,.$$

In the different representations of Chap. III, § 2, the components of $\boldsymbol{s}$ are:

(Pauli and Kramers) $\boldsymbol{s} = \boldsymbol{s}' \equiv \dfrac{\hbar}{2}(I_\varrho \times \sigma_1\,,\quad I_\varrho \times \sigma_2\,,\quad I_\varrho \times \sigma_3)\,,$

or, more explicitly,

$$s_i = \frac{\hbar}{2}\begin{pmatrix}\sigma_i & 0\\ 0 & \sigma_i\end{pmatrix};$$

(Majorana) $\boldsymbol{s}'' \equiv \dfrac{\hbar}{2}(\rho_2 \times \sigma_3\,,\quad I_\varrho \times \sigma_2\,,\quad -\rho_2 \times \sigma_1)\,.$

---

* The first of these relations is proved at once. For the second, notice that

$$[\gamma_i, \gamma_l\gamma_m] = \gamma_l[\gamma_i, \gamma_m] + [\gamma_i, \gamma_l]\,\gamma_m =$$
$$(\gamma_l\gamma_i + \gamma_i\gamma_l)\,\gamma_m - \gamma_l(\gamma_m\gamma_i + \gamma_i\gamma_m) = 2(\delta_{il}\gamma_m - \delta_{im}\gamma_l)\,.$$

Then

$$\gamma_i s_k - s_k\gamma_i = -\tfrac{1}{4}\,\mathrm{i}\,\hbar\,\varepsilon_{klm}\,[\gamma_i, \gamma_l\gamma_m] = \mathrm{i}\,\hbar\,\varepsilon_{ikm}\gamma_m\,.$$

3. Hamiltonian form of the Dirac equation

The Dirac equation (Chap. III, § 1, eq. (2)) may be written in a form analogous to that of the Schroedinger equation,

$$\hbar i \frac{\partial}{\partial t} \psi = H \psi ,$$

with the Hamiltonian operator

$$H = c\, \boldsymbol{\alpha} \cdot \frac{\hbar}{i} \nabla_x + \beta\, mc^2 .$$

In the spirit of this formulation, we may introduce a space of the four-component functions, in which the product is defined as

$$(\psi_1, \psi_2) = \int \psi_1^\dagger \psi_2 \, d^3x .$$

This is a constant of motion, as

$$\frac{d}{dt} (\psi_1, \psi_2) = \int \left(\frac{\partial \psi_1^\dagger}{\partial t} \psi_2 + \psi_1^\dagger \frac{\partial \psi_2}{\partial t} \right) d^3x =$$

$$= (i/\hbar) \int (\psi_1^\dagger H \psi_2 - \psi_1^\dagger H \psi_2) \, d^3x = 0 .$$

The norm of a wave function,

$$(\psi, \psi) = \int \psi^\dagger \psi \, d^3x ,$$

is positive-definite and coincides with the charge.

Finally, the mean value of an operator A is

$$\langle A \rangle = \int \psi^\dagger A \psi \, d^3x / \int \psi^\dagger \psi \, d^3x .$$

For the sake of brevity we will usually omit the denominator (ψ, ψ), which is equivalent to assuming the normalization $(\psi, \psi) = 1$ for the wave functions.

It is easy to see that the physical quantities found in the previous section can all be expressed as the mean values of Hermitian operators:

$$\boldsymbol{E} = \langle c\, \boldsymbol{\alpha} \cdot \boldsymbol{p} + \beta\, mc^2 \rangle ,$$

$$\boldsymbol{P} = \langle - i\hbar \nabla_x \rangle = \langle \boldsymbol{p} \rangle ,$$

$$\boldsymbol{L} = \langle \boldsymbol{x} \times \boldsymbol{p} \rangle ,$$

$$\boldsymbol{S} = \langle - \tfrac{1}{4} i \hbar \gamma \times \gamma \rangle = \langle - \tfrac{1}{4} i \hbar \boldsymbol{\alpha} \times \boldsymbol{\alpha} \rangle .$$

For the current one has

$$J = \int \boldsymbol{j}\, d^3x = \langle c\, \boldsymbol{\alpha} \rangle .$$

The time derivative of an operator A, denoted by $\dot{A}$, is defined by the equation

$$\langle \dot{A} \rangle = d \langle A \rangle / dt ,$$

Clearly

$$\dot{A} = (i/\hbar)\, [H, A] .$$

Thus, if an operator commutes with H, its mean value is a constant of motion.

Now we have

$$\begin{aligned}
\dot{x}_k &= c\alpha_k , \\
\dot{p}_k &= 0 , \\
\dot{\alpha}_k &= (2i/\hbar)(H\, \alpha_k - cp_k) = (2i/\hbar)(cp_k - \alpha_k\, H) , \\
\dot{\beta} &= (2i/\hbar)(H\, \beta - mc^2) = (2i/\hbar)(mc^2 - \beta\, H) = (2ic/\hbar)\, \boldsymbol{\alpha} \cdot \boldsymbol{p}\, \beta .
\end{aligned} \tag{7}$$

Since, as will be seen later, a matrix element of α_k between a positive and a negative frequency solution of the Dirac equation does not generally vanish, the first of eqs. (7) is closely related to the phenomenon of the Zitterbewegung, which will be discussed in the following chapter.

Although the orbital angular momentum operator $\boldsymbol{x} \times \boldsymbol{p}$ and the spin operator $\boldsymbol{s}$ do not commute with H, their sum does so, in accordance with $\boldsymbol{L} + \boldsymbol{S}$ being a constant of motion. On the other hand, the spin component in the direction of the momentum $\boldsymbol{s} \cdot \boldsymbol{p}/|\boldsymbol{p}|$ commutes with H, and is therefore a constant of motion. In fact, since

$$\dot{\boldsymbol{s}} = (i/\hbar)\, [H, \boldsymbol{s}] = c\, \boldsymbol{p} \times \boldsymbol{\alpha} ,$$

one has

$$\frac{d(\boldsymbol{s} \cdot \boldsymbol{p})}{dt} = c\, \boldsymbol{p} \cdot (\boldsymbol{p} \times \boldsymbol{\alpha}) = 0 .$$

CHAPTER V

PLANE WAVE SOLUTIONS OF THE DIRAC EQUATION

1. Eigenfunctions of H, p and s_p

Since

$$[p, H] = 0\,,$$

the momentum operator and the Hamiltonian have common eigenfunctions, which are of the type

$$u(\boldsymbol{k})\,\mathrm{e}^{\mathrm{i}\,(\boldsymbol{k}\cdot\boldsymbol{x}-\omega t)} \tag{1}$$

and

$$v(\boldsymbol{k})\,\mathrm{e}^{-\mathrm{i}\,(\boldsymbol{k}\cdot\boldsymbol{x}-\omega t)}\,. \tag{1'}$$

Inserting these in the Dirac equation, one finds the algebraic equations

$$\omega\,u(\boldsymbol{k}) = c(\boldsymbol{\alpha}\cdot\boldsymbol{k} + \beta\kappa)\,u(\boldsymbol{k})\,,\quad \omega\,v(\boldsymbol{k}) = c(\boldsymbol{\alpha}\cdot\boldsymbol{k} - \beta\kappa)\,v(\boldsymbol{k})\,,$$

where $\omega = c\,|(\boldsymbol{k}^2 + \kappa^2)^{\frac{1}{2}}|$, or, in a different form,

$$\begin{aligned} \left(\mathrm{i}\,\boldsymbol{\gamma}\cdot\boldsymbol{k} - \frac{\omega}{c}\,\gamma_4 + \kappa\right)u(\boldsymbol{k}) &= 0\,,\\ \left(-\,\mathrm{i}\,\boldsymbol{\gamma}\cdot\boldsymbol{k} + \frac{\omega}{c}\,\gamma_4 + \kappa\right)v(\boldsymbol{k}) &= 0\,. \end{aligned} \tag{2}$$

Obviously, (1) is an eigenfunction of $\boldsymbol{p}$ and H belonging to the eigenvalues $\hbar\boldsymbol{k}$ and $\hbar\omega$ respectively, while (1′) belongs to the eigenvalues $-\,\hbar\boldsymbol{k}$ and $-\,\hbar\omega$.

NOTE 1. It is interesting to consider the operators

$$\Lambda^{\pm} = \tfrac{1}{2}(I \pm \Lambda)\,,$$

with

$$\Lambda = (\boldsymbol{\alpha}\cdot\boldsymbol{p} + \beta\,mc)/(\boldsymbol{p}^2 + m^2c^2)^{\frac{1}{2}}\,.$$

These are projection operators for positive and negative frequencies respectively. In fact

$$\Lambda\,u(\boldsymbol{k})\,\mathrm{e}^{\mathrm{i}\,(\boldsymbol{k}\cdot\boldsymbol{x}-\omega t)} = \left[\omega/c(\boldsymbol{k}^2+\kappa^2)^{\frac{1}{2}}\right]u(\boldsymbol{k})\,\mathrm{e}^{\mathrm{i}\,(\boldsymbol{k}\cdot\boldsymbol{x}-\omega t)} = u(\boldsymbol{k})\,\mathrm{e}^{\mathrm{i}\,(\boldsymbol{k}\cdot\boldsymbol{x}-\omega t)}\,,$$

and similarly

$$\Lambda\, v(\boldsymbol{k})\, \mathrm{e}^{-\mathrm{i}(\boldsymbol{k}\cdot\boldsymbol{x}-\omega t)} = -\, v(\boldsymbol{k})\, \mathrm{e}^{-\mathrm{i}(\boldsymbol{k}\cdot\boldsymbol{x}-\omega t)}$$

Therefore

$$\Lambda^+ u(\boldsymbol{k})\, \mathrm{e}^{\mathrm{i}(\boldsymbol{k}\cdot\boldsymbol{x}-\omega t)} = u(\boldsymbol{k})\, \mathrm{e}^{\mathrm{i}(\boldsymbol{k}\cdot\boldsymbol{x}-\omega t)}\,,$$

$$\Lambda^+ v(\boldsymbol{k})\, \mathrm{e}^{-\mathrm{i}(\boldsymbol{k}\cdot\boldsymbol{x}-\omega t)} = 0\,,$$

$$\Lambda^- v(\boldsymbol{k})\, \mathrm{e}^{-\mathrm{i}(\boldsymbol{k}\cdot\boldsymbol{x}-\omega t)} = v(\boldsymbol{k})\, \mathrm{e}^{-\mathrm{i}(\boldsymbol{k}\cdot\boldsymbol{x}-\omega t)}\,,$$

$$\Lambda^- u(\boldsymbol{k})\, \mathrm{e}^{\mathrm{i}(\boldsymbol{k}\cdot\boldsymbol{x}-\omega t)} = 0\,.$$

Note the properties

$$\Lambda^2 = I\,, \quad (\Lambda^+)^2 = \Lambda^+\,, \quad (\Lambda^-)^2 = \Lambda^-\,,$$

$$\Lambda^+\Lambda^- = \Lambda^-\Lambda^+ = 0\,.$$

If a wave function is represented in the form

$$\psi(\boldsymbol{x}, t) = \psi^+(\boldsymbol{x}, t) + \psi^-(\boldsymbol{x}, t)$$

with

$$\psi^+(\boldsymbol{x}, t) = \int_{c\kappa}^{\infty} \phi^+(\omega, \boldsymbol{x})\, \mathrm{e}^{-\mathrm{i}\omega t}\, \mathrm{d}\omega\,,$$

$$\psi^-(\boldsymbol{x}, t) = \int_{c\kappa}^{\infty} \phi^-(\omega, \boldsymbol{x})\, \mathrm{e}^{\mathrm{i}\omega t}\, \mathrm{d}\omega\,,$$

one has

$$\Lambda^+ \psi = \psi^+\,, \quad \Lambda^- \psi = \psi^-\,.$$

The functions (1), (1′) may be chosen in such a way that they are eigenfunctions of the component of the spin along the direction of the momentum. (In fact, $\boldsymbol{s}\cdot\boldsymbol{p}$ commutes with $\boldsymbol{p}$ and with the Hamiltonian, see Chap. IV, § 3.)

Since, as follows from the anticommutation relations for the spin,

$$(\boldsymbol{s}\cdot\boldsymbol{p}/|\boldsymbol{p}|)^2 = \tfrac{1}{4}\hbar^2\,,$$

the operator $s_p \equiv \boldsymbol{s}\cdot\boldsymbol{p}/|\boldsymbol{p}|$ has the eigenvalues $\pm\frac{1}{2}\hbar$. Therefore we may define positive frequency solutions of the Dirac equation for momentum $\boldsymbol{p}$ and $s_p = \pm\frac{1}{2}\hbar$, which are denoted by

$$u(\boldsymbol{k}, \uparrow)\, \mathrm{e}^{\mathrm{i}(\boldsymbol{k}\cdot\boldsymbol{x}-\omega t)}\,, \quad u(\boldsymbol{k}, \downarrow)\, \mathrm{e}^{\mathrm{i}(\boldsymbol{k}\cdot\boldsymbol{x}-\omega t)}\,, \tag{3}$$

with $u(\boldsymbol{k}, \uparrow)$ and $u(\boldsymbol{k}, \downarrow)$ satisfying the equations

$$\begin{aligned} s_p\, u(\boldsymbol{k}, \uparrow) &= \tfrac{1}{2}\hbar\, u(\boldsymbol{k}, \uparrow)\,, \\ s_p\, u(\boldsymbol{k}, \downarrow) &= -\tfrac{1}{2}\hbar\, u(\boldsymbol{k}, \downarrow)\,. \end{aligned} \tag{4}$$

The solutions (3), being eigenfunctions of an Hermitian operator (s_p) belonging to different eigenvalues, are orthogonal *, namely

$$u^\dagger(\boldsymbol{k},\uparrow)\,u(\boldsymbol{k},\downarrow)=0\,,\qquad u^\dagger(\boldsymbol{k},\downarrow)\,u(\boldsymbol{k},\uparrow)=0\,. \tag{5}$$

Similarly the negative frequency solutions

$$v(\boldsymbol{k},\uparrow)\,\mathrm{e}^{-\mathrm{i}(\boldsymbol{k}\cdot\boldsymbol{x}-\omega t)}\,,\quad v(\boldsymbol{k},\downarrow)\,\mathrm{e}^{-\mathrm{i}(\boldsymbol{k}\cdot\boldsymbol{x}-\omega t)} \tag{3'}$$

are defined as eigenfunctions of s_p belonging to the eigenvalues $\pm\frac{1}{2}\hbar$

$$s_p\,v(\boldsymbol{k},\uparrow)=\tfrac{1}{2}\hbar\,v(\boldsymbol{k},\uparrow)\,,\qquad s_p\,v(\boldsymbol{k},\downarrow)=-\tfrac{1}{2}\hbar\,v(\boldsymbol{k},\downarrow)\,. \tag{4'}$$

Since for these solutions the momentum is $-\hbar\boldsymbol{k}$, they correspond to spin parallel and antiparallel to $-\boldsymbol{k}$, respectively, and are orthogonal

$$v^\dagger(\boldsymbol{k},\uparrow)\,v(\boldsymbol{k},\downarrow)=v^\dagger(\boldsymbol{k},\downarrow)\,v(\boldsymbol{k},\uparrow)=0\,, \tag{6}$$

A positive and a negative frequency solution of opposite momenta are also orthogonal, as they are eigenfunctions of the Hermitian operator $c(\boldsymbol{\alpha}\cdot\boldsymbol{k}+\beta\kappa)$ belonging to different eigenvalues. Thus we also have

$$u^\dagger(\boldsymbol{k},\uparrow)\,v(-\boldsymbol{k},\uparrow)=u^\dagger(\boldsymbol{k},\uparrow)\,v(-\boldsymbol{k},\downarrow)=$$
$$=u^\dagger(\boldsymbol{k},\downarrow)\,v(-\boldsymbol{k},\uparrow)=u^\dagger(\boldsymbol{k},\downarrow)\,v(-\boldsymbol{k},\downarrow)=\ldots=0\,. \tag{7}$$

For normalization we adopt a non-Lorentz-invariant one by requiring that $\rho=1$ (unit charge per unit volume). This gives

$$u^\dagger(\boldsymbol{k},\uparrow)\,u(\boldsymbol{k},\uparrow)=u^\dagger(\boldsymbol{k},\downarrow)\,u(\boldsymbol{k},\downarrow)=$$
$$=v^\dagger(\boldsymbol{k},\uparrow)\,v(\boldsymbol{k},\uparrow)=v^\dagger(\boldsymbol{k},\downarrow)\,v(\boldsymbol{k},\downarrow)=1\,. \tag{8}$$

* This is a general property of the eigenfunctions of an Hermitian operator. In fact, if

$$A\psi_1=\lambda_1\psi_1,\qquad A\psi_2=\lambda_2\psi_2\,,$$

the Hermiticity condition for A

$$(A\psi_1,\psi_2)=(\psi_1,A\psi_2)$$

gives

$$(\lambda_1\psi_1,\psi_2)=(\psi_1,\lambda_2\psi_2)\,.$$

If $\lambda_1\neq\lambda_2$, then

$$(\psi_1,\psi_2)=0$$

NOTE 1. The above normalization is that used in Heitler's book (W. Heitler, *The Quantum Theory of Radiation* (Oxford, 1954), p. 107).

Some authors (for instance S. Schweber, *Relativistic Quantum Field Theory* (Row, Peterson and Co., Evanston, 1961), p. 85) introduce a Lorentz invariant normalization as follows.

Let

$$\psi = u\, \mathrm{e}^{\mathrm{i}p_\mu x_\mu/\hbar}$$

be a solution of the Dirac equation. The function $u = u(\boldsymbol{k})$ and its adjoint obey the equations

$$(\mathrm{i}\, p_\mu \gamma_\mu + mc)\, u = 0\,,$$
$$\bar{u}\, (\mathrm{i}\, p_\mu \gamma_\mu + mc) = 0\,.$$

On multiplying the first by $\bar{u}\, \gamma_\nu$ from the left and the second by $\gamma_\nu\, u$ from the right, and summing, one has

$$2\, mc\, \bar{u}\, \gamma_\nu\, u + \mathrm{i}\, p_\mu\, \bar{u}(\gamma_\nu \gamma_\mu + \gamma_\mu \gamma_\nu)\, u = 0\,,$$

from which

$$\mathrm{i}\, mc\, \bar{u}\, \gamma_\nu\, u = p_\nu\, \bar{u}\, u\,. \tag{9}$$

Since the quantities $\bar{u}\, \gamma_\nu\, u$ are transformed as the components of a four-vector (as will be seen in Chap. VI, § 3) and $\bar{u}u$ is a scalar, the above relation expresses the equality of two four-vectors.

The normalization

$$\bar{u}\, u = E/|E|\,,$$

i.e.

$$\bar{u}\, u = 1 \quad \text{for positive frequency}\,,$$
$$\bar{u}\, u = -1 \quad \text{for negative frequency}\,,$$

is Lorentz invariant. Inserting this in eq. (9) one has

$$u^\dagger u = |E|/mc^2\,,$$

manifestly different from (8).

If the momentum is along the third axis ($\boldsymbol{k} \equiv (0, 0, |\boldsymbol{k}|)$), the plane wave solutions (1), (1′) may be expressed in the form

$$e^{i(k,x)}\, u\left(\boldsymbol{k}, {\uparrow \atop \downarrow}\right) =$$

$$= \tfrac{1}{2}[k_0(k_0 \pm k)]^{-\frac{1}{2}} \left(-\,i\,\boldsymbol{\alpha}\cdot\nabla_x + \frac{i}{c}\frac{\partial}{\partial t} + \beta\kappa\right) e^{i(k,x)} \begin{pmatrix} a_\pm \\ b_\pm \end{pmatrix},$$

$$e^{-i(k,x)}\, v\left(\boldsymbol{k}, {\uparrow \atop \downarrow}\right) =$$

$$= \pm\,\tfrac{1}{2}[k_0(k_0 \mp k)]^{-\frac{1}{2}} \left(-\,i\,\boldsymbol{\alpha}\cdot\nabla_x + \frac{i}{c}\frac{\partial}{\partial t} + \beta\kappa\right) e^{-i(k,x)} \begin{pmatrix} c_\pm \\ d_\pm \end{pmatrix},$$

($k = |\boldsymbol{k}|$), where, in the Pauli representation,

$$a_\pm = b_\pm = c_\mp = d_\mp = \chi_\pm\,, \quad \left(\chi_+ = \begin{pmatrix} 1 \\ 0 \end{pmatrix},\ \ \chi_- = \begin{pmatrix} 0 \\ 1 \end{pmatrix}\right),$$

in the Kramers representation,

$$a'_\pm = c'_\mp = \sqrt{2}\,\chi_\pm\,, \quad b'_\pm = d'_\mp = 0\,,$$

and, in the Majorana representation,

$$a''_\pm = c''_\mp = e^{\frac{1}{4}i\pi}\,\chi_\pm\,, \quad b''_\pm = d''_\mp = \pm\, e^{-\frac{1}{4}i\pi}\,\chi_\mp\,.$$

In fact

$$\left(-\,i\,\boldsymbol{\alpha}\cdot\nabla_x - \frac{i}{c}\frac{\partial}{\partial t} + \beta\kappa\right)\left(-\,i\,\boldsymbol{\alpha}\cdot\nabla_x + \frac{i}{c}\frac{\partial}{\partial t} + \beta\kappa\right) e^{\pm i(k,x)} =$$

$$= (\kappa^2 - \Box)\, e^{\pm i(k,x)} = 0\,,$$

while the orthonormality conditions and the validity of eqs. (4) and (4′) are readily verified by calculation.

2. Green's function

The plane wave solutions (3) and (3′) form a complete set. An arbitrary solution of the free particle Dirac equation may be represented in the form

$$\psi(x) = (2\pi)^{-\frac{3}{2}} \sum_{\lambda=\uparrow,\downarrow} \int \left[A(\boldsymbol{k},\lambda)\, u(\boldsymbol{k},\lambda)\, e^{i(\boldsymbol{k}\cdot x - \omega t)} + \right.$$
$$\left. + \, B(\boldsymbol{k},\lambda)\, v(-\,\boldsymbol{k},\lambda)\, e^{i(\boldsymbol{k}\cdot x + \omega t)}\right] d^3k \qquad (10)$$

(the two terms in square brackets are both eigenfunctions of the momentum belonging to the eigenvalue $\hbar\boldsymbol{k}$).

The coefficients of (10) may, conversely, be expressed in terms of the wave function

$$A(\boldsymbol{k}, \lambda) = (2\pi)^{-\frac{3}{2}} \int u^{\dagger}(\boldsymbol{k}, \lambda)\, \mathrm{e}^{-\mathrm{i}(\boldsymbol{k}\cdot\boldsymbol{x}-\omega t)}\, \psi(x)\, \mathrm{d}^3\mathrm{x}\,, \tag{11}$$

$$B(\boldsymbol{k}, \lambda) = (2\pi)^{-\frac{3}{2}} \int v^{\dagger}(-\boldsymbol{k}, \lambda)\, \mathrm{e}^{-\mathrm{i}(\boldsymbol{k}\cdot\boldsymbol{x}+\omega t)}\, \psi(x)\, \mathrm{d}^3\mathrm{x} \tag{11'}$$

(using the orthonormality relations for $u(\boldsymbol{k}, \lambda)$ and $v(-\boldsymbol{k}, \lambda)$).

Inserting (11) and (11′) in (10) one finds

$$(2\pi)^3\, \psi(x) =$$

$$= \int \left\{ \sum_{\lambda=\uparrow,\downarrow} \int \left[u(\boldsymbol{k},\lambda)\, u^{\dagger}(\boldsymbol{k},\lambda) + v(-\boldsymbol{k},\lambda)\, v^{\dagger}(-\boldsymbol{k},\lambda)\right] \mathrm{e}^{\mathrm{i}\boldsymbol{k}\cdot(x-x')}\, \mathrm{d}^3\mathrm{k} \right\} \psi(x')\, \mathrm{d}^3\mathrm{x}'.$$

Since this must be an identity, the expression in brackets must be equal to $(2\pi)^3\, \delta(\boldsymbol{x} - \boldsymbol{x}')$ (completeness relation). This implies that it must be *

$$\sum_{\lambda=\uparrow,\downarrow} \left[u(\boldsymbol{k}, \lambda)\, u^{\dagger}(\boldsymbol{k}, \lambda) + v(-\boldsymbol{k}, \lambda)\, v^{\dagger}(-\boldsymbol{k}, \lambda)\right] = I\,. \tag{12}$$

Since the Dirac equation is of the first order in the time, like the non-relativistic Schroedinger equation, the wave function at time t may be expressed in terms of its values at a different time t' as

$$\psi(\boldsymbol{x}, t) = \int \mathrm{G_D}(\boldsymbol{x} - \boldsymbol{x}', t - t')\, \psi(\boldsymbol{x}', t')\, \mathrm{d}^3\mathrm{x}'\,. \tag{13}$$

The Green's function $\mathrm{G_D}(\boldsymbol{x} - \boldsymbol{x}', t - t')$ has the properties

$$\begin{aligned} \left(\frac{1}{c}\frac{\partial}{\partial t} + \boldsymbol{\alpha}\cdot\nabla_x + \mathrm{i}\kappa\beta\right) \mathrm{G_D}(\boldsymbol{x} - \boldsymbol{x}', t - t') &= 0\,, \quad \text{for} \quad t \neq t'\,, \\ \mathrm{G_D}(\boldsymbol{x} - \boldsymbol{x}', 0) &= \delta(\boldsymbol{x} - \boldsymbol{x}')\, I\,, \end{aligned} \tag{14}$$

and may be constructed with the complete set of solutions (3) and (3′)

$$\begin{aligned} \mathrm{G_D}(\boldsymbol{x} - \boldsymbol{x}', t - t') = (2\pi)^{-3} \sum_{\lambda=\uparrow,\downarrow} \int \mathrm{e}^{\mathrm{i}\boldsymbol{k}\cdot(x-x')} \times \\ \times \left[u(\boldsymbol{k}, \lambda)\, u^{\dagger}(\boldsymbol{k}, \lambda)\, \mathrm{e}^{-\mathrm{i}\omega(t-t')} + v(-\boldsymbol{k}, \lambda)\, v^{\dagger}(-\boldsymbol{k}, \lambda)\, \mathrm{e}^{\mathrm{i}\omega(t-t')}\right] \mathrm{d}^3\mathrm{k}\,. \end{aligned} \tag{15}$$

This Green's function can also be written as

$$\mathrm{G_D}(\boldsymbol{x} - \boldsymbol{x}', t - t') = \left(-\frac{1}{c}\frac{\partial}{\partial t} + \boldsymbol{\alpha}\cdot\nabla_x + \mathrm{i}\kappa\beta\right) \mathrm{G_{KG}}(\boldsymbol{x} - \boldsymbol{x}', t - t')\,,$$

* The expressions (11) and (11′), and the completeness relation (12), would read differently, if a normalization other than (8) were chosen for u and v.

where G_{KG} is the relativistic Green's function for particles of spin zero (Part I, Chap. I, § 2, eq. (9)). We leave it to the reader to show that this expression satisfies the conditions (14).

It should be stressed that the Green's function G_D is that for the propagation of free waves in the one-particle theory. A wave packet involving only positive frequencies cannot, if propagating according to eq. (13) with the Green's function (15), acquire negative frequencies, as is easy to show. In Part III the Green's function G_D will be modified in the light of a description of antiparticles by means of negative frequency wave functions.

3. Zitterbewegung

Using the expression (10) for the wave function, the physical quantities of the theory may be written as integrals in $\boldsymbol{k}$ space.

For instance we have

$$\begin{aligned} Q &= \int \rho \, \mathrm{d}^3\mathrm{x} = \sum_{\lambda=\uparrow,\downarrow} \int (|A(\boldsymbol{k},\lambda)|^2 + |B(\boldsymbol{k},\lambda)|^2)\, \mathrm{d}^3\mathrm{k}\,, \\ \boldsymbol{P} &= \sum_{\lambda=\uparrow,\downarrow} \int (|A(\boldsymbol{k},\lambda)|^2 + |B(\boldsymbol{k},\lambda)|^2)\, \hbar \boldsymbol{k}\, \mathrm{d}^3\mathrm{k}\,, \\ E &= \sum_{\lambda=\uparrow,\downarrow} \int (|A(\boldsymbol{k},\lambda)|^2 - |B(\boldsymbol{k},\lambda)|^2)\, \hbar\omega\, \mathrm{d}^3\mathrm{k}\,. \end{aligned} \tag{16}$$

It is important to notice that E is the difference of two positive terms, arising from the positive and from the negative frequencies respectively (non-positive-definiteness, already noticed in Chap. IV, § 1).

The current may similarly be expressed in the form

$$\begin{aligned} \boldsymbol{J} = \int \boldsymbol{j}\, \mathrm{d}^3\mathrm{x} = \sum_{\lambda=\uparrow,\downarrow} \sum_{\lambda'=\uparrow,\downarrow} \int [&A^*(\boldsymbol{k},\lambda)\, A(\boldsymbol{k},\lambda')\, u^\dagger(\boldsymbol{k},\lambda)\, c\boldsymbol{\alpha}\, u(\boldsymbol{k},\lambda') + \\ &+ B^*(\boldsymbol{k},\lambda)\, B(\boldsymbol{k},\lambda')\, v^\dagger(-\boldsymbol{k},\lambda)\, c\boldsymbol{\alpha}\, v(-\boldsymbol{k},\lambda') + \\ &+ A^*(\boldsymbol{k},\lambda)\, B(\boldsymbol{k},\lambda')\, u^\dagger(\boldsymbol{k},\lambda)\, c\boldsymbol{\alpha}\, v(-\boldsymbol{k},\lambda')\, \mathrm{e}^{2\mathrm{i}\omega t} + \\ &+ B^*(\boldsymbol{k},\lambda)\, A(\boldsymbol{k},\lambda')\, v^\dagger(-\boldsymbol{k},\lambda)\, c\boldsymbol{\alpha}\, u(\boldsymbol{k},\lambda')\, \mathrm{e}^{-2\mathrm{i}\omega t}]\, \mathrm{d}^3\mathrm{k}\,. \end{aligned} \tag{17}$$

Since

$$\mathrm{d}\,\langle \boldsymbol{x} \rangle / \mathrm{d}t = \boldsymbol{J}/Q$$

(see Chap. IV, § 3), it follows from eq. (17) that $\langle \dot{\boldsymbol{x}} \rangle$ is not constant for a free particle wave packet, but is rather of the form

$$\langle \dot{\boldsymbol{x}} \rangle = \boldsymbol{a} + \langle \boldsymbol{b}(\omega)\, \mathrm{e}^{2\mathrm{i}\omega t} \rangle + \langle \boldsymbol{c}(\omega)\, \mathrm{e}^{-2\mathrm{i}\omega t} \rangle$$

(Zitterbewegung). The oscillating terms arise from interference of positive and negative frequency plane wave components, and are zero for a wave involving either only positive or only negative frequencies.

4. Plane wave solutions in various representations

The solutions of eqs. (2) depend, of course, on the representation of the γ's employed.

Tables I and II show the components of solutions of positive and negative frequency in the representations of Pauli, Kramers and Majorana.

TABLE I

	$\sqrt{2}\,u(\boldsymbol{k},\uparrow)$	$\sqrt{2}\,u(\boldsymbol{k},\downarrow)$	$u'(\boldsymbol{k},\uparrow)$	$u'(\boldsymbol{k},\downarrow)$	$\sqrt{2}\,\mathrm{e}^{-\frac{1}{4}\mathrm{i}\pi}\,u''(\boldsymbol{k},\uparrow)$	$\sqrt{2}\,\mathrm{e}^{-\frac{1}{4}\mathrm{i}\pi}\,u''(\boldsymbol{k},\downarrow)$
(1)	$\alpha(c+s)$	$-\beta^*(c+s)$	$c\alpha$	$-s\beta^*$	$c\alpha - \mathrm{i}s\beta$	$-s\beta^* - \mathrm{i}c\alpha^*$
(2)	$\beta(c+s)$	$\alpha^*(c+s)$	$c\beta$	$s\alpha^*$	$c\beta + \mathrm{i}s\alpha$	$s\alpha^* - \mathrm{i}c\beta^*$
(3)	$\alpha(c-s)$	$\beta^*(c-s)$	$s\alpha$	$-c\beta^*$	$\mathrm{i}c\beta + s\alpha$	$\mathrm{i}s\alpha^* - c\beta^*$
(4)	$\beta(c-s)$	$-\alpha^*(c-s)$	$s\beta$	$c\alpha^*$	$-\mathrm{i}c\alpha + s\beta$	$\mathrm{i}s\beta^* + c\alpha^*$

TABLE II

	$\sqrt{2}\,v(\boldsymbol{k},\uparrow)$	$\sqrt{2}\,v(\boldsymbol{k},\downarrow)$	$v'(\boldsymbol{k},\uparrow)$	$v'(\boldsymbol{k},\downarrow)$	$\sqrt{2}\,\mathrm{e}^{-\frac{1}{4}\mathrm{i}\pi}\,v''(\boldsymbol{k},\uparrow)$	$\sqrt{2}\,\mathrm{e}^{-\frac{1}{4}\mathrm{i}\pi}\,v''(\boldsymbol{k},\downarrow)$
(1)	$-\beta^*(c-s)$	$\alpha(c-s)$	$s\beta^*$	$c\alpha$	$s\beta^* - \mathrm{i}c\alpha^*$	$c\alpha + \mathrm{i}s\beta$
(2)	$\alpha^*(c-s)$	$\beta(c-s)$	$-s\alpha^*$	$c\beta$	$-s\alpha^* - \mathrm{i}c\beta^*$	$c\beta - \mathrm{i}s\alpha$
(3)	$\beta^*(c+s)$	$\alpha(c+s)$	$-c\beta^*$	$-s\alpha$	$-\mathrm{i}s\alpha^* - c\beta^*$	$\mathrm{i}c\beta - s\alpha$
(4)	$-\alpha^*(c+s)$	$\beta(c+s)$	$c\alpha^*$	$-s\beta$	$-\mathrm{i}s\beta^* + c\alpha^*$	$-\mathrm{i}c\alpha - s\beta$

The notation used is

$$c = \cos(\tfrac{1}{2}\chi)\,, \quad s = \sin(\tfrac{1}{2}\chi)\,, \quad \text{with} \quad \chi = \operatorname{ctg}^{-1}\left(\frac{|\boldsymbol{k}|}{\kappa}\right),$$

$$k_3 = |\boldsymbol{k}|\cos\theta\,, \quad k_1 \pm \mathrm{i}\,k_2 = |\boldsymbol{k}|\sin\theta\,\mathrm{e}^{\pm\mathrm{i}\phi}\,, \quad \alpha = \cos(\theta/2)\,\mathrm{e}^{-\mathrm{i}\phi/2}\,,$$

$$\beta = \sin(\theta/2)\,\mathrm{e}^{\mathrm{i}\phi/2}\,,$$

so that

$$c = \left(\frac{\sqrt{\boldsymbol{k}^2+\kappa^2} + |\boldsymbol{k}|}{2\sqrt{\boldsymbol{k}^2+\kappa^2}}\right)^{\frac{1}{2}}, \quad s = \left(\frac{\sqrt{\boldsymbol{k}^2+\kappa^2} - |\boldsymbol{k}|}{2\sqrt{\boldsymbol{k}^2+\kappa^2}}\right)^{\frac{1}{2}}.$$

(On this page c does not denote the speed of light.)

It is straightforward, though laborious, to verify that the solutions listed above form a complete set of orthonormal functions, i.e. that they satisfy the orthonormality conditions(5), (6), (7), (8) and the completeness relation (12).

In the Majorana representation the negative frequency solutions are the complex conjugates of the positive frequency solutions with the same spin

$$(u''_\alpha(\boldsymbol{k}, \uparrow))^* = v''_\alpha(\boldsymbol{k}, \uparrow), \quad (u''_\alpha(\boldsymbol{k}, \downarrow))^* = v''_\alpha(\boldsymbol{k}, \downarrow), \quad (\alpha = 1, 2, 3, 4),$$

as might have been expected, since, in the Majorana representation, the matrices multiplying $u(\boldsymbol{k})$ and $v(\boldsymbol{k})$ in eqs. (2) are complex conjugate (γ'' is real, γ''_4 is purely imaginary!).

A relation between the u's and the v's exists also in the other representations, though more complicated.

In Pauli's representation one has *

$$\begin{aligned} u(\boldsymbol{k}, \lambda) &= (v^\dagger(\boldsymbol{k}, \lambda)\, \gamma_2)^T = \gamma_2\, v^*(\boldsymbol{k}, \lambda), \\ v(\boldsymbol{k}, \lambda) &= (u^\dagger(\boldsymbol{k}, \lambda)\, \gamma_2)^T = \gamma_2\, u^*(\boldsymbol{k}, \lambda), \end{aligned}$$

and, in that of Kramers,

$$\begin{aligned} u'(\boldsymbol{k}, \lambda) &= (v'^\dagger(\boldsymbol{k}, \lambda)\, \gamma'_2)^T = \gamma'_2\, v'^*(\boldsymbol{k}, \lambda), \\ v'(\boldsymbol{k}, \lambda) &= (u'^\dagger(\boldsymbol{k}, \lambda)\, \gamma'_2)^T = \gamma'_2\, u'^*(\boldsymbol{k}, \lambda). \end{aligned}$$

5. Pauli's representation and the non-relativistic limit

For $\theta = 0$, $\alpha = 1$, $\beta = 0$ (momentum along the z axis), one has, in Pauli's representation

$$\begin{aligned} u(\boldsymbol{k}, \uparrow) &= \frac{1}{2\sqrt{\omega}} \begin{pmatrix} (\omega + c\,|\boldsymbol{k}|)^{\frac{1}{2}} + (\omega - c\,|\boldsymbol{k}|)^{\frac{1}{2}} \\ 0 \\ (\omega + c\,|\boldsymbol{k}|)^{\frac{1}{2}} - (\omega - c\,|\boldsymbol{k}|)^{\frac{1}{2}} \\ 0 \end{pmatrix}, \\ u(\boldsymbol{k}, \downarrow) &= \frac{1}{2\sqrt{\omega}} \begin{pmatrix} 0 \\ (\omega + c\,|\boldsymbol{k}|)^{\frac{1}{2}} + (\omega - c\,|\boldsymbol{k}|)^{\frac{1}{2}} \\ 0 \\ -(\omega + c\,|\boldsymbol{k}|)^{\frac{1}{2}} + (\omega - c\,|\boldsymbol{k}|)^{\frac{1}{2}} \end{pmatrix}, \end{aligned} \tag{18}$$

* Clearly from eqs. (2) we have

$$\left(-\,i\gamma^* \cdot \boldsymbol{k} - \frac{\omega}{c}\, \gamma_4^* + \kappa\right) u^*(\boldsymbol{k}) = 0.$$

Now as $\gamma_k^* = -\gamma_k$, $(k = 1, 3)$, $\gamma_m^* = \gamma_m$, $(m = 2, 4)$, multiplying by γ_2 from the left one has

$$\left(-\,i\gamma \cdot \boldsymbol{k} + \frac{\omega}{c}\, \gamma_4 + \kappa\right) \gamma_2 u^* = 0.$$

$$v(\boldsymbol{k}, \uparrow) = \frac{1}{2\sqrt{\omega}} \begin{pmatrix} 0 \\ (\omega + c\,|\boldsymbol{k}|)^{\frac{1}{2}} - (\omega - c\,|\boldsymbol{k}|)^{\frac{1}{2}} \\ 0 \\ -(\omega + c\,|\boldsymbol{k}|)^{\frac{1}{2}} - (\omega - c\,|\boldsymbol{k}|)^{\frac{1}{2}} \end{pmatrix},$$

$$v(\boldsymbol{k}, \downarrow) = \frac{1}{2\sqrt{\omega}} \begin{pmatrix} (\omega + c\,|\boldsymbol{k}|)^{\frac{1}{2}} - (\omega - c\,|\boldsymbol{k}|)^{\frac{1}{2}} \\ 0 \\ (\omega + c\,|\boldsymbol{k}|)^{\frac{1}{2}} + (\omega - c\,|\boldsymbol{k}|)^{\frac{1}{2}} \\ 0 \end{pmatrix}. \tag{18'}$$

These are essentially the solutions given in Heitler's book *. Note, however, that Heitler's negative energy solutions are for the momentum $\boldsymbol{p} \equiv (0, 0, \hbar\,|\boldsymbol{k}|)$, whereas (18′) are solutions for the momentum $\boldsymbol{p} \equiv (0, 0, -\hbar\,|\boldsymbol{k}|)$ and satisfy the relations

$$s_3\, v(\boldsymbol{k}, \uparrow) = -\tfrac{1}{2}\hbar\, v(\boldsymbol{k}, \uparrow),$$
$$s_3\, v(\boldsymbol{k}, \downarrow) = \tfrac{1}{2}\hbar\, v(\boldsymbol{k}, \downarrow),$$

as $s_p = -s_3$.

For $\boldsymbol{k} = 0$ the solutions (18), (18′) reduce to

$$u(\boldsymbol{k}, \uparrow) \to \begin{pmatrix} 1 \\ 0 \\ 0 \\ 0 \end{pmatrix}, \quad u(\boldsymbol{k}, \downarrow) \to \begin{pmatrix} 0 \\ 1 \\ 0 \\ 0 \end{pmatrix},$$

$$v(\boldsymbol{k}, \uparrow) \to \begin{pmatrix} 0 \\ 0 \\ 0 \\ -1 \end{pmatrix}, \quad v(\boldsymbol{k}, \downarrow) \to \begin{pmatrix} 0 \\ 0 \\ 1 \\ 0 \end{pmatrix}.$$

The vanishing of the last (first) two components of solutions of positive (negative) frequency for $\boldsymbol{k} = 0$ is a property of Pauli's representation, and may be proved directly without using Tables I and II. Thus, in the non-relativistic limit, the four-component wave function in the Pauli representation reduces to a two-component wave function like that of the Schroedinger-Pauli theory of particles of spin $\frac{1}{2}$.

For a wave function of the type

$$\psi(x) = \begin{pmatrix} \begin{pmatrix} u_1 \\ u_2 \end{pmatrix} \\ \begin{pmatrix} u_3 \\ u_4 \end{pmatrix} \end{pmatrix} e^{i(\boldsymbol{p}\cdot\boldsymbol{x} - Et)/\hbar} = \begin{pmatrix} U_1 \\ U_2 \end{pmatrix} e^{i(\boldsymbol{p}\cdot\boldsymbol{x} - Et)/\hbar},$$

* W. Heitler, *The Quantum Theory of Radiation* (Oxford, 1954), p. 107.

the Dirac equation, in Pauli's representation, reads

$$E\begin{pmatrix}U_1\\U_2\end{pmatrix} = c\,\boldsymbol{p}\cdot\begin{pmatrix}0 & \boldsymbol{\sigma}\\ \boldsymbol{\sigma} & 0\end{pmatrix}\begin{pmatrix}U_1\\U_2\end{pmatrix} + mc^2\begin{pmatrix}I_\sigma & 0\\ 0 & -I_\sigma\end{pmatrix}\begin{pmatrix}U_1\\U_2\end{pmatrix},$$

namely

$$\begin{aligned}(E-mc^2)\,U_1 &= c\,\boldsymbol{\sigma}\cdot\boldsymbol{p}\,U_2\,,\\ (E+mc^2)\,U_2 &= c\,\boldsymbol{\sigma}\cdot\boldsymbol{p}\,U_1\,.\end{aligned} \tag{19}$$

We assume that $E > 0$. An arbitrary choice can be made for U_1, while U_2, according to eqs. (19), is fully determined by U_1

$$U_2 = c\,\boldsymbol{\sigma}\cdot\boldsymbol{p}\,U_1/(E+mc^2)\,.$$

The positive energy solutions (18) (Pauli representation) correspond to the two independent choices

$$U_1(\uparrow) = \begin{pmatrix}1\\0\end{pmatrix},\quad U_1(\downarrow) = \begin{pmatrix}0\\1\end{pmatrix}.$$

We now show that, for $E > 0$, the "ratio" of U_2 and U_1 is of the order of v/c. In fact

$$U_2^\dagger U_2 = c^2\, U_1^\dagger(\boldsymbol{\sigma}\cdot\boldsymbol{p})^2\, U_1/(E+mc^2)^2 = (E-mc^2)\, U_1^\dagger U_1/(E+mc^2)\,, \tag{20}$$

and, since in the non-relativistic limit $E \simeq mc^2 + \frac{1}{2}mv^2$, one has

$$U_2^\dagger U_2 \simeq (v/2c)^2\, U_1^\dagger U_1\,.$$

6. The Foldy-Wouthuysen transformation

As has been shown in the previous section, the first (last) two components of the wave function in the Pauli representation become negligible in the non-relativistic limit for negative (positive) frequency, respectively. This property arises from the fact that, when the momentum of the particle is small compared with mc, the term involving the *odd* matrices α_1, α_2, α_3, which have matrix elements connecting upper and lower components, becomes negligible, and the Hamiltonian reduces to the term in β, an *even* matrix without such matrix elements *.

* The matrices

$$I,\ \beta,\ \boldsymbol{\sigma} = \frac{\boldsymbol{\alpha}\times\boldsymbol{\alpha}}{2\mathrm{i}}$$

and $\beta\boldsymbol{\sigma}$, which do not mix the first two with the last two components of the wave function, are said to be even, in contrast to the odd matrices $\boldsymbol{\alpha}$, $\beta\boldsymbol{\alpha}_1$ $\gamma_5 = -\mathrm{i}\alpha_1\alpha_2\alpha_3$ and $\beta\gamma_5$, which do so.

This suggests the question: "Can a representation of the Dirac equation for arbitrary momentum be found, such that the lower (upper) components of the wave function are identically zero?" *.

Foldy and Wouthuysen have shown ** that this is achieved by the unitary transformation

$$U_{\text{F.W.}} = e^{i S_{\text{F.W.}}}$$

with

$$S_{\text{F.W.}} = \frac{1}{2i} \beta \frac{\boldsymbol{\alpha} \cdot \boldsymbol{p}}{|\boldsymbol{p}|} \operatorname{tg}^{-1} \left(\frac{|\boldsymbol{p}|}{mc} \right).$$

Since

$$e^{\pm i S_{\text{F.W.}}} = \frac{|H| + mc^2 \pm c\beta\, \boldsymbol{\alpha} \cdot \boldsymbol{p}}{[2\,|H|(|H| + mc^2)]^{\frac{1}{2}}},$$

where

$$|H| = c\sqrt{p^2 + m^2c^2},$$

the transformed Hamiltonian

$$H_{\text{F.W.}} = U_{\text{F.W.}} H\, U_{\text{F.W.}}^{-1} = U_{\text{F.W.}} (c\, \boldsymbol{\alpha} \cdot \boldsymbol{p} + \beta mc^2)\, U_{\text{F.W.}}^{-1}$$

is easily evaluated and found to be of the form

$$H_{\text{F.W.}} = \beta\, |H| = \beta c \sqrt{p^2 + m^2c^2},$$

while the transformed wave function is

$$\psi_{\text{F.W.}} = \frac{|H| + \beta H}{[2\,|H|(|H| + mc^2)]^{\frac{1}{2}}} \psi.$$

Clearly a (positive/negative) frequency solution $\psi^{\pm}$ of the Dirac equation is transformed into

$$\psi_{\text{F.W.}}^{\pm} = \left[\frac{|H|}{2(|H| + mc^2)} \right]^{\frac{1}{2}} (I \pm \beta)\, \psi^{\pm},$$

having vanishing (lower/upper) components if the Dirac matrices and $\psi^{\pm}$ are in the Pauli representation.

Since $\beta\, \psi_{\text{F.W.}}^{\pm} = \pm\, \psi_{\text{F.W.}}^{\pm}$, the Dirac equations

$$\left(H_{\text{F.W.}} - \hbar i \frac{\partial}{\partial t} \right) \psi_{\text{F.W.}} = 0$$

* Note that the functions $(I \pm \beta)\, \psi\, (x)/2$, having this property, do not, however, satisfy the Dirac equation for $\boldsymbol{p} \neq 0$, even if $\psi(x)$ does.

** L. L. Foldy and S. A. Wouthuysen, Phys. Rev. **78** (1950), 29.

for free particles of (positive/negative) energy reduce, in the non-relativistic limit, to the Schroedinger equations

$$\pm\left(mc^2 + \frac{p^2}{2m}\right)\psi^{\pm}_{\mathrm{F.W.}} = \hbar i \frac{\partial \psi^{\pm}_{\mathrm{F.W.}}}{\partial t}$$

for particles of (positive/negative) mass. This shows that the F.-W. representation is most suitable for the treatment of the non-relativistic approximation.

It must be stressed that the F.-W. transformation, unlike the transformations (12), (13), (14), Chap. III, § 2, is *not* a point transformation.

The wave function

$$\psi(x) = (2\pi)^{-\frac{3}{2}} \times \\ \times \sum_{\lambda} \int \left[A(\boldsymbol{k}, \lambda)\, u(\boldsymbol{k}, \lambda)\, e^{i(k, x)} + B(-\boldsymbol{k}, \lambda)\, v(\boldsymbol{k}, \lambda)\, e^{-i(k, x)}\right] d^3k \tag{21}$$

is transformed into

$$\psi_{\mathrm{F.W.}}(x) = (2\pi)^{-\frac{3}{2}} \sum_{\lambda} \int \sqrt{\frac{(k^2+\kappa^2)^{\frac{1}{2}}}{2[(k^2+\kappa^2)^{\frac{1}{2}}+\kappa]}}\,(I + \beta\Lambda(\boldsymbol{p})) \times \\ \times \left[A(\boldsymbol{k}, \lambda)\, u(\boldsymbol{k}, \lambda)\, e^{i(k, x)} + B(-\boldsymbol{k}, \lambda)\, v(\boldsymbol{k}, \lambda)\, e^{-i(k, x)}\right] d^3k,$$

where $\Lambda(\boldsymbol{p})$ is the operator defined in *Note 1*, p. 146.

Substituting

$$A(\boldsymbol{k}, \lambda) = u^{\dagger}(\boldsymbol{k}, \lambda)\,(2\pi)^{-\frac{3}{2}} \int \psi(x)\, e^{-i(k, x)}\, d^3x\,,$$
$$B(-\boldsymbol{k}, \lambda) = v^{\dagger}(\boldsymbol{k}, \lambda)\,(2\pi)^{-\frac{3}{2}} \int \psi(x)\, e^{i(k, x)}\, d^3x\,,$$

one has

$$\psi_{\mathrm{F.W.}}(x) = \int U_{\mathrm{F.W.}}(\boldsymbol{x}, \boldsymbol{x}')\, \psi(x')\, d^3x'\,, \quad (t = t')\,,$$

with

$$U_{\mathrm{F.W.}}(\boldsymbol{x}, \boldsymbol{x}') = (2\pi)^{-3} \sum_{\lambda} \int \sqrt{\frac{(k^2+\kappa^2)^{\frac{1}{2}}}{2[(k^2+\kappa^2)^{\frac{1}{2}}+\kappa]}}\,(I + \beta\Lambda(\boldsymbol{p})) \times \\ \times \left[u(\boldsymbol{k}, \lambda)\, u^{\dagger}(\boldsymbol{k}, \lambda)\, e^{i\boldsymbol{k}\cdot(\boldsymbol{x}-\boldsymbol{x}')} + v(\boldsymbol{k}, \lambda)\, v^{\dagger}(\boldsymbol{k}, \lambda)\, e^{-i\boldsymbol{k}\cdot(\boldsymbol{x}-\boldsymbol{x}')}\right] d^3k\,,$$

showing that $\psi_{\mathrm{F.W.}}$ at a given point consists of contributions depending on ψ over a region, about the point, with an extension of the order of the Compton wavelength of the particle.

Whereas the momentum operator $\boldsymbol{p}$ is the same in the conventional and in the F.-W. representation *, $(\boldsymbol{p})_{\mathrm{F.W.}} = \boldsymbol{p}$, this is not the case for the position, the velocity, the orbital angular momentum and the spin.

---

* Since the transformation does not involve $\boldsymbol{x}$.

In fact

$$(\boldsymbol{x})_{\text{F.W.}} = U_{\text{F.W.}}\, \boldsymbol{x}\, U^{-1}_{\text{F.W.}} =$$

$$= \boldsymbol{x} - \frac{\mathrm{i}\,\hbar c\,\beta\boldsymbol{\alpha}}{2\,|H|} + \frac{\hbar c^2[\mathrm{i}c\beta(\boldsymbol{\alpha}\cdot\boldsymbol{p})\boldsymbol{p} - (\boldsymbol{\sigma}\times\boldsymbol{p})\,|H|]}{2\,|H|^2\,(|H| + mc^2)}, \tag{22}$$

$$(\dot{\boldsymbol{x}})_{\text{F.W.}} = U_{\text{F.W.}}\, c\boldsymbol{\alpha}\, U^{-1}_{\text{F.W.}} = c\boldsymbol{\alpha} + \frac{c^2\,\beta\boldsymbol{p}}{|H|} - \frac{c^3(\boldsymbol{\alpha}\cdot\boldsymbol{p})\,\boldsymbol{p}}{|H|\,(|H| + mc^2)}, \tag{22'}$$

$$(\boldsymbol{L})_{\text{F.W.}} = U_{\text{F.W.}}\,(\boldsymbol{x}\times\boldsymbol{p})\,U^{-1}_{\text{F.W.}} = (\boldsymbol{x})_{\text{F.W.}}\times\boldsymbol{p}, \tag{22''}$$

$$(\boldsymbol{s})_{\text{F.W.}} = U_{\text{F.W.}}\,\frac{\hbar\,\boldsymbol{\alpha}\times\boldsymbol{\alpha}}{4\mathrm{i}}\,U^{-1}_{\text{F.W.}} =$$

$$= \frac{\hbar}{2}\left[\boldsymbol{\sigma} + \frac{\mathrm{i}\,c\beta(\boldsymbol{\alpha}\times\boldsymbol{p})}{|H|} - \frac{c^2\,\boldsymbol{p}\times(\boldsymbol{\sigma}\times\boldsymbol{p})}{|H|\,(|H| + mc^2)}\right], \tag{22'''}$$

as the reader may verify as an excercise.

On the other hand, one may define the operators

mean position $\boldsymbol{x}_{\text{av.}}$,
mean velocity $\boldsymbol{v}_{\text{av.}}$,
mean orbital angular momentum $\boldsymbol{L}_{\text{av.}} = \boldsymbol{x}_{\text{av.}}\times\boldsymbol{p}$,
mean spin $\boldsymbol{s}_{\text{av.}}$,

which, *in the* F.-W. *representation*, are given by

$$(\boldsymbol{x}_{\text{av.}})_{\text{F.W.}} = \boldsymbol{x},\quad (\boldsymbol{v}_{\text{av.}})_{\text{F.W.}} = \frac{c^2\beta\boldsymbol{p}}{|H|},\quad (\boldsymbol{L}_{\text{av.}})_{\text{F.W.}} = \boldsymbol{x}\times\boldsymbol{p},\quad (\boldsymbol{s}_{\text{av.}})_{\text{F.W.}} = \frac{\hbar}{2}\,\boldsymbol{\sigma}.$$

In contrast to $\boldsymbol{x}$ and $\boldsymbol{v} = c\boldsymbol{\alpha}$ *, the position and velocity operators in the conventional representation, the operators $\boldsymbol{x}_{\text{av.}}$ and $\boldsymbol{v}_{\text{av.}}$ have the properties of the position and the velocity of a classical particle.

In fact, $\boldsymbol{v}_{\text{av.}}$ has the correct momentum dependence and

$$\dot{\boldsymbol{x}}_{\text{av.}} = \frac{\mathrm{i}}{\hbar}[H, \boldsymbol{x}_{\text{av.}}] = \boldsymbol{v}_{\text{av.}}$$

as is easy to show by working in the F.-W. representation. Using this one also sees at once that, in contrast to $\boldsymbol{L}$ and $\boldsymbol{s}$, $\boldsymbol{L}_{\text{av.}}$ and $\boldsymbol{s}_{\text{av.}}$ are constants of the motion.

* The eigenvalues of v_i, $(i = 1, 2, 3)$, are $\pm\, c$

Of course, the operators $\boldsymbol{x}_{\text{av.}}$, $\boldsymbol{v}_{\text{av.}}$ etc. are given, in the old representation, by $U_{\text{F.W.}}^{-1}\, \boldsymbol{x}\, U_{\text{F.W.}}$ etc., which, when worked out, are expressions of the same degree of complication as the right hand sides of eqs. (22) etc.

The mean value of the operator $\boldsymbol{v}_{\text{av.}}$, for the wave function (21), involving positive and negative frequencies, is

$$\langle \boldsymbol{v}_{\text{av.}} \rangle = \left\langle U_{\text{F.W.}}^{-1} \frac{c\beta \boldsymbol{p}}{(p^2 + m^2c^2)^{\frac{1}{2}}} U_{\text{F.W.}} \right\rangle = \left\langle \frac{\boldsymbol{p}}{p^2 + m^2c^2} H \right\rangle =$$

$$= \left\langle \frac{c\boldsymbol{p}}{(p^2 + m^2c^2)^{\frac{1}{2}}} \Lambda(\boldsymbol{p}) \right\rangle = \left\langle \frac{c\boldsymbol{p}}{(p^2 + m^2c^2)^{\frac{1}{2}}} (\Lambda_+(\boldsymbol{p}) - \Lambda_-(\boldsymbol{p})) \right\rangle =$$

$$= \sum_{\lambda} \int \frac{c\boldsymbol{k}}{(k^2 + \kappa^2)^{\frac{1}{2}}} (|A(\boldsymbol{k}, \lambda)|^2 + |B(-\boldsymbol{k}, \lambda)|^2)\, \mathrm{d}^3k$$

(*no* Zitterbewegung!) *.

In the conventional representation, the mean-spin operator is

$$\boldsymbol{s}_{\text{av.}} = \frac{\hbar}{2}\left[\boldsymbol{\sigma} - \frac{\mathrm{i}\, c\beta(\boldsymbol{\alpha} \times \boldsymbol{p})}{|H|} - \frac{c^2(\boldsymbol{p} \times (\boldsymbol{\sigma} \times \boldsymbol{p}))}{|H|(|H| + mc^2)}\right].$$

The expression in square brackets differs by a factor $H/|H|$ from the three-vector polarization operator, which was first introduced by Stech **.

The components $s_{\text{av.},i}$, $(i = 1, 2, 3)$, obey the same commutation and anticommutation relations as the Pauli matrices

$$s_{\text{av.},i}\, s_{\text{av.},j} = \tfrac{1}{4}\, \hbar^2\, \delta_{ij} + \tfrac{1}{2}\, \mathrm{i}\, \hbar\, \varepsilon_{ijk}\, s_{\text{av.},k}\,.$$

They commute with the Hamiltonian, and their expectation values are, therefore, constants of motion.

If $\boldsymbol{\varepsilon}$ is a unit vector, a free-particle solution of the Dirac equation $\psi_{p,\varepsilon}$ (momentum $\boldsymbol{p}$) which is an eigenfunction of the operator $(\boldsymbol{s}_{\text{av.}} \cdot \boldsymbol{\varepsilon})$ belonging to the eigenvalue $\frac{1}{2}\hbar$, describes a particle polarized along $\boldsymbol{\varepsilon}$. Since $(\boldsymbol{s}_{\text{av.}}\, \boldsymbol{p}) = \frac{1}{2}\hbar(\boldsymbol{\sigma} \cdot \boldsymbol{p})$, the free-particle solutions (3) and (3′) describe particles polarized in the direction of the momentum or in the opposite direction. $\psi_{p,\varepsilon}$ may be constructed as a linear combination of such solutions.

---

* The "velocity" of a positive (negative) frequency component is parallel (antiparallel) to the momentum. (Remember that $A(\boldsymbol{k}, \lambda)$ and $B(-\boldsymbol{k}, \lambda)$ correspond to momentum $\hbar\boldsymbol{k}$ and $-\hbar\boldsymbol{k}$ respectively.)

** B. Stech, Z. f. Phys. **144** (1956), 214.

CHAPTER VI

INVARIANCE PROPERTIES OF THE DIRAC EQUATION

1. General coordinate transformations

The Dirac equation is invariant under a coordinate transformation

$$x'_\mu = x'_\mu(x)\,, \quad (\mu = 1, 2, 3, 4)\,, \tag{1}$$

if the transformed wave function

$$\psi'(x') = S\,\psi(x) \tag{2}$$

obeys the equation

$$\left(\gamma_\mu \frac{\partial}{\partial x'_\mu} + \kappa\right) \psi'(x') = 0 \tag{3}$$

identical in form with the equation for $\psi(x)$.

Multiplying eq. (3) by S^{-1}, one sees that it reduces to the original equation for $\psi(x)$ if *

$$S^{-1}\,\gamma_\mu\, S \frac{\partial}{\partial x'_\mu} = \gamma_\nu \frac{\partial}{\partial x_\nu}\,. \tag{4}$$

Thus the proof of the invariance of the Dirac equation under a coordinate transformation consists essentially in finding a matrix satisfying eq. (4).

Notice that the adjoint Dirac equation

$$\frac{\partial \bar{\psi}}{\partial x_\mu}\,\gamma_\mu - \kappa\,\bar{\psi} = 0$$

is also invariant if the adjoint wave function is transformed according to

$$\bar{\psi}' = \bar{\psi}\, S^{-1}\,. \tag{5}$$

* In this formula S is assumed to be independent of x, which is not always the case, but is true for the transformations studied in § 2.

As an exercise the reader may show that, for $m = 0$, the Dirac equation is invariant under the reciprocal radii transformation

$$x_\mu{}' = x_\mu / x^2, \qquad x^2 = x_\mu x_\mu\,.$$

In this case

$$S = x^2 \gamma_\nu x_\nu, \qquad S^{-1} = \gamma_\nu x_\nu / (x^2)^2\,.$$

In order that $\bar{\psi}' = \overline{\psi'}$, it must be

$$S^{-1} = \gamma_4 \, S^\dagger \, \gamma_4 \, . \tag{6}$$

It is readily seen that the Lagrangian density is invariant under the above coordinate transformations (1), (2) with (4) and (6), as, of course, could be expected:

$$\mathfrak{L}' = -\frac{\hbar c}{2}\left[\bar{\psi}'\left(\gamma_\mu \frac{\partial \psi'}{\partial x'_\mu} + \kappa\,\psi'\right) - \left(\frac{\partial \bar{\psi}'}{\partial x'_\mu}\gamma_\mu - \kappa\,\bar{\psi}'\right)\psi'\right] =$$

$$= -\frac{\hbar c}{2}\left[\bar{\psi}\left(S^{-1}\gamma_\mu\, S\frac{\partial \psi}{\partial x'_\mu} + \kappa\,\psi\right) - \left(\frac{\partial \bar{\psi}}{\partial x'_\mu} S^{-1}\gamma_\mu\, S - \kappa\,\bar{\psi}\right)\psi\right] = \mathfrak{L}.$$

2. Linear orthogonal transformations

In the case of the linear transformations

$$x'_\mu = a_{\mu\nu}\, x_\nu \tag{7}$$

whose coefficients satisfy the orthogonality relations

$$a_{\mu\varrho}\, a_{\mu\sigma} = \delta_{\varrho\sigma}\,, \quad a_{\mu\varrho}\, a_{\nu\varrho} = \delta_{\mu\nu}\,, \tag{8}$$

eq. (4) becomes

$$S^{-1}\,\gamma_\mu\, S = a_{\mu\nu}\,\gamma_\nu\, . \tag{9}$$

2.1. Space rotations

For an infinitesimal space rotation of angle $\delta\phi$ about the unit vector $\boldsymbol{n}$, neglecting terms of second order, one has

$$S = 1 + (\mathrm{i}\,\boldsymbol{s}\cdot\boldsymbol{n}\,\delta\phi/\hbar) = 1 + \tfrac{1}{2}(n_1\,\gamma_2\,\gamma_3 + n_2\,\gamma_3\,\gamma_1 + n_3\,\gamma_1\,\gamma_2)\,\delta\phi\,,$$

$$S^{-1} = 1 - (\mathrm{i}\,\boldsymbol{s}\cdot\boldsymbol{n}\,\delta\phi/\hbar)\,.$$

In fact (see p. 143)

$$S^{-1}\,\gamma_i\, S = \gamma_i + (\mathrm{i}/\hbar)(\gamma_i\, s_j - s_j\,\gamma_i)\, n_j\,\delta\phi = \gamma_i + \varepsilon_{ilm}\,\gamma_l\, n_m\,\delta\phi\,,$$

or concisely

$$S^{-1}\,\boldsymbol{\gamma}\, S = \boldsymbol{\gamma} + \boldsymbol{\gamma}\times\boldsymbol{n}\,\delta\phi,$$

in accordance with (9).

For a finite rotation S is

$$S = \mathrm{e}^{\mathrm{i}\boldsymbol{s}\cdot\boldsymbol{n}\phi/\hbar} = \cos(\tfrac{1}{2}\,\phi) + (2\mathrm{i}/\hbar)\,\boldsymbol{s}\cdot\boldsymbol{n}\,\sin(\tfrac{1}{2}\,\phi)\,, \tag{10}$$

as is easily shown using the commutation relations for the spin components*.

According to eq. (10) the generator of infinitesimal rotations is $\boldsymbol{s}$, and not $\boldsymbol{s} + \boldsymbol{L} = \boldsymbol{M}$, since we express the transformed components $\psi'_\alpha(x')$ in terms of the original components $\psi_\alpha(x)$. On the other hand, if $\psi'(x)$ were to be expressed in terms of $\psi(x)$ (the argument is the same!), we would have

$$\psi'(x) = e^{i(\boldsymbol{s}+\boldsymbol{L})\cdot\boldsymbol{n}\phi/\hbar}\,\psi(x)\,,$$

$\boldsymbol{L}$ being the orbital angular momentum.

For rotations around the first, the second and the third axis one has respectively

$$S_1 = e^{\gamma_2\gamma_3\phi/2} = \cos(\tfrac{1}{2}\phi) + \gamma_2\gamma_3 \sin(\tfrac{1}{2}\phi)\,,$$
$$S_2 = e^{\gamma_3\gamma_1\phi/2} = \cos(\tfrac{1}{2}\phi) + \gamma_3\gamma_1 \sin(\tfrac{1}{2}\phi)\,,$$
$$S_3 = e^{\gamma_1\gamma_2\phi/2} = \cos(\tfrac{1}{2}\phi) + \gamma_1\gamma_2 \sin(\tfrac{1}{2}\phi)\,.$$

Because of the occurrence of $\phi/2$ rather than ϕ in the exponent, a wave function changes sign under a rotation of 2π. Like the two-component wave functions of Chap. I, § 3, the Dirac wave functions provide a two-valued representation of the rotation group.

Since $\boldsymbol{s}$ is Hermitian, the matrix S is unitary

$$S^{-1} = e^{-i\boldsymbol{s}\cdot\boldsymbol{n}\phi/\hbar} = S^\dagger\,.$$

As γ_4 commutes with the spin, this equation is equivalent to eq. (6).

From the invariance of the action under space rotations, the components of the angular momentum given in Chap. IV, § 2, may be deduced by Noether's theorem, as outlined in that section.

2.2. Lorentz transformations

For a Lorentz transformation of velocity v along the first axis, S is given by**

$$S = e^{i\gamma_1\gamma_4\chi/2} = \cosh(\tfrac{1}{2}\chi) + i\,\gamma_1\,\gamma_4 \sinh(\tfrac{1}{2}\chi)$$

---

* The reduction of the exponential form to a linear combination of trigonometric functions is done, as in Chap. I, § 3, by series expansion using the formulae,

$$(\gamma_2\gamma_3)^2 = (\gamma_3\gamma_1)^2 = (\gamma_1\gamma_2)^2 = -I\,, \qquad (i\boldsymbol{s}\cdot\boldsymbol{n}/\hbar)^2 = -\tfrac{1}{4}I\,.$$

** This is shown by the following calculations:

$$S^{-1}\gamma_1 S = \gamma_1\cosh^2(\tfrac{1}{2}\chi) + 2i\gamma_4 \sinh(\tfrac{1}{2}\chi)\cosh(\tfrac{1}{2}\chi) + \gamma_1 \sinh^2(\tfrac{1}{2}\chi) = \gamma_1\cosh\chi + i\gamma_4\sinh\chi = (1 - v^2/c^2)^{-\frac{1}{2}}(\gamma_1 + iv\gamma_4/c)$$

(use the formulae $\cosh^2(\tfrac{1}{2}\chi) + \sinh^2(\tfrac{1}{2}\chi) = \cosh\chi$, $2\sinh(\tfrac{1}{2}\chi)\cosh(\tfrac{1}{2}\chi) = \sinh\chi$),

$$S^{-1}\gamma_2 S = \gamma_2, \qquad S^{-1}\gamma_3 S = \gamma_3, \qquad S^{-1}\gamma_4 S = (1 - v^2/c^2)^{-\frac{1}{2}}(\gamma_4 - iv\gamma_1/c)$$

in agreement with eq. (9).

with

$$\cosh \chi = (1 - \beta^2)^{-\frac{1}{2}}, \quad \sinh \chi = \beta(1 - \beta^2)^{-\frac{1}{2}}, \quad (\beta = v/c).$$

This matrix, which does not commute with γ_4, is Hermitian rather than unitary *. Its inverse is

$$S^{-1} = e^{-i\gamma_1\gamma_4\chi/2} = \cosh(\tfrac{1}{2}\chi) - i\,\gamma_1\,\gamma_4 \sinh(\tfrac{1}{2}\chi).$$

For Lorentz transformations along the second and the third axis one has

$$S = e^{i\gamma_2\gamma_4\chi/2} = \cosh(\tfrac{1}{2}\chi) + i\,\gamma_2\gamma_4 \sinh(\tfrac{1}{2}\chi),$$
$$S = e^{i\gamma_3\gamma_4\chi/2} = \cosh(\tfrac{1}{2}\chi) + i\,\gamma_3\gamma_4 \sinh(\tfrac{1}{2}\chi),$$

respectively.

Bargmann and Wigner ** have studied the transformation properties of the wave function $\psi(x)$ for a particle having the four-momentum p_μ, under the *little group* consisting of the Lorentz transformations which leave the components of the four-momentum of the particle unchanged, so that

$$p_\mu' = a_{\mu\nu}\, p_\nu = p_\mu .$$

For an infinitesimal transformation with coefficients $a_{\mu\nu} = \delta_{\mu\nu} + \xi_{\mu\nu}$, the infinitesimal parameters $\xi_{\mu\nu}$ must satisfy the condition

$$\xi_{\mu\nu}\, p_\nu = 0 .$$

The variation in form of the wavefunction is

$$\delta\psi(x) = \psi'(x) - \psi(x) = \xi_{\mu\nu}\left(\tfrac{1}{4}\,\gamma_\mu\,\gamma_\nu + x_\mu \frac{\partial}{\partial x_\nu}\right)\psi(x),$$

which, since

$$\frac{\hbar}{i}\frac{\partial}{\partial x_\mu}\,\psi = p_\mu\,\psi,$$

reduces to

$$\delta\psi(x) = \xi_{\mu\nu}\left(\tfrac{1}{4}\,\gamma_\mu\,\gamma_\nu + \frac{i}{\hbar}\,x_\mu\,p_\nu\right)\psi(x) = \tfrac{1}{4}\,\xi_{\mu\nu}\,\gamma_\mu\,\gamma_\nu\,\psi(x).$$

The right side may be expressed in terms of the four-vector polarization operator $T_\mu = \gamma_5\left[i\,\gamma_\mu - (mc)^{-1}\,p_\mu\right]$ (see § 3). In fact, using the Dirac equation and the conditions $\xi_{\mu\nu}\,p_\nu = 0$, one has (no convention of equal indices!)

$$\sum_{\mu\varrho\lambda} \varepsilon_{\nu\mu\varrho\lambda}\,\xi_{\mu\varrho}\,T_\lambda\,\psi(x) = \sum_{\mu\varrho\lambda} \varepsilon_{\nu\mu\varrho\lambda}\,\xi_{\mu\varrho}\,\varepsilon_{\nu\mu\varrho\lambda}\,\gamma_\nu\,\gamma_\mu\,\gamma_\varrho\,\gamma_\lambda\left[i\,\gamma_\lambda - (mc)^{-1}\,p_\lambda\right]\psi(x)$$

* See, however, the remark on p. 128.

** V. Bargmann and E. P. Wigner, Proc. Natl. Acad. Sci. U.S. **34** (1948), 211.

$$= \sum_{\mu(\neq\nu)} \sum_{\varrho(\neq\nu)} \sum_{\lambda(\neq\nu,\mu,\varrho)} \xi_{\mu\varrho} \left[\mathrm{i}\, \gamma_\nu \gamma_\mu \gamma_\varrho - (mc)^{-1} \gamma_\nu \gamma_\mu \gamma_\varrho \gamma_\lambda p_\lambda \right] \psi(x)$$

$$= \sum_{\mu(\neq\nu)} \sum_{\varrho(\neq\nu)} \xi_{\mu\varrho} \{ \mathrm{i}\, \gamma_\nu \gamma_\mu \gamma_\varrho - (mc)^{-1} \gamma_\nu \gamma_\mu \gamma_\varrho [- p_\nu \gamma_\nu - p_\varrho \gamma_\varrho - p_\mu \gamma_\mu + \mathrm{i}\, mc] \} \psi(x)$$

$$= (mc)^{-1} \sum_{\mu(\neq\nu)} \sum_{\varrho(\neq\nu)} \xi_{\mu\varrho} (p_\nu \gamma_\mu \gamma_\varrho + p_\varrho \gamma_\nu \gamma_\mu - p_\mu \gamma_\nu \gamma_\varrho) \psi(x)$$

$$= (mc)^{-1} p_\nu \sum_{\mu(\neq\nu)} \left[\sum_{\varrho(\neq\nu)} \xi_{\mu\varrho} \gamma_\mu \gamma_\varrho - 2\, \xi_{\mu\nu} \gamma_\nu \gamma_\mu \right] \psi(x)$$

$$= (mc)^{-1} p_\nu \sum_\mu \sum_\varrho \xi_{\mu\varrho} \gamma_\mu \gamma_\varrho \psi(x) .$$

By this relation, the change in form of the wave function may be written

$$\delta\psi(x) = \frac{mc}{4p_\nu} \sum_{\mu\varrho\lambda} \varepsilon_{\nu\mu\varrho\lambda} \xi_{\mu\varrho} T_\lambda \psi(x)$$

(no sum convention for ν; p_ν is assumed to be different from zero). This shows that the components of the polarization four-vector are the *generators* of the *little group*.

2.3. *Space inversion*

The coefficients $a_{\mu\nu}$ in eqs. (7) and (9) are, in this case,

$$a_{\mu\nu} = 0 \quad \text{for} \quad \mu \neq \nu , \quad a_{11} = a_{22} = a_{33} = -1 , \quad a_{44} = 1 .$$

Eq. (9) therefore becomes

$$\gamma_i S = - S \gamma_i , \quad (i = 1, 2, 3) , \quad \gamma_4 S = S \gamma_4 .$$

Since S must anticommute with γ_i, $(i = 1, 2, 3)$, and commute with γ_4, it may be written as

$$S = \eta_{\mathrm{P}} \gamma_4 ,$$

where η_{P} is a constant.

On requiring that two space inversions should lead back to the original wave function up to a sign (since the wave functions are two-valued), we find

$$\eta_{\mathrm{P}}^2 = \pm 1 .$$

Thus four values are possible for the intrinsic parity of Dirac particles *,

---

* By operating with

$$\frac{1}{2} \left[1 - \frac{\mathrm{i}}{4!} \gamma_5 \varepsilon_{\mu\nu\varrho\sigma} \gamma_\mu \gamma_\nu \gamma_\varrho \gamma_\sigma \right]$$

on wave functions with intrinsic parity ± 1, one gets wave functions with intrinsic parity $\pm$ i.

as has already been found in the non-relativistic theory of particles of spin $\frac{1}{2}$.

2.4. Transformation properties of wave functions in Kramers' representation

In the Kramers representation the γ matrices are given by eqs. (10), and the Dirac equation has the form (11), p. 135. This last splits into the two equations

$$\boldsymbol{\sigma} \cdot \nabla_x \Psi_2 - \frac{1}{c}\frac{\partial}{\partial t} \Psi_1 = \mathrm{i}\frac{mc}{\hbar} \Psi_1 ,$$

$$\boldsymbol{\sigma} \cdot \nabla_x \Psi_1 + \frac{1}{c}\frac{\partial}{\partial t} \Psi_2 = -\mathrm{i}\frac{mc}{\hbar} \Psi_2 .$$

Since

$$\gamma' \gamma_4' = \mathrm{i}\, \rho_3 \times \boldsymbol{\sigma} = \begin{pmatrix} \mathrm{i}\,\boldsymbol{\sigma} & 0 \\ 0 & -\mathrm{i}\,\boldsymbol{\sigma} \end{pmatrix}, \quad \boldsymbol{s}' = \frac{\hbar}{2}\begin{pmatrix} \boldsymbol{\sigma} & 0 \\ 0 & \boldsymbol{\sigma} \end{pmatrix},$$

it is apparent that the Lorentz transformation matrices

$$\mathrm{e}^{\mathrm{i}\gamma_1'\gamma_4'\chi/2}, \quad \mathrm{e}^{\mathrm{i}\gamma_2'\gamma_4'\chi/2}, \quad \mathrm{e}^{\mathrm{i}\gamma_3'\gamma_4'\chi/2}$$

and the space rotation matrix $\mathrm{e}^{\mathrm{i}\boldsymbol{s}' \cdot \boldsymbol{n}\phi/\hbar}$, do not mix Ψ_1 and Ψ_2 with each other, but only the components of Ψ_1 among themselves, and those of Ψ_2 also among themselves.

On the other hand, for a space inversion

$$S\begin{pmatrix} \Psi_1 \\ \Psi_2 \end{pmatrix} = \eta_{\mathrm{P}}\, \gamma_4' \begin{pmatrix} \Psi_1 \\ \Psi_2 \end{pmatrix} = \eta_{\mathrm{P}} \begin{pmatrix} \Psi_2 \\ \Psi_1 \end{pmatrix},$$

showing that Ψ_1 and Ψ_2 are interchanged, namely the first component of the wave function is interchanged with the third and the second with the fourth. It is precisely by this interchange of Ψ_1 with Ψ_2, that the invariance of the Dirac equation under space inversion is made possible.

At this point it may be interesting to remark that, if ψ_{K}' is a solution of the Dirac equation for massless particles ($m = 0$) in the Kramers representation, so is also $\gamma_5' \, \psi_{\mathrm{K}}'$. The wave functions $(1 \pm \gamma_5')\, \psi_{\mathrm{K}}'$ are solutions of the Dirac equation for $m = 0$, characterized by the fact that their (first/last) two components are zero. The (last/first) two components obey the Weyl equation

$$\left(\boldsymbol{\sigma} \cdot \nabla_x \mp \frac{1}{c}\frac{\partial}{\partial t}\right) \Psi = 0,$$

respectively.

2.5. The Cini-Touschek transformation

In § 2.4. it has been noted that, if $\psi'_{\mathrm{K}}(x)$ is the wave function of a massless particle in the Kramers representation,

$$\chi'_{\pm} = \frac{1 \mp \gamma'_5}{2} \psi'_{\mathrm{K}}(x)$$

are two-component wave functions satisfying the Weyl equation. If $\psi'_{\mathrm{K}}(x)$ describes a particle of momentum $\boldsymbol{p}$, then

$$(\boldsymbol{\sigma} \cdot \boldsymbol{p})\, \chi'_{\pm} = \pm\, \varepsilon\, |\boldsymbol{p}|\, \chi'_{\pm}\,,$$

where ε is the sign of the energy, i.e. $\chi'_{\pm}$ for $\varepsilon = +1$ correspond to spin parallel and antiparallel to the momentum, respectively, and vice versa for $\varepsilon = -1$.

Cini and Touschek * have found a representation which is suitable for the treatment of the high-energy limit of the Dirac equation, in the same way as the F.-W. representation is for the non-relativistic limit.

The transformation leading from the conventional Dirac theory to the C.-T. representation is

$$U_{\mathrm{C.T.}} = \mathrm{e}^{\mathrm{i}S_{\mathrm{C.T.}}}$$

with **

$$S_{\mathrm{C.T.}} = \frac{\mathrm{i}}{2} \beta' \frac{\boldsymbol{\alpha}' \cdot \boldsymbol{p}}{|\boldsymbol{p}|} \,\mathrm{tg}^{-1} \left(\frac{mc}{|\boldsymbol{p}|} \right).$$

This can also be expressed in the form

$$\mathrm{e}^{\pm \mathrm{i}S_{\mathrm{C.T.}}} = \frac{|H| + c\,|\boldsymbol{p}| \mp \beta' \dfrac{\boldsymbol{\alpha}' \cdot \boldsymbol{p}}{|\boldsymbol{p}|} mc^2}{[2\,|H|\,(|H| + c\,|\boldsymbol{p}|)]^{\frac{1}{2}}}\,,$$

by which the transformed Hamiltonian is found to be

$$H_{\mathrm{C.T.}} = |H| \frac{\boldsymbol{\alpha}' \cdot \boldsymbol{p}}{|\boldsymbol{p}|}\,.$$

Thus the Dirac equation in the C.-T. representation is

$$(\boldsymbol{\alpha}' \cdot \boldsymbol{p})\, \psi'_{\mathrm{C.T.}} = \varepsilon\, |\boldsymbol{p}|\, \psi'_{\mathrm{C.T.}}\,, \tag{11}$$

* M. Cini and B. Touschek, Nuovo Cimento **7** (1958), 422. See also P. M. Mathews and A. Sankaranarayanan, Progr. of Theor. Phys. **26** (1961), 1 for the definition of the position operator in the C.-T. representation.

** Here it is convenient to assume that the Dirac theory is formulated in the Kramers representation for the matrices.

where ε is the sign of the energy and $\psi'_{\text{C.T.}} = U_{\text{C.T.}}\, \psi'_{\text{K}}$.

The wave functions

$$(\chi'_{\text{C.T.}})_\pm = \frac{1 \mp \gamma'_5}{2}\, \psi'_{\text{C.T.}}$$

are also solutions of eq. (11), which reduces to

$$(\boldsymbol{\sigma} \cdot \boldsymbol{p})(\chi'_{\text{C.T.}})_\pm = \pm\, \varepsilon\, |\boldsymbol{p}|\, (\chi'_{\text{C.T.}})_\pm\,,$$

identical in form with the equations for massless neutrinos.

3. Transformation properties of bilinear forms

By using eqs. (2), (5) and (9), one can study the behaviour, under linear coordinate transformations, of quantities of the type

$$\bar{\psi}\, A\, \psi\,,$$

where A is a 4×4 matrix.

Thus one finds that $\bar{\psi}\,\psi$ is transformed as a scalar. In fact

$$\bar{\psi}'\, \psi' = \bar{\psi}\, S^{-1}\, S\, \psi = \bar{\psi}\, \psi\,.$$

For the four components of the four-current

$$j_\mu = \mathrm{i}c\, \bar{\psi}\, \gamma_\mu\, \psi\,,$$

one has

$$j'_\mu = \mathrm{i}c\, \bar{\psi}'\, \gamma_\mu\, \psi' = \mathrm{i}c\, \bar{\psi}\, S^{-1}\, \gamma_\mu\, S\, \psi = a_{\mu\nu}\, j_\nu\,.$$

Therefore they behave like the components of a four-vector under Lorentz transformations. Under space inversion the first three components change sign (i.e. they form a polar three-vector), while the fourth component remains unchanged.

The quantities

$$\bar{\psi}\, \gamma_\mu\, \gamma_\nu\, \psi$$

are transformed like the components of a tensor of the second rank

$$\bar{\psi}'\, \gamma_\mu\, \gamma_\nu\, \psi' = \bar{\psi}\, S^{-1}\, \gamma_\mu\, S\, S^{-1}\, \gamma_\nu\, S\, \psi = a_{\mu\varrho}\, a_{\nu\sigma}\, \bar{\psi}\, \gamma_\varrho\, \gamma_\sigma\, \psi\,.$$

Instead of $\bar{\psi}\, \gamma_\mu\, \gamma_\nu\, \psi$ it is often convenient to introduce the skew-symmetric tensor of the second rank (hexavector)

$$m_{\mu\nu} = \bar{\psi}\, \sigma_{\mu\nu}\, \psi\,,$$

with

$$\sigma_{\mu\nu} = (\gamma_\mu\, \gamma_\nu - \gamma_\nu\, \gamma_\mu)/2\,\mathrm{i}\,.$$

Note that the quantities

$$\int m_{23}\, d^3x\,, \quad \int m_{31}\, d^3x\,, \quad \int m_{12}\, d^3x\,,$$

similar to, but not identical with the components of the spin, are not constants of the motion *.

Under space inversion the components m_{23}, m_{31} and m_{12} are transformed like those of an axial vector.

Finally, using the Dirac equation, one may show that the bilinear forms

$$n_{\mu\nu} = \bar{\psi}\, \gamma_\mu\, \gamma_5 \left[\mathrm{i}\, \gamma_\nu - (mc)^{-1}\, p_\nu\right] \psi$$

obey the continuity equations

$$\frac{\partial n_{\mu\nu}}{\partial x_\mu} = 0\,, \quad (\nu = 1, \dots 4)\,.$$

Therefore, the quantities

$$t_\nu = \int n_{4\nu}\, d^3x = \int \psi^\dagger\, \gamma_5 \left[\mathrm{i}\, \gamma_\nu - (mc)^{-1}\, p_\nu\right] \psi\, d^3x\,, \qquad (\nu = 1, \dots 4)\,,$$

are constants of motion. Under a Lorentz transformation they behave like the components of a four-vector, while, under a space inversion,

$$t_i' = t_i\,, \quad (i = 1, 2, 3)\,, \quad t_4' = -\,t_4\,,$$

and, under time reversal (see § 4),

$$t_i' = -\,t_i\,, \quad t_4' = t_4\,.$$

Such quantities may be regarded as the mean values of the components of the four-vector polarization operator **

$$T_\nu = \gamma_5 \left[\mathrm{i}\, \gamma_\nu - (mc)^{-1}\, p_\nu\right],$$

* They may be written in the form

$$(2/\hbar) \int \psi^\dagger \beta s_1 \psi\, d^3x\,, \qquad (2/\hbar) \int \psi^\dagger \beta s_2 \psi\, d^3x\,, \qquad (2/\hbar) \int \psi^\dagger \beta s_3 \psi\, d^3x\,,$$

where s_1, s_2, s_3 are the spin matrices.

Notice that

$$d(\beta s_1)/dt = \tfrac{1}{4}\, c\varepsilon_{1mn} p_k (\alpha_k \alpha_m \alpha_n + \alpha_m \alpha_n \alpha_k)\, \beta =$$

$$= -\tfrac{1}{2}\, c\, \varepsilon_{1mn} p_k \alpha_m \alpha_k \alpha_n \beta = c p_1 \alpha_1 \alpha_2 \alpha_3 \beta \neq 0\,.$$

Similarly

$$d(\beta s_2)/dt = c p_2 \alpha_2 \alpha_3 \alpha_1 \beta, \qquad d(\beta s_3)/dt = c p_3 \alpha_3 \alpha_1 \alpha_2 \beta\,.$$

** See the review: D. M. Fradkin and R. H. Good, Jr., Revs. Mod. Phys. **33** (1961), 343; see also J. Hilgevoord and S. A. Wouthuysen, Nucl. Phys. **40** (1963), 1.

which commute with the Hamiltonian, and are closely related with the mean-spin operator $\boldsymbol{s}_{\text{av}}$ discussed in Chap. V, § 6. One has

$$\frac{\hbar}{2}\boldsymbol{T} = \frac{H}{|H|}\left[\boldsymbol{s}_{\text{av.}} + \frac{(\boldsymbol{s}_{\text{av.}}\cdot\boldsymbol{p})\,\boldsymbol{p}}{m(|H| + mc^2)}\right],$$

$$\frac{\hbar}{2}T_4 = \frac{\mathrm{i}}{mc}\,(\boldsymbol{s}_{\text{av.}}\cdot\boldsymbol{p})\,.$$

The important tensor of the fourth rank

$$\frac{1}{4!}\sum_P (-1)^{\mathrm{P}}\, P\,\bar{\psi}\,\gamma_\mu\,\gamma_\nu\,\gamma_\rho\,\gamma_\sigma\,\psi$$

is skew-symmetric (P is a permutation of the indices μ, ν, ρ, σ; the exponent P is zero or one according as the permutation is even or odd). The components having any two indices equal vanish, while those for which μ, ν, ρ, σ are a permutation of 1, 2, 3, 4 equal $\bar{\psi}\,\gamma_5\,\psi$ or $-\,\bar{\psi}\,\gamma_5\,\psi$, according as such a permutation is even or odd.

Performing a linear orthogonal transformation (7), (8), (9), one has, for the non-vanishing components

$$\frac{1}{4!}\sum_P (-1)^{\mathrm{P}}\, P\,\bar{\psi}'\,\gamma_\mu\,\gamma_\nu\,\gamma_\rho\,\gamma_\sigma\,\psi' =$$

$$= \frac{1}{4!}\sum_P (-1)^{\mathrm{P}}\, P\,\bar{\psi}\,S^{-1}\,\gamma_\mu\,SS^{-1}\,\gamma_\nu\,SS^{-1}\,\gamma_\rho\,SS^{-1}\,\gamma_\sigma\,S\,\psi$$

$$= \frac{1}{4!}\sum_P (-1)^{\mathrm{P}}\, P\,a_{\mu\mu'}\,a_{\nu\nu'}\,a_{\rho\rho'}\,a_{\sigma\sigma'}\,\bar{\psi}\,\gamma_{\mu'}\,\gamma_{\nu'}\,\gamma_{\rho'}\,\gamma_{\sigma'}\,\psi.$$

It is easily seen that only terms for which μ', ν', ρ', σ' are permutations of 1, 2, 3, 4, and, therefore, also of μ, ν, ρ, σ, contribute to the last expression. Thus $\bar{\psi}\,\gamma_{\mu'}\,\gamma_{\nu'}\,\gamma_{\rho'}\,\gamma_{\sigma'}\,\psi$ may be replaced by $(-1)^{\mathrm{P}'}\,\bar{\psi}\,\gamma_\mu\,\gamma_\nu\,\gamma_\rho\,\gamma_\sigma\,\psi$, where P′ is zero or one, according as μ', ν', ρ', σ' is an even or an odd permutation of μ, ν, ρ, σ. Hence we have

$$\frac{1}{4!}\sum_P (-1)^{\mathrm{P}}\, P\,\bar{\psi}'\,\gamma_\mu\,\gamma_\nu\,\gamma_\rho\,\gamma_\sigma\,\psi' =$$

$$= \bar{\psi}\, \gamma_\mu\, \gamma_\nu\, \gamma_\rho\, \gamma_\sigma\, \psi\, \frac{1}{4!} \sum_P (-1)^P P \sum_{\mu', \nu', \rho', \sigma'} (-1)^{P'} a_{\mu\mu'}\, a_{\nu\nu'}\, a_{\rho\rho'}\, a_{\sigma\sigma'}$$

$$= (\operatorname{Det} a)\, \bar{\psi}\, \gamma_\mu\, \gamma_\nu\, \gamma_\rho\, \gamma_\sigma\, \psi,$$

where Det a is the determinant of the 4×4 matrix $\|a_{\mu\nu}\|$.

For a proper Lorentz transformation, Det $a = 1$. For a space inversion, Det $a = -1$. Therefore, under this latter transformation,

$$\bar{\psi}'\, \gamma_5\, \psi' = -\, \bar{\psi}\, \gamma_5\, \psi\,,$$

which follows also, directly, from the fact that the transformation matrix $S = \eta_P\, \gamma_4$ for space inversions anticommutes with γ_5.

4. Time reversal

Since a time inversion is a linear transformation of the type (7), following the procedure outlined in § 2 one might try to determine a matrix S such that

$$S^{-1}\, \gamma_i\, S = \gamma_i\,, \quad (i = 1, 2, 3)\,, \quad S^{-1}\, \gamma_4\, S = -\, \gamma_4\,, \tag{12}$$

which would transform the wave functions according to (2). The invariance of the Dirac equation under time reversal would thus have been proved *.

Nevertheless, on transforming the components of the four-momentum, one would find

$$\boldsymbol{P}' = \boldsymbol{P}\,, \quad E' = -\, E\,,$$

which is unacceptable on physical grounds.

A transformation law of a type different from (2) must therefore be assumed for the wave functions. In looking for it, we are helped by the remark that time reversal is equivalent to the complex conjugation of x_μ. This points to the possibility that the transformed wave function may have to be linear in the complex conjugate of the original wave function.

In fact, the Dirac equation is invariant under time reversal if, in the Pauli representation, the transformed wave function is assumed to be

$$\psi' = (\eta_T\, \gamma_1\, \gamma_3\, \psi)^*\,,$$

* The matrix which satisfies eqs. (12) is $S = \gamma_1\gamma_2\gamma_3$, $(S^\dagger = S^{-1} = -\, S)$, which, however does not satisfy eq. (6), and so $\bar{\psi}' = \psi'^\dagger\, \gamma_4 = \psi^\dagger\; \gamma_3\, \gamma_2\, \gamma_1\, \gamma_4 = -\, \bar{\psi}\, S^{-1} = \bar{\psi} S$, $\bar{\psi}' = \psi^\dagger\, \gamma_4\, S^{-1} = \bar{\psi} S^{-1}$.

Note that, since there is no 2×2 matrix which anticommutes with σ_1, σ_2 and σ_3, the analogue of (12) for the Weyl equation does not exist.

or

$$\psi'(\boldsymbol{x}, t') = \eta_T^* \gamma_1^T \gamma_3^T \psi^*(\boldsymbol{x}, t), \tag{13}$$

with $|\eta_T| = 1$.

Since, in the Pauli representation,

$$\gamma_1^* = -\gamma_1 = \gamma_1^T, \quad \gamma_3^* = -\gamma_3 = \gamma_3^T, \quad \gamma_2^* = \gamma_2 = \gamma_2^T, \quad \gamma_4^* = \gamma_4 = \gamma_4^T,$$

eq. (13) may be written as

$$\psi' = \eta_T^* \gamma_1 \gamma_3 \psi^* .$$

Now, in the Pauli representation,

$$(\gamma_\mu \gamma_1 \gamma_3)^* = \gamma_\mu^* \gamma_1 \gamma_3 = \gamma_1 \gamma_3 \gamma_\mu .$$

Thus the equation

$$\gamma_\mu \frac{\partial \psi'}{\partial x'_\mu} + \kappa \psi' = 0$$

in the "time reversed system" leads to

$$\eta_T^* \left(\gamma_\mu \gamma_1 \gamma_3 \frac{\partial \psi^*}{\partial x^*_\mu} + \kappa \gamma_1 \gamma_3 \psi^* \right) = \left[\eta_T \gamma_1 \gamma_3 \left(\gamma_\mu \frac{\partial}{\partial x_\mu} + \kappa \right) \psi \right]^* = 0,$$

i.e. to the Dirac equation in the original system.

Moreover,

$$\boldsymbol{j}' = (\boldsymbol{j}')^* = -\mathrm{i}c\, \bar{\psi} \gamma_3 \gamma_1 (\boldsymbol{\gamma} \gamma_1 \gamma_3)^* \psi = -\mathrm{i}c\, \bar{\psi} \boldsymbol{\gamma} \psi = -\boldsymbol{j}$$

and

$$\rho' = \psi'^\dagger \psi' = \psi^\dagger \psi = \rho .$$

The invariance of the Lagrangian is easily proved. In fact, it is easy to see that $\mathfrak{L} = \mathfrak{L}^*$. Now

$$\gamma_\mu \frac{\partial \psi'}{\partial x'_\mu} + \kappa \psi' = \eta_T^* \gamma_1 \gamma_3 \left(\gamma_\mu \frac{\partial \psi}{\partial x_\mu} + \kappa \psi \right)^*$$

and

$$\bar{\psi}' \left(\gamma_\mu \frac{\partial \psi'}{\partial x'_\mu} + \kappa \psi' \right) = \left[\bar{\psi} \left(\gamma_\mu \frac{\partial \psi}{\partial x_\mu} + \kappa \psi \right) \right]^* .$$

Similarly

$$\left(\frac{\partial \bar{\psi}'}{\partial x'_\mu} \gamma_\mu - \kappa \bar{\psi}' \right) \psi' = \left[\left(\frac{\partial \bar{\psi}}{\partial x_\mu} \gamma_\mu - \kappa \bar{\psi} \right) \psi \right]^* .$$

Thus $\mathfrak{L}'(x') = (\mathfrak{L}(x))^* = \mathfrak{L}(x)$. Likewise, for the components of the energy-momentum tensor (eq. (3), Chap. IV, § 1) one has $T'_{ik} = T_{ik}, T'_{44} = T_{44}$,

$T'_{4i} = -T_{4i}$ under time reversal, and so the energy and the momentum have the correct transformation properties, $E' = E$, $\boldsymbol{P}' = -\boldsymbol{P}$.

As an exercise, the reader may verify that the following expressions are invariant under space inversion and time reversal:

$$\bar{\psi}\left(\frac{1}{4!}\gamma_\mu\,\gamma_\nu\,\gamma_\rho\,\gamma_\sigma\right)\psi\cdot\phi_{\mu\nu\rho\sigma} = \varepsilon_{1234}\,\bar{\psi}\,\gamma_5\,\psi\cdot\phi$$

and

$$\bar{\psi}\left(\frac{1}{4!}\gamma_\mu\,\gamma_\nu\,\gamma_\rho\,\gamma_\sigma\right)\gamma_\lambda\,\psi\cdot\frac{\partial\phi_{\mu\nu\rho\sigma}}{\partial x_\lambda} = \varepsilon_{1234}\,\bar{\psi}\,\gamma_5\,\gamma_\lambda\,\psi\cdot\frac{\partial\phi}{\partial x_\lambda}\,.$$

Here ψ is a Dirac wave function and $\phi_{\mu\nu\rho\sigma} = \varepsilon_{\mu\nu\rho\sigma}\,\phi$ is the pseudoscalar wave function of a spin zero particle of intrinsic parity -1, ϕ being a real scalar. Note that $\phi_{\mu\nu\rho\sigma}$ changes sign under time reversal. The above expressions occur in the theory of pion-nucleon interactions.

We suggest also the following exercise. Show that the β-decay interaction

$$\sum_i C_i(\bar{\psi}_{\mathrm{N}}\,O_i\,\psi_{\mathrm{P}})\left(\bar{\psi}_\nu\,O_i\,\frac{1+\gamma_5}{2}\,\psi_{\mathrm{e}}\right) + \text{complex conjugate}\,,$$

where

$$O_1 = I\,,\quad O_2 = \gamma_\mu\,,\quad O_3 = \sigma_{\mu\nu}\,,\quad O_4 = \mathrm{i}\,\gamma_\mu\,\gamma_5\,,\quad O_5 = \gamma_5\,,$$

ψ_{N}, ψ_{P}, ψ_{e}, ψ_ν are the neutron, proton, electron and neutrino wave functions and the C_i's are coupling constants, is invariant under Lorentz transformations, but not under space inversions.

The probability for the reaction $\mathrm{N} \to \mathrm{P}$ + electron + antineutrino is a bilinear form of the matrix elements

$$u^\dagger(\boldsymbol{k}_{\mathrm{e}}, \lambda_{\mathrm{e}})\,\frac{1+\gamma_5}{2}\,O_i\,\gamma_4\,v(\boldsymbol{k}_\nu, \lambda_\nu)\,,$$

where $\boldsymbol{k}_{\mathrm{e}}$, $\boldsymbol{k}_\nu$ denote the momenta of the electron and of the antineutrino, and λ_{e}, λ_ν their spin orientations. Show that, if the electron mass is neglected, such matrix elements vanish when the electron spin is parallel to $\boldsymbol{k}_{\mathrm{e}}$ (i.e. if $\lambda_{\mathrm{e}} = \uparrow$).

Chapter VII

THE DIRAC EQUATION FOR A CHARGED PARTICLE IN AN ELECTROMAGNETIC FIELD

1. Gauge invariance

The equation for the wave function of a particle of spin $\frac{1}{2}$ and charge $-e(e > 0)$, interacting with an electromagnetic field of four-potential A_μ, is obtained from the free particle Dirac equation by making the replacement

$$-i\hbar\, \partial/\partial x_\mu \to -i\hbar\, \partial/\partial x_\mu + (e/c)\, A_\mu\,, \tag{1}$$

as has been done for particles of spin zero.

This yields the equation

$$\gamma_\mu \left(\frac{\partial}{\partial x_\mu} + \frac{ie}{\hbar c} A_\mu\right) \psi + \kappa\, \psi = 0\,, \tag{2}$$

and the adjoint equation

$$\left(\frac{\partial}{\partial x_\mu} - \frac{ie}{\hbar c} A_\mu\right) \bar{\psi}\, \gamma_\mu - \kappa\, \bar{\psi} = 0\,. \tag{3}$$

All operations used in the previous chapter to prove the invariance of the Dirac equation under space rotations, Lorentz transformations etc. are also applicable to eqs. (2) and (3). In fact the transformation of the four-potential under a linear coordinate transformation is

$$A'_\mu = a_{\mu\nu}\, A_\nu\,,$$

the coefficients $a_{\mu\nu}$ being the same as those for the transformation

$$\partial/\partial x'_\mu = a_{\mu\nu}\, \partial/\partial x_\nu$$

of the derivatives.

Moreover we recall that the substitution (1) automatically leads to gauge-invariant equations. In fact eqs. (2) and (3) are invariant under the combined gauge transformations of the first and second kind

$$\psi \to e^{-ie\Lambda/\hbar c}\, \psi\,, \quad \bar{\psi} \to e^{ie\Lambda/\hbar c}\, \bar{\psi}\,, \quad A_\mu \to A_\mu + \partial\Lambda/\partial x_\mu$$

($\Lambda = \Lambda(x)$ being an arbitrary function of space-time).

Writing eq. (2) for ψ_2 and multiplying from the left by $\bar{\psi}_1$, and writing (3) for $\bar{\psi}_1$ and multiplying from the right by ψ_2, we obtain on summing

$$\frac{\partial}{\partial x_\mu}(\bar{\psi}_1 \gamma_\mu \psi_2) = 0\,. \tag{4}$$

Thus the quantity

$$\int \bar{\psi}_1 \gamma_4 \psi_2 \, \mathrm{d}^3\mathrm{x} = \int \psi_1^\dagger \psi_2 \, \mathrm{d}^3\mathrm{x}$$

is constant in time and may be defined as the product of the wave functions ψ_1 and ψ_2. In contrast to the case of spin zero, the product of two wave functions is given by the same formula both for free particles and for particles in an electromagnetic field. In particular, for $\psi_1 = \psi_2$ one finds the expression for the four-current

$$j_\mu = \mathrm{i}c\, \bar{\psi} \gamma_\mu \psi\,. \tag{5}$$

Therefore, the charge density is

$$\rho = \psi^\dagger \psi$$

as for free particles.

The four-current (5) is obviously invariant under gauge transformations of the first kind $\psi \to \mathrm{e}^{-\mathrm{i}e\Lambda/\hbar c}\, \psi$ and $\bar{\psi} \to \mathrm{e}^{\mathrm{i}e\Lambda/\hbar c}\, \bar{\psi}$. (Remember that in the case of spinless particles, the expression for the four-current of free particles, $j_\mu = (\hbar/2m\mathrm{i})(\psi^* \, \partial\psi/\partial x_\mu - \partial\psi^*/\partial x_\mu \, \psi)$, was not invariant under gauge transformations of the first kind, owing to the derivatives. On replacing $\partial/\partial x_\mu$ by $\partial/\partial x_\mu + \mathrm{i}eA_\mu/\hbar c$ the four-current in the presence of an electromagnetic field, $j_\mu = (\hbar/2m\mathrm{i})(\psi^* \, \partial\psi/\partial x_\mu - \partial\psi^*/\partial x_\mu \, \psi) + (e/mc) A_\mu \psi^* \psi$, was obtained.)

Eqs. (2) and (3) may be derived from the Lagrangian density

$$\begin{aligned}\mathfrak{L} = -\frac{\hbar c}{2}\Bigg[\bar{\psi}\left(\gamma_\mu\left(\frac{\partial}{\partial x_\mu} + \frac{\mathrm{i}e}{\hbar c}A_\mu\right) + \kappa\right)\psi - \\ -\left(\left(\frac{\partial}{\partial x_\mu} - \frac{\mathrm{i}e}{\hbar c}A_\mu\right)\bar{\psi}\gamma_\mu - \kappa\,\bar{\psi}\right)\psi\Bigg]\end{aligned} \tag{6}$$

by stipulating the invariance of the action under independent variations of $\bar{\psi}$ and ψ.

The Lagrangian density (6) differs from that for free particles by the interaction Lagrangian density

$$\mathfrak{L}_{\mathrm{int}} = -\,\mathrm{i}e\, A_\mu \bar{\psi} \gamma_\mu \psi = -\,(e/c)\, A_\mu j_\mu\,.$$

Note that

$$\frac{\partial \mathfrak{L}}{\partial A_\mu} = -\frac{e}{c} j_\mu ,$$

as for particles of spin zero (Part I, Chap. V, § 4).

The Lagrangian density (6) is invariant under combined gauge transformations. The continuity equation for the four-current can be derived by requiring that the action is invariant either under gauge transformations of the first kind or under those of the second kind (with $\Lambda(x) = 0$ on the boundary of the domain of integration, in both cases).

2. Charge conjugation

The Dirac equations (2) and (3) are invariant under the transformation

$$\psi' = \eta_C^* S_C^{-1} \bar{\psi}^T , \quad \bar{\psi}' = -\eta_C \psi^T S_C \tag{7}$$

accompanied by either

$$A_\mu \to -A_\mu , \tag{8}$$

or

$$e \to -e . \tag{8'}$$

In eqs. (7), η_C is a constant of unit modulus, $|\eta_C| = 1$, and S_C is a unitary matrix satisfying

$$\gamma_\mu^T = -S_C \gamma_\mu S_C^{-1} \tag{9}$$

(in other words, S_C induces a transformation which transposes and changes the sign of all the γ's).

In fact, the equation

$$\left(\frac{\partial}{\partial x_\mu} + \frac{ie}{\hbar c} A_\mu\right) \gamma_\mu \psi' + \kappa \psi' = 0$$

gives

$$\left(\frac{\partial}{\partial x_\mu} + \frac{ie}{\hbar c} A_\mu\right) S_C \gamma_\mu S_C^{-1} S_C \psi' + \kappa S_C \psi' = 0 .$$

Using the first of eqs. (7), one has

$$-\left(\frac{\partial}{\partial x_\mu} + \frac{ie}{\hbar c} A_\mu\right) \gamma_\mu^T \bar{\psi}^T + \kappa \bar{\psi}^T = 0 .$$

Transposing this equation, and effecting one of the replacements (8), (8′) one obtains

$$\left(\frac{\partial}{\partial x_\mu} - \frac{\mathrm{i}e}{\hbar c} A_\mu\right) \bar{\psi}\, \gamma_\mu - \kappa\, \bar{\psi} = 0\,,$$

i.e. eq. (3).

The group of transformations (7) and (8) or (8′), is called charge conjugation. Its analogue for spin zero particles has been discussed in Part I, Chap. VII, § 4.

The matrices S_C satisfying (9) are, in the different representations,

$$\text{(Pauli)}\quad S_C = \gamma_2\,\gamma_4\,, \qquad \text{(Kramers)}\quad S_C = \gamma_2'\,\gamma_4'\,, \qquad \text{(Majorana)}\quad S_C = \gamma_4''\,.$$

Particularly interesting is the case of the Majorana representation, in which the transformation (7) amounts to the replacement of the wave function by its *complex conjugate*. In this representation a steady wave function for a particle of charge e and energy E obeys the same equation as the wave function for a particle of charge $-\,e$ and energy $-\,E$. This is an extension of the property $u'' = v''^*$ for the free particle wave functions (Chap. V, § 4).

It is important to stress that, under the transformations (7) and (8) or (8′), the Lagrangian density (6) changes sign, whereas the four-current (5) remains unchanged. On the other hand $\mathfrak{L}' = \mathfrak{L}$ and $j_\mu = -\,j_\mu$ in quantum field theory, where ψ and $\psi^\dagger$ are anticommuting operators, as is briefly explained at the end of this book. The extra minus sign results from bringing $\psi^\dagger$ to the left of ψ. For instance,

$$j_4' = \mathrm{i}c\,\psi'^\dagger\,\psi' = \mathrm{i}c\,\psi_\alpha\,\psi_\alpha^* = -\,\mathrm{i}c\,\psi_\alpha^*\,\psi_\alpha = -\,j_4\,.$$

3. Spin magnetic moment

In order to exhibit the difference between the electromagnetic properties of particles of spin $\frac{1}{2}$ and those of spin zero, it is convenient to establish a wave equation of the second order for the former, and to compare it with the Klein-Gordon equation.

We introduce the operators

$$\pi_\mu = -\,\mathrm{i}\,\hbar\,\partial/\partial x_\mu + (e/c)\,A_\mu\,,$$

which satisy the relations *

$$[\pi_\mu, \pi_\nu] = (\hbar e/\mathrm{i}c)\,F_{\mu\nu}\,, \quad (\gamma_\mu \pi_\mu)^2 = \pi_\mu \pi_\mu + (\hbar e/2c)\,\sigma_{\mu\nu}\,F_{\mu\nu}$$

with

$$\sigma_{\mu\nu} = (\gamma_\mu\,\gamma_\nu - \gamma_\nu\,\gamma_\mu)/2\mathrm{i}\,.$$

---

* In fact, $(\gamma_\mu\pi_\mu)^2 = \gamma_\mu\gamma_\nu\pi_\mu\pi_\nu = (\delta_{\mu\nu} + \mathrm{i}\sigma_{\mu\nu})\,\pi_\mu\pi_\nu = \pi_\mu\pi_\mu + \frac{1}{2}\,\mathrm{i}\sigma_{\mu\nu}\,[\pi_\mu, \pi_\nu]$.

But for a factor $-$ ic, the fourth component is the operator "kinetic energy + rest energy"

$$-\mathrm{i}c\,\pi_4 = \hbar \mathrm{i}\,\partial/\partial t + e\,A_0\,.$$

The Dirac equation, in the presence of an electromagnetic field, now reads

$$(\gamma_\mu\,\pi_\mu - \mathrm{i}\,mc)\,\psi = 0\,.$$

Multiplication from the left by $(\gamma_\mu\,\pi_\mu + \mathrm{i}\,mc)$ gives the equation

$$(\pi_\mu\,\pi_\mu + m^2c^2)\,\psi + (\hbar e/2c)\,\sigma_{\mu\nu}\,F_{\mu\nu}\,\psi = 0\,, \tag{10}$$

which differs from the corresponding Klein-Gordon equation for particles of spin zero,

$$(\pi_\mu\,\pi_\mu + m^2c^2)\,\psi = 0\,,$$

by the term

$$\frac{\hbar e}{2c}\,\sigma_{\mu\nu}\,F_{\mu\nu} = \frac{\hbar e}{2\mathrm{i}c}\,\gamma_\mu\,\gamma_\nu\,F_{\mu\nu} = (\hbar e/c)(\boldsymbol{\sigma}\cdot\boldsymbol{H} - \mathrm{i}\,\boldsymbol{\alpha}\cdot\boldsymbol{E})\,.$$

Eq. (10) may also be put into the form

$$\left[-\frac{\hbar^2}{2m}\,\Delta - \frac{\mathrm{i}e\hbar}{mc}\,\boldsymbol{A}\cdot\nabla_x + \frac{e^2}{2mc^2}\,A^2 + \frac{\hbar e}{2mc}\,(\boldsymbol{\sigma}\cdot\boldsymbol{H} - \mathrm{i}\,\boldsymbol{\alpha}\cdot\boldsymbol{E})\right]\psi =$$

$$= \frac{1}{2mc^2}\left[\left(\hbar\mathrm{i}\,\frac{\partial}{\partial t} + eA_0\right)^2 - (mc^2)^2\right]\psi\,,$$

which will be useful in treating the non-relativistic limit. The spin is therefore responsible for an interaction with the magnetic field, which can be interpreted as that of a magnetic dipole of moment

$$\boldsymbol{\mu} = -\frac{e\hbar}{2mc}\,\boldsymbol{\sigma}$$

(*spin magnetic moment*).

For the electron, $\boldsymbol{\mu}$ is antiparallel to the spin and has the value,

$$\frac{e\hbar}{2m_e c} = 0.9273 \times 10^{-20}\ \text{erg gauss}^{-1}$$

(*Bohr magneton*).

Besides the interaction with the magnetic field, there is an electric dipole interaction with the electric field. The electric dipole moment

$$\boldsymbol{m} = \frac{\mathrm{i}e\hbar}{2mc}\,\boldsymbol{\alpha}$$

appears together with the magnetic dipole moment in accordance with the requirements of relativity. However, since it is of the order of v/c, it is negligible in the non-relativistic approximation. In order to show this, eq. (2) may be written in the Hamiltonian form

$$\hbar \mathrm{i} \frac{\partial}{\partial t} \psi = H \psi$$

with

$$H = c\, \boldsymbol{\alpha} \cdot \left(\frac{\hbar}{\mathrm{i}} \nabla_x + \frac{e}{c} \boldsymbol{A} \right) + \beta\, mc^2 - e\, A_0 \,.$$

Thus, also in the presence of an electromagnetic field *,

$$\dot{\boldsymbol{x}} = (\mathrm{i}/\hbar)\,[H, \boldsymbol{x}] = c\, \boldsymbol{\alpha} \,.$$

The operator $\boldsymbol{m}$ may therefore be written as

$$\boldsymbol{m} = \frac{\mathrm{i}e\hbar}{2mc^2}\, \dot{\boldsymbol{x}} \,,$$

from which its order of magnitude is apparent.

The four-current can be split into a "Klein-Gordon part" and a spin contribution. Using the Dirac equation one has in fact **

$$j_\mu = \frac{\hbar}{2m\mathrm{i}} \left(\bar{\psi} \frac{\partial \psi}{\partial x_\mu} - \frac{\partial \bar{\psi}}{\partial x_\mu} \psi \right) + \frac{e}{mc} A_\mu \bar{\psi} \psi + \frac{\hbar}{2m\mathrm{i}} \sum_{\nu (\neq \mu)} \frac{\partial}{\partial x_\nu} (\bar{\psi}\, \gamma_\mu\, \gamma_\nu\, \psi) \,. \quad (11)$$

The spin contribution is therefore

$$j'_\mu = \frac{\hbar}{2m\mathrm{i}} \sum_{\nu (\neq \mu)} \frac{\partial}{\partial x_\nu} (\bar{\psi}\, \gamma_\mu\, \gamma_\nu\, \psi) = \frac{\hbar}{2m} \frac{\partial}{\partial x_\nu} (\bar{\psi}\, \sigma_{\mu\nu}\, \psi) \,.$$

---

* It is also easy to show that

$$\dot{p}_i = (\mathrm{i}/\hbar)\,[H, p_i] = e\, \partial A_0/\partial x_i - e \alpha_k\, \partial A_k/\partial x_i.$$

This can be written as $\dot{\boldsymbol{p}} + (e/c)\, \dot{\boldsymbol{A}} = -e\boldsymbol{E} - (e/c)\, \dot{\boldsymbol{x}} \times \boldsymbol{H}$, just as in the Introduction, § 3.

** Using the Dirac equation, one replaces alternatively ψ with $-(1/\kappa)\, \gamma_\mu [\partial/\partial x_\mu + (\mathrm{i}e/\hbar c)\, A_\mu]\, \psi$ or $\bar{\psi}$ with $(1/\kappa) [\partial/\partial x_\mu - (\mathrm{i}e/\hbar c) A_\mu]\, \bar{\psi} \gamma_\mu$ in the expression $j_\mu = \mathrm{i}c \bar{\psi} \gamma_\mu \psi$. Summing the two expressions obtained in this way, and dividing by two, one has

$$j_\mu = \frac{\hbar}{2m\mathrm{i}} \left[\bar{\psi} \gamma_\mu \gamma_\nu \left(\frac{\partial}{\partial x_\nu} + \frac{\mathrm{i}e}{\hbar c} A_\nu \right) \psi - \left(\left(\frac{\partial}{\partial x_\nu} - \frac{\mathrm{i}e}{\hbar c} A_\nu \right) \bar{\psi} \right) \gamma_\nu \gamma_\mu \psi \right] = \frac{\hbar}{2m\mathrm{i}} \left(\bar{\psi} \frac{\partial \psi}{\partial x_\mu} - \frac{\partial \bar{\psi}}{\partial x_\mu} \psi \right) +$$

$$+ \frac{e}{2mc} A_\nu \bar{\psi} (\gamma_\mu \gamma_\nu + \gamma_\nu \gamma_\mu)\, \psi + \frac{\hbar}{2m\mathrm{i}} \sum_{\nu (\neq \mu)} \left(\bar{\psi} \gamma_\mu \gamma_\nu \frac{\partial \psi}{\partial x_\nu} - \frac{\partial \bar{\psi}}{\partial x_\nu} \gamma_\nu \gamma_\mu \psi \right),$$

i.e. the expression (11).

Remembering that $\sigma_{\mu\nu}$ is skew-symmetric while $\partial^2/\partial x_\mu \, \partial x_\nu$ is symmetric, it is readily shown that j'_μ itself obeys the continuity equation.

The tensor

$$m_{\mu\nu} = \bar{\psi} \, \sigma_{\mu\nu} \, \psi$$

is called the polarization and magnetization tensor *.

The spatial components of j'_μ are

$$\boldsymbol{j}' = \frac{\hbar}{2m} \nabla_x \times (\bar{\psi} \, \boldsymbol{\sigma} \, \psi) - \frac{i\hbar}{2mc} \frac{\partial}{\partial t} (\bar{\psi} \, \boldsymbol{\alpha} \, \psi) =$$

$$= -\frac{1}{e} \frac{\partial}{\partial t} (\bar{\psi} \, \boldsymbol{m} \, \psi) - \frac{c}{e} \nabla_x \times (\bar{\psi} \, \boldsymbol{\mu} \, \psi) \, .$$

In the extreme non-relativistic approximation $v/c \to 0$, these reduce to

$$\boldsymbol{j}' = (\hbar/2m) \, \nabla_x \times (\Psi_1^\dagger \, \boldsymbol{\sigma} \, \Psi_1) \, , \tag{12}$$

(Ψ_1 is the "large" component of the wave function in the Pauli representation).

The time component of j'_μ is

$$j'_4 = (ic/e) \, \nabla_x (\bar{\psi} \, \boldsymbol{m} \, \psi) \, .$$

It vanishes as $v/c \to 0$. Notice that its integral over all space is zero.

The polarization of a particle in an electromagnetic field $F_{\mu\nu} \equiv (\boldsymbol{E}, \boldsymbol{H})$ may be described by the four-vector polarization operator

$$T_\mu = \gamma_5 \left[\mathrm{i} \, \gamma_\mu - (mc)^{-1} \, \pi_\mu \right] .$$

The components of this are not constants of motion, but obey the equations **

---

* The reader may find it interesting to compare the angular momentum tensor M_{ik} (see Chap. IV, § 2) with the "moment" of the current defined as

$$\mathfrak{M}_{ik} = (-\,e/2c) \int (x_i j_k - x_k j_i) \, \mathrm{d}^3 x \, .$$

In particular, it may be shown that the spin magnetic moment tensor

$$(-e/mc) \, S_{ik} = (-e\hbar/2mc) \int \bar{\psi} \, \gamma_4 \sigma_{ik} \psi \, \mathrm{d}^3 x$$

and the spin part of $\mathfrak{M}_{ik}$

$$\mathfrak{S}_{ik} = (-\,e/2c) \int (x_i j_k' - x_k j_i') \, \mathrm{d}^3 x$$

coincide but for terms of the order v^2/c^2.

** Fradkin and Good, *loc. cit.*

$$\frac{\mathrm{d}\boldsymbol{T}}{\mathrm{d}t} = -\frac{e}{mc}(\boldsymbol{\sigma} \times \boldsymbol{H} - \gamma_5 \boldsymbol{E}),$$

$$\frac{\mathrm{d}T_4}{\mathrm{d}t} = -\frac{\mathrm{i}e}{mc}(\boldsymbol{\sigma} \cdot \boldsymbol{E}),$$

or, in covariant form,

$$\frac{\mathrm{d}T_\mu}{\mathrm{d}t} = -\mathrm{i}\frac{e}{mc}\gamma_4 \gamma_5 \gamma_\nu F_{\mu\nu}.$$

Using these equations, one may calculate the rate at which an initially polarized particle is depolarized by the interaction with the electromagnetic field.

NOTE 1. Feynman and Gell-Mann (Phys. Rev. **109** (1958), 193) have cast the theory of spin $\frac{1}{2}$ particles in an interesting form, based on eq. (10),

$$(\pi_\mu \pi_\mu + m^2c^2)\psi + (\hbar e/c)(\boldsymbol{\sigma} \cdot \boldsymbol{H} - \mathrm{i}\boldsymbol{\alpha} \cdot \boldsymbol{E})\psi = 0.$$

Remembering that γ_5 commutes with $\boldsymbol{\sigma}$ and $\boldsymbol{\alpha}$, it is clear that

$$\chi_+ = \tfrac{1}{2}(1 + \gamma_5)\psi$$

obeys the equation

$$(\pi_\mu \pi_\mu + m^2c^2)\chi_+ + (\hbar e/c)(\boldsymbol{\sigma} \cdot \boldsymbol{H} - \mathrm{i}\boldsymbol{\alpha} \cdot \boldsymbol{E})\chi_+ = 0.$$

If

$$\psi = \begin{pmatrix} \Psi_1 \\ \Psi_2 \end{pmatrix},$$

then (Pauli's representation for the γ's!)

$$\chi_+ = \tfrac{1}{2}\begin{pmatrix} \Psi_1 - \Psi_2 \\ \Psi_2 - \Psi_1 \end{pmatrix}.$$

Thus χ_+ is of the form

$$\chi_+ = \begin{pmatrix} \Psi \\ -\Psi \end{pmatrix}.$$

Inserting this in the equation for χ_+, we have

$$(\pi_\mu \pi_\mu + m^2c^2)\Psi + (\hbar e/c)\boldsymbol{\sigma} \cdot (\boldsymbol{H} + \mathrm{i}\boldsymbol{E})\Psi = 0. \qquad (\alpha)$$

Gell-Mann and Feynman have proposed that this second order equation for the two-component wave function Ψ should be regarded as the basic equation of the theory, and have given an interesting application of this idea to β–decay. Within the framework of the Dirac theory, eq. (α) is entirely equivalent to the Dirac equation. In fact, retracing the derivation of (α), one sees that, for any stationary solution Ψ of (α), a solution of the Dirac equation can be constructed,

$$\psi = \frac{1}{E + eA_0 + mc^2 - c\boldsymbol{\sigma} \cdot \boldsymbol{\pi}} \begin{pmatrix} (E + eA_0 + mc^2)\,\Psi \\ c\,\boldsymbol{\sigma} \cdot \boldsymbol{\pi}\,\Psi \end{pmatrix}.$$

Furthermore, a solution of eq. (α) is determined by the initial values of the two components of Ψ and of those of $\partial\Psi/\partial t$, namely by the initial values of four functions, as is the case for the Dirac equation.

A variant of the above formulation employs $\chi_- = \frac{1}{2}(1 - \gamma_5)\,\psi$ etc.

4. Anomalous magnetic moments

According to Dirac's theory, one would expect that the proton, a particle of spin $\frac{1}{2}$, charge e and mass $m_p = 1836\, m_e$, should be described by a wave function satisfying the equation

$$\gamma_\mu \left(\frac{\partial}{\partial x_\mu} - \frac{ie}{\hbar c} A_\mu \right) \psi + \frac{m_p c}{\hbar} \psi = 0\,. \tag{13}$$

The proton would thus possess a magnetic moment

$$\boldsymbol{\mu} = \frac{e\hbar}{2m_p c} \boldsymbol{\sigma}\,,$$

parallel to the spin and of magnitude equal to a nuclear magneton $e\hbar/2m_p c$. Experiment, however, shows that the magnitude of the proton magnetic moment is 2.79 nuclear magnetons. Similarly, the neutron, a neutral particle of spin $\frac{1}{2}$, would have no magnetic moment according to Dirac's theory. Again experiment has shown that it possesses a magnetic moment, antiparallel to the spin, of magnitude 1.91 nuclear magnetons.

These anomalies may be attributed to the interaction of protons and neutrons with the *pion field,* and calculations based on quantum field theory have supported this view. Here we confine ourselves to a phenomenological description of the nucleon magnetic moments.

Pauli showed that this can be achieved by replacing eq. (13) by the equation *

$$\left[\gamma_\mu\left(\frac{\partial}{\partial x_\mu}-\frac{ie}{\hbar c}A_\mu\right)+\frac{\lambda i}{2\hbar c}F_{\mu\nu}\gamma_\mu\gamma_\nu+\frac{m_p c}{\hbar}\right]\psi=0\,, \tag{14}$$

where λ is the "anomaly" of the magnetic moment of the proton, i.e.

$$\lambda = 1.79\,\frac{e\hbar}{2m_p c}\,.$$

"Pauli's term" $F_{\mu\nu}\gamma_\mu\gamma_\nu$ is manifestly gauge invariant, and eq. (14) is clearly Lorentz invariant.

It will now be shown that the magnetic moment of a particle, whose wave function obeys eq. (14), is indeed

$$\boldsymbol{\mu}=\left(\frac{e\hbar}{2m_p c}+\lambda\right)\boldsymbol{\sigma}\,.$$

For this purpose the Dirac-Pauli eq. (14), $\left(\pi_\mu = -\,i\hbar\partial/\partial x_\mu-(e/c)A_\mu\right)$

$$\left[\gamma_\mu\pi_\mu+(\lambda/2c)\,F_{\mu\nu}\gamma_\mu\gamma_\nu-im_p c\right]\psi=0\,,$$

is multiplied by $(1/2m_p)[\gamma_\mu\pi_\mu-(\lambda/2c)\,F_{\mu\nu}\gamma_\mu\gamma_\nu+im_p c]$ from the left. This gives the equation

$$\left\{\frac{1}{2m_p}\left(\frac{\hbar}{i}\frac{\partial}{\partial x_\mu}-\frac{e}{c}A_\mu\right)^2-\left(\frac{e\hbar}{2m_p c}+\lambda\right)(\boldsymbol{\sigma}\cdot\boldsymbol{H}-i\,\boldsymbol{\alpha}\cdot\boldsymbol{E})+\tfrac{1}{2}\,m_p c^2+\right.$$

$$\left.+\frac{1}{2m_p}\left[\frac{\lambda}{2c}(\gamma_\varrho\pi_\varrho F_{\mu\nu}\gamma_\mu\gamma_\nu-F_{\varrho\sigma}\gamma_\varrho\gamma_\sigma\gamma_\mu\pi_\mu)-\frac{\lambda^2}{4c^2}F_{\varrho\sigma}F_{\mu\nu}\gamma_\varrho\gamma_\sigma\gamma_\mu\gamma_\nu\right]\right\}\psi=0$$

from which our assertion is apparent.

Later it will be useful to have the Hamiltonian form of eq. (14) for a particle of charge $-\,e$. Putting

$$\psi = e^{-imc^2t/\hbar}\,\psi_0 \tag{15}$$

and denoting by $H_{\rm kin}$ the kinetic energy operator ** defined by

* Similarly a neutron may be described by the equation

$$\left(\gamma_\mu\frac{\partial}{\partial x_\mu}+\frac{\lambda i}{2\hbar c}F_{\mu\nu}\gamma_\mu\gamma_\nu+\frac{m_n c}{\hbar}\right)\psi=0$$

with $\lambda = -\,1.91$ nuclear magnetons.

** Not including the rest energy, whose contributions are taken care of by the exponential in (15).

$$H_{\text{kin}} = \hbar i \frac{\partial}{\partial t} + e\, A_0\,,$$

one has

$$H_{\text{kin}}\, \psi_0 + mc^2(1 - \beta)\, \psi_0 = \left[c\, \boldsymbol{\alpha} \cdot \boldsymbol{\pi} - \lambda\, \beta(\boldsymbol{\sigma} \cdot \boldsymbol{H} - \mathrm{i}\, \boldsymbol{\alpha} \cdot \boldsymbol{E})\right] \psi_0\,. \quad (16)$$

5. The Foldy-Wouthuysen transformation

The non-relativistic approximation is most conveniently treated by formulating the theory in a representation where the Hamiltonian does not involve odd matrices. For free particles this has been done in Chap. V, § 6, by means of the Foldy-Wouthuysen transformation. However, for particles subject to forces it does not appear possible to achieve this purpose by a single unitary transformation.

The case of a charged particle in a static magnetic field is an exception. The transformation

$$\mathrm{e}^{\pm \mathrm{i} S_{\text{F.W.}}} = \frac{|H| + mc^2 \pm \beta\, \boldsymbol{\alpha} \cdot (c\boldsymbol{p} + e\boldsymbol{A})}{[2\,|H|\,(|H| + mc^2)]^{\frac{1}{2}}}\,,$$

with

$$|H| = \sqrt{H^2} = \sqrt{[\boldsymbol{\alpha} \cdot (c\boldsymbol{p} + e\boldsymbol{A})]^2 + m^2c^4}$$
$$= \sqrt{(c\boldsymbol{p} + e\boldsymbol{A})^2 + m^2c^4 + e\hbar c\, \boldsymbol{\sigma} \cdot \boldsymbol{H}}\,,$$

gives the new Hamiltonian

$$H_{\text{F.W.}} = \beta\, |H|\,.$$

Nevertheless, if the potentials which describe the forces acting on the particle have a Fourier representation not involving wavevectors $|\boldsymbol{k}| \geqslant \kappa$ and frequencies $\geqslant c\,\kappa$ (so that they cannot induce transitions between free-particle states with momentum and energy changes $\geqslant mc$ and $\geqslant mc^2$ respectively), one can always make a sequence of transformations, each of which eliminates odd operators from the Hamiltonian to one higher order in the expansion parameter $1/m$.

Following Foldy and Wouthuysen *, we consider a Dirac equation of the form

$$H\, \psi = (\beta\, mc^2 + \mathfrak{E}(x) + \mathfrak{O}(x))\, \psi = \hbar i \frac{\partial \psi}{\partial t}\,, \quad (17)$$

where $\mathfrak{E}(x)$ is an even matrix and $\mathfrak{O}(x)$ an odd matrix, both, in general, functions of space-time. We assume that $\mathfrak{E}$ and $\mathfrak{O}$ are of no lower order in $1/m$ than $(1/m)^0$.

* L. L. Foldy and S. A. Wouthuysen, Phys. Rev. **78** (1950), 29.

By the unitary transformation

$$U_{\mathrm{F.W.}} = \mathrm{e}^{\mathrm{i} S_{\mathrm{F.W.}}},$$

with

$$S_{\mathrm{F.W.}} = -\frac{\mathrm{i}}{2mc^2}\beta\, \mathfrak{O}, \tag{18}$$

the Dirac equation takes the form

$$H_{\mathrm{F.W.}}\, \psi_{\mathrm{F.W.}} = \hbar \mathrm{i} \frac{\partial \psi_{\mathrm{F.W.}}}{\partial t},$$

where the new Hamiltonian $H_{\mathrm{F.W.}}$ does not involve any odd matrix of order $1/m$.

In fact, on substituting $\psi = U^{-1}_{\mathrm{F.W.}}\, \psi_{\mathrm{F.W.}}$, eq. (17) becomes

$$H_{\mathrm{F.W.}}\, \psi_{\mathrm{F.W.}} = \left(U_{\mathrm{F.W.}}\, H\, U^{-1}_{\mathrm{F.W.}} - \mathrm{i}\hbar\, U_{\mathrm{F.W.}} \frac{\partial U^{-1}_{\mathrm{F.W.}}}{\partial t} \right) \psi_{\mathrm{F.W.}} = \hbar \mathrm{i} \frac{\partial \psi_{\mathrm{F.W.}}}{\partial t}.$$

For simplicity we assume that the forces acting on the particle are static, so that the time derivative of $U^{-1}_{\mathrm{F.W.}}$ is zero.

Using the expansions

$$U_{\mathrm{F.W.}} = \sum_{\nu=0}^{\infty} \frac{(\mathrm{i}\, S_{\mathrm{F.W.}})^{\nu}}{\nu!}, \quad U^{-1}_{\mathrm{F.W.}} = \sum_{\nu=0}^{\infty} \frac{(-\,\mathrm{i}\, S_{\mathrm{F.W.}})^{\nu}}{\nu!},$$

the new Hamiltonian can also be expanded as follows

$$H_{\mathrm{F.W.}} = H + \mathrm{i}\, [S_{\mathrm{F.W.}}, H] + \frac{\mathrm{i}^2}{2!} [S_{\mathrm{F.W.}}, [S_{\mathrm{F.W.}}, H]] + \ldots$$

Inserting the expressions (17) for the old Hamiltonian and (18) for $S_{\mathrm{F.W.}}$, and remembering that β commutes with all even matrices and anticommutes with all odd matrices, we have

$$H_{\mathrm{F.W.}} = \beta mc^2 + \mathfrak{E} + \frac{\beta}{2mc^2} (\mathfrak{O}^2 + [\mathfrak{O}, \mathfrak{E}]) -$$

$$- \frac{1}{(mc^2)^2} \left(\tfrac{1}{3}\, \mathfrak{O}^3 + \tfrac{1}{8} [\mathfrak{O}, [\mathfrak{O}, \mathfrak{E}]]\right) + \ldots,$$

which is free from odd matrices to the order $1/m$.

A further transformation,

$$U'_{\mathrm{F.W.}} = \mathrm{e}^{\mathrm{i}\, S'_{\mathrm{F.W.}}}$$

with

$$S'_{\mathrm{F.W.}} = -\frac{\mathrm{i}}{(2mc^2)^2}\,[\mathfrak{O}, \mathfrak{E}]\,,$$

leads to the Hamiltonian

$$H'_{\mathrm{F.W.}} = U'_{\mathrm{F.W.}}\, H_{\mathrm{F.W.}}\, U'^{-1}_{\mathrm{F.W.}} = \beta\, mc^2 + \mathfrak{E} + \frac{\beta\,\mathfrak{O}^2}{2mc^2}$$

$$-\frac{1}{(mc^2)^2}\left(\tfrac{1}{3}\,\mathfrak{O}^3 + \tfrac{1}{8}\,[\mathfrak{O}, [\mathfrak{O}, \mathfrak{E}]] + \tfrac{1}{4}\,[\mathfrak{E}, [\mathfrak{O}, \mathfrak{E}]]\right) + \ldots,$$

free from odd matrices to the order $(1/m)^2$.

Proceeding one step further, odd matrices may be eliminated to the order $(1/m)^3$ by yet another transformation,

$$U''_{\mathrm{F.W.}} = \mathrm{e}^{\mathrm{i}\,S''_{\mathrm{F.W.}}}$$

with

$$S''_{\mathrm{F.W.}} = \frac{\mathrm{i}}{2(mc^2)^3}\,\beta\left(\tfrac{1}{3}\,\mathfrak{O}^3 + \tfrac{1}{4}\,[\mathfrak{E}, [\mathfrak{O}, \mathfrak{E}]]\right).$$

This gives

$$H''_{\mathrm{F.W.}} = U''_{\mathrm{F.W.}}\, H'_{\mathrm{F.W.}}\, U''^{-1}_{\mathrm{F.W.}} = \beta\, mc^2 + \mathfrak{E} + \frac{\beta\,\mathfrak{O}^2}{2mc^2}$$

$$-\frac{1}{8(mc^2)^2}\,[\mathfrak{O}, [\mathfrak{O}, \mathfrak{E}]] + \ldots$$

The Hamiltonian

$$H'''_{\mathrm{F.W.}} = H''_{\mathrm{F.W.}} - \frac{\beta}{8(mc^2)^3}\left(\mathfrak{O}^4 + [\mathfrak{O}, \mathfrak{E}]^2\right), \tag{19}$$

free from odd matrices to the order $(1/m)^4$, is obtained by the above transformations followed by yet another, retaining terms in $(1/m)^4$ at each step. For a charged particle in a static electromagnetic field

$$\mathfrak{E} = -\,cA_0(\boldsymbol{x})\,, \quad \mathfrak{O} = c\,\boldsymbol{\alpha}\cdot\boldsymbol{\pi}\,,$$

and, therefore,

$$\mathfrak{O}^2 = c^2\,(\boldsymbol{\sigma}\cdot\boldsymbol{\pi})^2 = c^2\boldsymbol{\pi}^2 + ec\hbar\,\boldsymbol{\sigma}\cdot\boldsymbol{H}\,,$$

$$[\mathfrak{O}, [\mathfrak{O}, \mathfrak{E}]] = 2e^2c\hbar\,\boldsymbol{\sigma}\cdot(\boldsymbol{A}\times\boldsymbol{E}) - 2ec^2\hbar\,\boldsymbol{\sigma}\cdot(\boldsymbol{E}\times\boldsymbol{p}) - ec^2\hbar^2\,\mathrm{div}\,\boldsymbol{E}\,.$$

Inserting these expressions in eq. (19), one has

$$H'''_{\text{F.W.}} = \beta mc^2 - e A_0 + \frac{\beta}{2m} (\boldsymbol{\sigma} \cdot \boldsymbol{\pi})^2 + \frac{e\hbar}{4m^2c^2} \boldsymbol{\sigma} \cdot (\boldsymbol{E} \times \boldsymbol{\pi}) +$$

$$+ \frac{e\hbar^2}{8m^2c^2} \operatorname{div} \boldsymbol{E} - \frac{\beta}{8m^3c^2} (\boldsymbol{\sigma} \cdot \boldsymbol{\pi})^4 + \frac{e^2\hbar^2\beta}{8m^3c^4} (\boldsymbol{\sigma} \cdot \boldsymbol{E})^2 .$$

Here $\boldsymbol{E}$, $\boldsymbol{H}$, A_0 and $\boldsymbol{A}$ are evaluated at the point $\boldsymbol{x}$, which, to the approximation considered, is the *mean position* of the particle (cf. the case of free particles, Chap. V, § 6).

The local interactions involved in $H'''_{\text{F.W.}}$ correspond to non-local ones (Zitterbewegung) in the two-component theory obtained by expressing the original Dirac equation only in terms of the first two components of the wavefunction.

Since the Hamiltonian $H'''_{\text{F.W.}}$ is free of odd matrices, its eigenfunctions have the last (first) two components equal to zero for positive (negative) energy. For a stationary solution of energy $E > 0$, the equation

$$\left(H'''_{\text{F.W.}} - \hbar i \frac{\partial}{\partial t} \right) \psi'''_{\text{F.W.}} = 0$$

is identical with the second approximation Schroedinger-Pauli equation shown on p. 191, but for the last term ($\boldsymbol{E}^2$) which is a higher order correction.

CHAPTER VIII

NON-RELATIVISTIC LIMIT OF THE DIRAC EQUATION

1. The Schroedinger-Pauli equation

Henceforth we shall use Pauli's representation of the γ's, which has already been found suitable for the treatment of the non-relativistic limit in the case of free particles. In this representation two of the four components of the Dirac wave function become negligible for small velocities, and the Dirac equation reduces to the Schroedinger-Pauli equation for a two-component wave function.

The wave function ψ_0 in eq. (16) of the preceding chapter may be represented in the form

$$\psi_0 = \begin{pmatrix} \Psi_1 \\ \Psi_2 \end{pmatrix},$$

Ψ_1 and Ψ_2 being two-element column matrices. Then eq. (16) separates into the two equations

$$\begin{aligned} H_{\text{kin}}\, \Psi_1 &= c\, \boldsymbol{\sigma} \cdot \boldsymbol{\pi}\, \Psi_2 - \lambda\, \boldsymbol{\pi} \cdot \boldsymbol{H}\, \Psi_1 + \mathrm{i}\lambda\, \boldsymbol{\sigma} \cdot \boldsymbol{E}\, \Psi_2\,, \\ H_{\text{kin}}\, \Psi_2 + 2mc^2\, \Psi_2 &= c\, \boldsymbol{\sigma} \cdot \boldsymbol{\pi}\, \Psi_1 + \lambda\, \boldsymbol{\sigma} \cdot \boldsymbol{H}\, \Psi_2 - \mathrm{i}\lambda\, \boldsymbol{\sigma} \cdot \boldsymbol{E}\, \Psi_1\,. \end{aligned} \tag{1}$$

We expand the equations in powers of the operator $H_{\text{kin}}/2mc^2$. Of course, since H_{kin} depends on A_0, such expansions are allowed only if

$$|e\, A_0|/2mc^2 \ll 1\,.$$

For instance they would not be permissible for a Coulomb field, as in the case of the hydrogen atom, since A_0 would then diverge for $r = 0$.

The first approximation is obtained at once on neglecting $H_{\text{kin}}/2mc^2$ as compared with one. This yields the equations

$$\begin{aligned} H_{\text{kin}}\, \Psi_1 &= (c\, \boldsymbol{\sigma} \cdot \boldsymbol{\pi} + \mathrm{i}\, \lambda\, \boldsymbol{\sigma} \cdot \boldsymbol{E})\, \Psi_2 - \lambda\, \boldsymbol{\sigma} \cdot \boldsymbol{H}\, \Psi_1\,, \\ \Psi_2 &= \left[(c\, \boldsymbol{\sigma} \cdot \boldsymbol{\pi} - \mathrm{i}\, \lambda\, \boldsymbol{\sigma} \cdot \boldsymbol{E})\, \Psi_1 + \lambda\, \boldsymbol{\sigma} \cdot \boldsymbol{H}\, \Psi_2\right]/2mc^2\,. \end{aligned}$$

From the second of these, Ψ_2 may be expressed in terms of Ψ_1:

$$\Psi_2 = (2mc^2 - \lambda\, \boldsymbol{\sigma} \cdot \boldsymbol{H})^{-1} (c\, \boldsymbol{\sigma} \cdot \boldsymbol{\pi} - \mathrm{i}\, \lambda\, \boldsymbol{\sigma} \cdot \boldsymbol{E})\, \Psi_1 \,. \tag{2}$$

Substitution in the first equation gives

$$(H_{\mathrm{kin}} + \lambda\, \boldsymbol{\sigma} \cdot \boldsymbol{H})\, \Psi_1 = \tag{3}$$
$$= (c\, \boldsymbol{\sigma} \cdot \boldsymbol{\pi} + \mathrm{i}\, \lambda\, \boldsymbol{\sigma} \cdot \boldsymbol{E})\, (2mc^2 - \lambda\, \boldsymbol{\sigma} \cdot \boldsymbol{H})^{-1} (c\, \boldsymbol{\sigma} \cdot \boldsymbol{\pi} - \mathrm{i}\, \lambda\, \boldsymbol{\sigma} \cdot \boldsymbol{E})\, \Psi_1 \,.$$

For simplicity we will assume that

$$|\lambda\, \boldsymbol{\sigma} \cdot \boldsymbol{H}| \ll 2mc^2 \,,$$
$$|\lambda\, \boldsymbol{\sigma} \cdot \boldsymbol{E}| \simeq 0 \,, \quad |\boldsymbol{\nabla}_x \cdot \boldsymbol{H}| \simeq 0 \,.$$

Eq. (2) then becomes

$$\Psi_2 = \boldsymbol{\sigma} \cdot \boldsymbol{\pi}\, \Psi_1 / 2mc \,,$$

showing that $\Psi_2^\dagger \Psi_2$ is of the order of v^2/c^2 compared with $\Psi_1^\dagger \Psi_1$, and is therefore negligible in the extreme non-relativistic limit.

Eq. (3) now becomes

$$(H_{\mathrm{kin}} + \lambda\, \boldsymbol{\sigma} \cdot \boldsymbol{H})\, \Psi_1 = (\boldsymbol{\sigma} \cdot \boldsymbol{\pi})(\boldsymbol{\sigma} \cdot \boldsymbol{\pi})\, \Psi_1 / 2m$$

and, taking into account the equation

$$(\boldsymbol{\sigma} \cdot \boldsymbol{\pi})(\boldsymbol{\sigma} \cdot \boldsymbol{\pi}) = \boldsymbol{\pi} \cdot \boldsymbol{\pi} + (\hbar e/c)\, \boldsymbol{\sigma} \cdot \boldsymbol{H} \,, \tag{4}$$

this may be written in the form

$$\left(\hbar \mathrm{i} \frac{\partial}{\partial t} + e\, A_0\right) \Psi_1 = \left[\frac{1}{2m} \left(\frac{\hbar}{\mathrm{i}} \boldsymbol{\nabla}_x + \frac{e}{c} \boldsymbol{A}\right)^2 + \left(\frac{e\hbar}{2mc} - \lambda\right) \boldsymbol{\sigma} \cdot \boldsymbol{H}\right] \Psi_1 \tag{5}$$

(*Schroedinger-Pauli equation,* first approximation).
From this equation one sees (once more) that the magnetic moment of the particle described by eq. (5) is

$$\boldsymbol{\mu} = \left(- \frac{\hbar e}{2mc} + \lambda\right) \boldsymbol{\sigma} \,.$$

To the approximation considered here, the charge and current densities are

$$\rho = \psi^\dagger \psi = \Psi_1^\dagger \Psi_1 + \Psi_1^\dagger\, (\boldsymbol{\sigma} \cdot \boldsymbol{\pi})^2 \Psi_1 / 4m^2c^2 \simeq \Psi_1^\dagger \Psi_1 \,,$$
$$\boldsymbol{j} = c\, \psi^\dagger\, \boldsymbol{\alpha}\, \psi = c\, \Psi_1^\dagger\, \boldsymbol{\sigma}\, \Psi_2 + c \Psi_2^\dagger\, \boldsymbol{\sigma}\, \Psi_1 \simeq \left[\Psi_1^\dagger\, \boldsymbol{\sigma} (\boldsymbol{\sigma} \cdot \boldsymbol{\pi})\, \Psi_1 + (\boldsymbol{\sigma} \cdot \boldsymbol{\pi}\, \Psi_1)^\dagger\, \boldsymbol{\sigma}\, \Psi_1\right] / 2m \,.$$

While the charge density is that of the non-relativistic theory of particles of spin $\frac{1}{2}$, the current density exhibits the additional term *

$$j' = \frac{\hbar}{2m} \nabla_x \times (\Psi_1^\dagger \boldsymbol{\sigma} \Psi_1),$$

already encountered in eq. (12) of the preceding chapter.

In order to simplify the discussion of the second approximation we will assume that the electromagnetic field is static and that there is no Pauli term ($\lambda = 0$).

For a stationary wave function eqs. (1) become

$$\begin{aligned} (E + e A_0)\, \Psi_1 &= c\, \boldsymbol{\sigma} \cdot \boldsymbol{\pi}\, \Psi_2 , \\ (E + e A_0 + 2mc^2)\, \Psi_2 &= c\, \boldsymbol{\sigma} \cdot \boldsymbol{\pi}\, \Psi_1 . \end{aligned} \tag{6}$$

Here E is the total energy (minus the rest energy), and $E + e A_0 = E_{\text{kin}}$ is the kinetic energy. From the second of eqs. (6) one has **

$$\Psi_2 = \left(1 + \frac{E_{\text{kin}}}{2mc^2}\right)^{-1} \frac{\boldsymbol{\sigma} \cdot \boldsymbol{\pi}}{2mc} \Psi_1 . \tag{7}$$

Substituting this in the first equation, one finds the *exact* equation

$$E_{\text{kin}}\, \Psi_1 = \frac{1}{2m}\, \boldsymbol{\sigma} \cdot \boldsymbol{\pi} \left(1 + \frac{E_{\text{kin}}}{2mc^2}\right)^{-1} \boldsymbol{\sigma} \cdot \boldsymbol{\pi}\, \Psi_1 .$$

For regions in space where $E_{\text{kin}} \ll 2mc^2$, $[1 + (E_{\text{kin}}/2mc^2)]^{-1}$ may be expanded in a power series of $E_{\text{kin}}/2mc^2$. Putting

$$\left(1 + \frac{E_{\text{kin}}}{2mc^2}\right)^{-1} \simeq 1 - \frac{E_{\text{kin}}}{2mc^2},$$

* For the component j_k one has

$$j_k = \frac{e}{mc} A_k \Psi_1^\dagger \Psi_1 + \frac{\hbar}{2mi} \left(\Psi_1^\dagger \frac{\partial \Psi_1}{\partial x_k} - \frac{\partial \Psi_1^\dagger}{\partial x_k} \Psi_1 \right) + \\ + \frac{\hbar}{2mi} \sum_{l(\neq k)} \left(\Psi_1^\dagger \sigma_k \sigma_l \frac{\partial \Psi_1}{\partial x_l} - \frac{\partial \Psi_1^\dagger}{\partial x_l} \sigma_l \sigma_k \Psi_1 \right).$$

The last term may be written as

$$j_k' = \frac{\hbar}{2m} \varepsilon_{klm} \left(\Psi_1^\dagger \sigma_m \frac{\partial \Psi_1}{\partial x_l} + \frac{\partial \Psi_1^\dagger}{\partial x_l} \sigma_m \Psi_1 \right) = \frac{\hbar}{2m} \varepsilon_{klm} \frac{\partial}{\partial x_l} (\Psi_1^\dagger \sigma_m \Psi_1) .$$

** Here $E_{\text{kin}} = E + eA_0$ is a function of x. Therefore the order of the factors on the right side may not be changed.

one obtains the second order approximation of the Dirac equation

$$E_{\mathrm{kin}}\,\Psi_1 = \left[\frac{(\boldsymbol{\sigma}\cdot\boldsymbol{\pi})^2}{2m} - \frac{(\boldsymbol{\sigma}\cdot\boldsymbol{\pi})\,E_{\mathrm{kin}}(\boldsymbol{\sigma}\cdot\boldsymbol{\pi})}{(2mc)^2}\right]\Psi_1\,. \tag{8}$$

Still working to second order, eq. (7) may be written in a more significant form. In fact, one has

$$\begin{aligned}2(\boldsymbol{\sigma}\cdot\boldsymbol{\pi})\,E_{\mathrm{kin}}(\boldsymbol{\sigma}\cdot\boldsymbol{\pi}) &= [\boldsymbol{\sigma}\cdot\boldsymbol{\pi},\ E_{\mathrm{kin}}](\boldsymbol{\sigma}\cdot\boldsymbol{\pi}) + E_{\mathrm{kin}}(\boldsymbol{\sigma}\cdot\boldsymbol{\pi})^2 + \text{h.c.} =\\ &= \mathrm{i}e\hbar(\boldsymbol{\sigma}\cdot\boldsymbol{E})(\boldsymbol{\sigma}\cdot\boldsymbol{\pi}) + E_{\mathrm{kin}}(\boldsymbol{\sigma}\cdot\boldsymbol{\pi})^2 + \text{h.c.} =\\ &= E_{\mathrm{kin}}(\boldsymbol{\sigma}\cdot\boldsymbol{\pi})^2 + (\boldsymbol{\sigma}\cdot\boldsymbol{\pi})^2 E_{\mathrm{kin}} - e\hbar^2 \operatorname{div}\boldsymbol{E} - 2e\hbar\,\boldsymbol{\sigma}\cdot(\boldsymbol{E}\times\boldsymbol{\pi}).\end{aligned}$$

The Dirac equations (7), (8) reduce to the following

$$\Psi_2 = \left(\frac{\boldsymbol{\sigma}\cdot\boldsymbol{\pi}}{2mc} - \frac{c\,E_{\mathrm{kin}}\,\boldsymbol{\sigma}\cdot\boldsymbol{\pi}}{(2mc^2)^2}\right)\Psi_1\,,$$

$$\begin{aligned}E_{\mathrm{kin}}\,\Psi_1 = \bigg[&\frac{1}{2m}(\boldsymbol{\sigma}\cdot\boldsymbol{\pi})^2 - \frac{1}{8m^2c^2}\left(E_{\mathrm{kin}}(\boldsymbol{\sigma}\cdot\boldsymbol{\pi})^2 + (\boldsymbol{\sigma}\cdot\boldsymbol{\pi})^2 E_{\mathrm{kin}}\right) +\\ &+ \frac{e\hbar^2}{8m^2c^2}\operatorname{div}\boldsymbol{E} + \frac{e\hbar}{4m^2c^2}\boldsymbol{\sigma}\cdot(\boldsymbol{E}\times\boldsymbol{\pi})\bigg]\Psi_1\,.\end{aligned} \tag{9}$$

The second of these is the second approximation Schroedinger-Pauli equation for the two-component wave function Ψ_1.

2. The hydrogen atom

We shall study the case when there is no magnetic field ($\boldsymbol{A} = 0$) and the field A_0 is central ($A_0 = A_0(r)$). Then, denoting the orbital angular momentum by $\boldsymbol{L}$, one has

$$(\boldsymbol{\sigma}\cdot\boldsymbol{\pi})^2 = \boldsymbol{p}^2 \simeq 2m\,E_{\mathrm{kin}}\,,$$

$$\boldsymbol{E}\times\boldsymbol{\pi} = -\frac{1}{r}\frac{\mathrm{d}A_0}{\mathrm{d}r}\boldsymbol{L}, \quad \operatorname{div}\boldsymbol{E} = -\Delta A_0,$$

and eq. (9) becomes

$$E_{\mathrm{kin}}\,\Psi_1 = \left(\frac{\boldsymbol{p}^2}{2m} - \frac{(\boldsymbol{p}^2)^2}{8m^3c^2} - \frac{e}{2m^2c^2}\frac{1}{r}\frac{\mathrm{d}A_0}{\mathrm{d}r}\boldsymbol{L}\cdot\boldsymbol{s} - \frac{e\hbar^2}{8m^2c^2}\Delta A_0\right)\Psi_1. \tag{10}$$

In order to interpret the terms on the right-hand side of this equation, we notice that the first two arise from the expansion of the kinetic energy

$$mc^2\left[\left(1+\frac{p^2}{m^2c^2}\right)^{\frac{1}{2}}-1\right]\simeq\left(\frac{p^2}{2m^2c^2}-\frac{(p^2)^2}{8m^4c^4}\right)mc^2+\ldots,$$

while the spin-orbit coupling $\boldsymbol{L}\cdot\boldsymbol{s}$ is due to the fact that the electron in motion "sees" the electrostatic field as a magnetic field. The last term on the right-hand side of eq. (10) has no classical analogue.

For a stationary wave function of energy E (E = total energy minus rest energy as in eq. (6)), eq. (10) reads

$$E\Psi_1=\left(-eA_0+\frac{p^2}{2m}-\frac{(p^2)^2}{8m^3c^2}-\frac{e}{2m^2c^2}\frac{1}{r}\frac{dA_0}{dr}\boldsymbol{L}\cdot\boldsymbol{s}-\frac{e\hbar^2}{8m^2c^2}\Delta A_0\right)\Psi_1.$$

This is of the form

$$E\Psi_1=H\Psi_1,$$

the Hamiltonian being the sum of the zeroth order Hamiltonian

$$H^{(0)}=-eA_0+(p^2/2m)$$

and of the perturbation

$$H^{(1)}=-\frac{(p^2)^2}{8m^3c^2}-\frac{e}{2m^2c^2}\frac{1}{r}\frac{dA_0}{dr}\boldsymbol{L}\cdot\boldsymbol{s}-\frac{e\hbar^2}{8m^2c^2}\Delta A_0.$$

If the Hamiltonian were only $H^{(0)}$, the components of the orbital angular momentum L_i, ($i=1,2,3$), and those of the spin s_i, ($i=1,2,3$), would commute with it, and would separately be constants of motion.

The eigenfunctions of $H^{(0)}$ can be classified by the eigenvalues of $\boldsymbol{L}^2$, L_3 and s_3, i.e. with the quantum numbers l, m_l (with $-l\leqslant m_l\leqslant l$) and m_s ($m_s=\pm\frac{1}{2}$).

The perturbation $H^{(1)}$, on the other hand, does not separately commute with L_i and s_i, ($i=1,2,3$), but only with $\boldsymbol{L}^2$ and $\boldsymbol{s}^2$ and with the components of the total angular momentum

$$\boldsymbol{M}=\boldsymbol{L}+\boldsymbol{s}.$$

As is known, from the commutation relations of an angular momentum it follows that, if $\hbar^2 j(j+1)$ are the eigenvalues of $\boldsymbol{M}^2$, those of M_3 are $\hbar m_j$ with $-j\leqslant m_j\leqslant j$. The eigenfunctions of the total Hamiltonian may therefore be classified by the quantum numbers j, l and m_j.

Now, from

$$\boldsymbol{M}^2=\boldsymbol{L}^2+2\boldsymbol{L}\cdot\boldsymbol{s}+\boldsymbol{s}^2,$$

it follows that the spin-orbit coupling $\boldsymbol{L}\cdot\boldsymbol{s}$ commutes with $\boldsymbol{M}^2$. Hence an eigenfunction of $\boldsymbol{M}^2$ and $\boldsymbol{L}^2$, characterized by the quantum numbers j and l, is also an eigenfunction of $2\boldsymbol{L}\cdot\boldsymbol{s}$ belonging to the eigenvalue $\hbar^2[j(j+1) - l(l+1) - s(s+1)]$, $(s=\frac{1}{2})$.

For a given l, j may have the values $l\pm\frac{1}{2}$. Similarly, for a given j, $l=j\mp\frac{1}{2}$. In order to distinguish between the two possibilities, one may give the eigenvalues of the operator $\boldsymbol{L}\cdot\boldsymbol{s}$, or, more conveniently, those of the operator

$$K = \boldsymbol{\sigma}\cdot\boldsymbol{L} + \hbar\,.$$

The eigenvalues of this operator may be used, in place of j or l, for the classification of the eigenfunctions of H.

It is easy to see that *

$$\boldsymbol{M}^2 = K^2 - \tfrac{1}{4}\hbar^2\,,\quad \boldsymbol{L}^2 = K^2 - \hbar K\,, \tag{11}$$

showing that the eigenfunctions of K are also eigenfunctions of $\boldsymbol{M}^2$ and $\boldsymbol{L}^2$. Moreover, denoting the eigenvalues of K by $\hbar k$, it follows from eqs. (11) that

$$l(l+1) = k^2 - k = k(k-1)\,,$$
$$j(j+1) = k^2 - \tfrac{1}{4} = (k+\tfrac{1}{2})(k-\tfrac{1}{2})\,,\quad \text{i.e.}\quad k = \pm\,(j+\tfrac{1}{2})\,.$$

According to these relations one has

$$\begin{aligned} &\alpha)\quad k>0\,,\quad l=k-1\,,\quad j=l+\tfrac{1}{2}\,,\\ &\beta)\quad k<0\,,\quad l=-k\,,\quad\;\; j=l-\tfrac{1}{2}\,. \end{aligned} \tag{12}$$

The eigenvalues of K are, therefore, 0, $\pm\,\hbar$, $\pm\,2\hbar$, ... etc.

Let us now determine the eigenvalues of H. For an energy eigenfunction of quantum numbers j, l, m_j, the perturbation Hamiltonian is

$$H^{(1)} = -\frac{1}{2mc^2}(E+e\,A_0)^2 + $$
$$-\frac{e\hbar^2}{4m^2c^2}\frac{\mathrm{d}A_0}{\mathrm{d}r}\frac{j(j+1)-l(l+1)-s(s+1)}{r} - \frac{e\hbar^2}{8m^2c^2}\Delta A_0\,.$$

If A_0 is the Coulomb field of a nucleus of charge Ze

$$A_0 = \frac{Ze}{4\pi r}\,,$$

* In fact,

$$K^2 = \hbar^2 + 2\hbar(\boldsymbol{\sigma}\cdot\boldsymbol{L}) + (\delta_{ik} + \mathrm{i}\varepsilon_{ikl}\sigma_l)\,L_iL_k = \hbar^2 + \hbar\boldsymbol{\sigma}\cdot\boldsymbol{L} + \boldsymbol{L}^2\,.$$

the eigenvalues of the "unperturbed" Hamiltonian are

$$E_n^{(0)} = -\, mc^2\alpha^2 Z^2/2n^2 \,.$$

These are shifted by an amount ΔE, which, according to perturbation theory, is the mean value of $H^{(1)}$ evaluated by using the unperturbed wave functions. Measuring the energy in units $mc^2\alpha^2$ and r in units $\hbar/mc\alpha$ (atomic system of units, see e.g. Bethe and Salpeter, *loc. cit.*, p. 89), we have

$$\Delta E = -\frac{1}{2}\alpha^2\left(\frac{1}{4}\frac{Z^4}{n^4} - \frac{Z^3}{n^2}\overline{r^{-1}} + Z^2\,\overline{r^{-2}}\right) + \tag{13}$$
$$+ \tfrac{1}{4} Z\alpha^2\left[j(j+1) - l(l+1) - s(s+1)\right]\overline{r^{-3}} + \frac{Z\alpha^2\pi}{2}\int \delta(\boldsymbol{x})\,|\Psi_{nl}(\boldsymbol{x})|^2\,\mathrm{d}^3\mathrm{x}.$$

The last term in (13) vanishes for $l > 0$, and gives

$$\frac{Z\alpha^2\pi}{2}\,|\Psi_{n0}(0)|^2 = \frac{Z\alpha^2}{8}\,(R_{n0}(0))^2\,,$$

for $l = 0$. Actually l must be different from zero for our procedure to be consistent. For instance, the third of the mean values

$$\overline{r^{-1}} = Z/n^2\,,\quad \overline{r^{-2}} = Z^2/n^3(l+\tfrac{1}{2})\,,\quad \overline{r^{-3}} = Z^3/n^3(l+1)(l+\tfrac{1}{2})\,l$$

is infinite for $l = 0$.

Using these mean values, and noticing that

$$j(j+1) - l(l+1) - s(s+1) = \begin{cases} l\,, & \text{for } j = l + \tfrac{1}{2}\,,\\ -(l+1)\,, & \text{for } j = l - \tfrac{1}{2}\,,\end{cases}$$

we have, in units of $mc^2\alpha^2$,

$$\Delta E = -\frac{\alpha^2 Z^4}{2n^3}\left(\frac{1}{j+\frac{1}{2}} - \frac{3}{4n}\right). \tag{14}$$

This formula is valid for both $j = l + \frac{1}{2}$ and $j = l - \frac{1}{2}$.

The result (14) agrees with the expansion of Dirac's formula for the fine structure of the levels of hydrogen-like atoms in a power series of $\alpha^2 Z^2$ (see Chap. IX, § 3) and with the old Bohr-Sommerfeld relativistic theory. It is not in accord with the result found for π-mesic atoms. In fact the integral denominator $j + \frac{1}{2}$ occurs in (14) instead of the half-integral $l + \frac{1}{2}$ of eq. (12), Part I, Chap. VI, § 3.

3. Eigenfunctions of M^2, L^2, K and M_3

A two-component function

$$\Psi = \begin{pmatrix} \psi_1 \\ \psi_2 \end{pmatrix}$$

is an eigenfunction of $\boldsymbol{L}^2$ belonging to the eigenvalue $\hbar^2 \, l(l+1)$ if

$$\psi_1 = f_1(r) \, Y_l^{(m_l)} , \quad \psi_2 = f_2(r) \, Y_l^{(m_l')} .$$

If Ψ is also an eigenfunction of M_3 belonging to the eigenvalue $\hbar m_j$, one has

$$M_3 \, \Psi = \begin{pmatrix} L_3 + \frac{1}{2}\hbar & 0 \\ 0 & L_3 - \frac{1}{2}\hbar \end{pmatrix} \begin{pmatrix} \psi_1 \\ \psi_2 \end{pmatrix} = \begin{pmatrix} (L_3 + \frac{1}{2}\hbar)\, \psi_1 \\ (L_3 - \frac{1}{2}\hbar)\, \psi_2 \end{pmatrix} = \hbar \, m_j \begin{pmatrix} \psi_1 \\ \psi_2 \end{pmatrix} .$$

Thus one easily recognizes that the angular parts of ψ_1 and ψ_2 are spherical harmonics whose order m_l differs by one,

$$\psi_1 = f_1(r) \, Y_l^{(m_l)} , \quad \psi_2 = f_2(r) \, Y_l^{(m_l+1)} .$$

A relation between the functions f_1 and f_2 may be found by using the operator $K = \boldsymbol{L} \cdot \boldsymbol{\sigma} + \hbar$, of which Ψ is assumed to be an eigenfunction belonging to the eigenvalue $\hbar k$. One has

$$(L_3 + \hbar)\, \psi_1 + (L_1 - \mathrm{i}\, L_2)\, \psi_2 = \hbar k \, \psi_1 ,$$
$$(L_1 + \mathrm{i}\, L_2)\, \psi_1 + (\hbar - L_3)\, \psi_2 = \hbar k \psi_2 .$$

These two equations are equivalent and one may, therefore, employ only the first. Remembering the properties of the spherical harmonics, Part I, Chap. IV, § 2, eqs. (6), we have

$$f_2 \, [(l + m_l + 1)(l - m_l)]^{\frac{1}{2}} \, Y_l^{(m_l)} = f_1 \, (k - m_l - 1) \, Y_l^{(m_l)} .$$

Thus,

if $k > 0$, i.e. $k = l + 1$,

$$(l + m_l + 1)^{\frac{1}{2}} f_2 = (l - m_l)^{\frac{1}{2}} f_1 , \tag{15}$$

if $k < 0$, i.e. $k = -l$,

$$(l - m_l)^{\frac{1}{2}} f_2 = -\,(l + m_l + 1)^{\frac{1}{2}} f_1 . \tag{15'}$$

In both cases f_1 and f_2 differ by a constant only, and Ψ may be written as $\Psi = f(r)\, X$ with

$$X = \begin{pmatrix} c_1 \, Y_l^{(m_l)} \\ c_2 \, Y_l^{(m_l+1)} \end{pmatrix} .$$

Normalizing f so that

$$\int_0^\infty |f(r)|^2 r^2 \, dr = 1 ,$$

the normalization condition for Ψ gives

$$|c_1|^2 + |c_2|^2 = 1 .$$

Combining this condition with eqs. (15), (15′) and remembering the relations (12) we finally obtain *

$k > 0$

$$k = l + 1 , \quad m_l = m_j - \tfrac{1}{2} , \quad j = l + \tfrac{1}{2} ,$$

$$c_1^+ = \sqrt{\frac{l + m_l + 1}{2l + 1}} = \sqrt{\frac{j + m_j}{2j}} , \quad c_2^+ = \sqrt{\frac{l - m_l}{2l + 1}} = \sqrt{\frac{j - m_j}{2j}} , \tag{16}$$

$k < 0$

$$k = - l , \quad m_l = m_j - \tfrac{1}{2} , \quad j = l - \tfrac{1}{2} ,$$

$$c_1^- = \sqrt{\frac{l - m_l}{2l + 1}} = \sqrt{\frac{j - m_j + 1}{2(j + 1)}} ,$$

$$c_2^- = - \sqrt{\frac{l + m_l + 1}{2l + 1}} = - \sqrt{\frac{j + m_j + 1}{2(j + 1)}} . \tag{16′}$$

Note that the functions

$$X_j^{(m_j)+} = \begin{pmatrix} c_1^+ \, Y_{j-\frac{1}{2}}^{(m_j - \frac{1}{2})} \\ c_2^+ \, Y_{j-\frac{1}{2}}^{(m_j + \frac{1}{2})} \end{pmatrix} , \tag{17}$$

$$X_j^{(m_j)-} = \begin{pmatrix} c_1^- \, Y_{j+\frac{1}{2}}^{(m_j - \frac{1}{2})} \\ c_2^- \, Y_{j+\frac{1}{2}}^{(m_j + \frac{1}{2})} \end{pmatrix} \tag{17′}$$

are both eigenfunctions of M_3 and $\boldsymbol{M}^2$ belonging to the eigenvalues $\hbar m_j$ and $\hbar^2 j(j + 1)$. They are also eigenfunctions of K belonging to the eigenvalues $k = j + \frac{1}{2}$ and $k = - (j + \frac{1}{2})$, respectively.

---

* In fact, for $k > 0$,

$$f_1/f_2 = c_1^+/c_2^+ = [(l + m_l + 1)/(l - m_l)]^{\frac{1}{2}} .$$

Then

$$| c_2^+ |^2 \left(1 + \frac{l + m_l + 1}{l - m_l} \right) = 1 ,$$

or $c_2^+ = [(l - m_l)/(2l + 1)]^{\frac{1}{2}}$. Similarly for c_1^+ and for the case $k < 0$.

One has

$$(M_1 - \mathrm{i}\, M_2)\, X_j^{(m_j)\pm} = \hbar[(j + m_j)(j - m_j + 1)]^{\frac{1}{2}}\, X_j^{(m_j-1)\pm}\,,$$

corresponding to the second equation (6) on p.76, and similarly for $M_1 + \mathrm{i}\, M_2$.

It is also readily seen that (17), (17′) are eigenfunctions of the space inversion operator belonging to eigenvalues of opposite sign. In fact, under space inversion

$$\mathrm{P}\, X_j^{(m_j)\pm} = (-1)^{j\mp\frac{1}{2}}\, X_j^{(m_j)\pm}\,.$$

The parity of $X_j^{(m_j)\pm}$ is therefore determined by the orbital angular momentum, which is the same for both components.

Chapter IX

DIRAC PARTICLE IN A CENTRAL ELECTROSTATIC FIELD

1. Dirac equation in polar coordinates

The Dirac equation may be written in polar coordinates

$$\mathrm{i}\,\hbar\frac{\partial}{\partial t}\psi = \left(-e\,A_0 - \mathrm{i}\,\hbar c\,\boldsymbol{\alpha}\cdot\boldsymbol{n}\frac{\partial}{\partial r} - c\,\boldsymbol{\alpha}\cdot\frac{\boldsymbol{n}\times\boldsymbol{L}}{r} + \beta mc^2\right)\psi\,. \quad (1)$$

In deriving this formula the momentum $\boldsymbol{p}$ has been decomposed into a component along the unit vector $\boldsymbol{n} = \boldsymbol{x}/r$ and one perpendicular to it,

$$\boldsymbol{p} = \boldsymbol{n}(\boldsymbol{n}\cdot\boldsymbol{p}) - \boldsymbol{n}\times(\boldsymbol{n}\times\boldsymbol{p}) = \boldsymbol{n}(\boldsymbol{n}\cdot\boldsymbol{p}) - \boldsymbol{n}\times\boldsymbol{L}/r\,.$$

In Pauli's representation for the α and β matrices, eq. (1) reads *

$$\mathrm{i}\,\hbar\frac{\partial}{\partial t}\psi = \left(-e\,A_0 - \mathrm{i}\,\hbar c\,\rho_1(\boldsymbol{\sigma}\cdot\boldsymbol{n})\frac{\partial}{\partial r} + \frac{1}{r}\,\mathrm{i}\,c\,\rho_1(\boldsymbol{\sigma}\cdot\boldsymbol{n})(\boldsymbol{\sigma}\cdot\boldsymbol{L}) + \rho_3\,mc^2\right)\psi\,. \quad (2)$$

If A_0 is spherically symmetric, it is easy to show that the operator

$$\boldsymbol{M} = \boldsymbol{L} + \boldsymbol{s}$$

commutes with H, and is therefore a constant of motion. However, in contrast to what was found in § 2 of the preceding chapter (approximate theory), now $\boldsymbol{L}^2$ is not a constant of the motion. On the other hand, the properties of the operator

$$K = \rho_3(\boldsymbol{L}\cdot\boldsymbol{\sigma} + \hbar) \quad (3)$$

are similar to those of the operator K of the approximate theory.

---

* In fact

$$\boldsymbol{\alpha}\cdot(\boldsymbol{n}\times\boldsymbol{L}) = \rho_1\varepsilon_{ikl}\sigma_i n_k L_l = -\mathrm{i}\rho_1\sigma_k\sigma_l n_k L_l = -\mathrm{i}\,\rho_1(\boldsymbol{\sigma}\cdot\boldsymbol{n})(\boldsymbol{\sigma}\cdot\boldsymbol{L})\,.$$

In the non-relativistic case, instead of $\boldsymbol{\alpha}\cdot\boldsymbol{p}$ one has $\boldsymbol{p}^2 = (\boldsymbol{p}\cdot\boldsymbol{n})(\boldsymbol{n}\cdot\boldsymbol{p}) + (L^2/r^2)$, corresponding to the classical formula $\boldsymbol{p}^2 = p_r^2 + (L^2/r^2)$.

Since

$$K^2 = \boldsymbol{M}^2 + \tfrac{1}{4}\hbar^2 ,$$

K^2 commutes with the Hamiltonian. It is not difficult to verify that also K commutes with H *. Its eigenvalues $\hbar k$, together with the eigenvalues $\hbar^2 j(j+1)$ of $\boldsymbol{M}^2$ and $\hbar m_j$ of M_3, may therefore be chosen as quantum numbers for classifying the eigenfunctions of the Hamiltonian.

The latter may therefore be written in the form

$$H = -e A_0 - \mathrm{i}\,\hbar c\,\rho_1\,\boldsymbol{\sigma}\cdot\boldsymbol{n}\frac{\partial}{\partial r} + \frac{1}{r}\mathrm{i}\,c\,\rho_1\,\boldsymbol{\sigma}\cdot\boldsymbol{n}(\rho_3 K - \hbar) + \rho_3 mc^2 . \quad (4)$$

Using the results obtained for the two-component wave functions in the previous chapter, one sees that the eigenfunctions of K are

$$\psi^{(1)} = \begin{pmatrix} f\,X_j^{(m_j)+} \\ g\,X_j^{(m_j)-} \end{pmatrix}, \quad \psi^{(2)} = \begin{pmatrix} f\,X_j^{(m_j)-} \\ g\,X_j^{(m_j)+} \end{pmatrix},$$

corresponding to the eigenvalues $k = j + \frac{1}{2}$ and $k = -(j+\frac{1}{2})$ respectively.

More explicitly one has

$$\psi^{(1)} = \begin{pmatrix} f\,X_j^{(m_j)+} \\ g\,X_j^{(m_j)-} \end{pmatrix} = \begin{bmatrix} \left(\dfrac{l+m_j+\frac{1}{2}}{2l+1}\right)^{\frac{1}{2}} f\ Y_l^{(m_j-\frac{1}{2})} \\[2ex] \left(\dfrac{l-m_j+\frac{1}{2}}{2l+1}\right)^{\frac{1}{2}} f\ Y_l^{(m_j+\frac{1}{2})} \\[2ex] \left(\dfrac{l-m_j+\frac{3}{2}}{2l+3}\right)^{\frac{1}{2}} g\ Y_{l+1}^{(m_j-\frac{1}{2})} \\[2ex] -\left(\dfrac{l+m_j+\frac{3}{2}}{2l+3}\right)^{\frac{1}{2}} g\ Y_{l+1}^{(m_j+\frac{1}{2})} \end{bmatrix}, \quad (5)$$

---

* Clearly K commutes with $A_0(r)$ and with ρ_3. We show that it commutes also with $\rho_1(\boldsymbol{\sigma}\cdot\boldsymbol{n})$. One has, in fact,

$$(\boldsymbol{\sigma}\cdot\boldsymbol{n})(\boldsymbol{\sigma}\cdot\boldsymbol{L}) + (\boldsymbol{\sigma}\cdot\boldsymbol{L})(\boldsymbol{\sigma}\cdot\boldsymbol{n}) = \mathrm{i}\boldsymbol{\sigma}\cdot[\boldsymbol{n}\times\boldsymbol{L} + \boldsymbol{L}\times\boldsymbol{n}] = \mathrm{i}\varepsilon_{ikl}\sigma_i\,[n_k, L_l] = -2\hbar\boldsymbol{\sigma}\cdot\boldsymbol{n} .$$

Note that

$$\boldsymbol{L}^2 = K^2 - \hbar(\boldsymbol{\sigma}\cdot\boldsymbol{L} + \hbar) ,$$

from which one sees that $\boldsymbol{L}^2$ does not commute with H. In fact the second term on the right does not commute with the Hamiltonian. For this reason the factor ρ_3 had to be introduced in the definition (3) of K.

where we have put $l = j - \frac{1}{2}$, and

$$\psi^{(2)} = \begin{pmatrix} f\ X_j^{(m_j)-} \\ g\ X_j^{(m_j)+} \end{pmatrix} = \begin{bmatrix} \left(\dfrac{l - m_j + \frac{1}{2}}{2l+1}\right)^{\frac{1}{2}} f\ Y_l^{(m_j - \frac{1}{2})} \\ -\left(\dfrac{l + m_j + \frac{1}{2}}{2l+1}\right)^{\frac{1}{2}} f\ Y_l^{(m_j + \frac{1}{2})} \\ \left(\dfrac{l + m_j - \frac{1}{2}}{2l-1}\right)^{\frac{1}{2}} g\ Y_{l-1}^{(m_j - \frac{1}{2})} \\ \left(\dfrac{l - m_j - \frac{1}{2}}{2l-1}\right)^{\frac{1}{2}} g\ Y_{l-1}^{(m_j + \frac{1}{2})} \end{bmatrix}, \tag{5'}$$

where $l = j + \frac{1}{2}$.

Since

$$\mathrm{P}\begin{pmatrix} f\ X_j^{(m_j)\pm}(\boldsymbol{x}) \\ g\ X_j^{(m_j)\mp}(\boldsymbol{x}) \end{pmatrix} = \gamma_4 \begin{pmatrix} f\ X_j^{(m_j)\pm}(-\boldsymbol{x}) \\ g\ X_j^{(m_j)\mp}(-\boldsymbol{x}) \end{pmatrix} = (-1)^{j\mp\frac{1}{2}} \begin{pmatrix} f\ X_j^{(m_j)\pm}(\boldsymbol{x}) \\ g\ X_j^{(m_j)\mp}(\boldsymbol{x}) \end{pmatrix},$$

such functions are eigenfunctions of the space inversion operaror belonging to the eigenvalues $(-1)^{j\mp\frac{1}{2}} = (-1)^l$ (l is the index appearing in eqs. (5), (5′), and has a different value for $\psi^{(1)}$ and $\psi^{(2)}$, $l = j - \frac{1}{2}$ and $l = j + \frac{1}{2}$ respectively).

The operator $\rho_1\, \boldsymbol{\sigma} \cdot \boldsymbol{n}$ acts on the wave functions (5), (5′) as follows:

$$\rho_1\, \boldsymbol{\sigma} \cdot \boldsymbol{n} \begin{pmatrix} f\ X_j^{(m_j)\pm} \\ g\ X_j^{(m_j)\mp} \end{pmatrix} = \begin{pmatrix} \boldsymbol{\sigma} \cdot \boldsymbol{n}\, g(r)\, X_j^{(m_j)\mp} \\ \boldsymbol{\sigma} \cdot \boldsymbol{n}\, f(r)\, X_j^{(m_j)\pm} \end{pmatrix} = \begin{pmatrix} g(r)\, X_j^{(m_j)\pm} \\ f(r)\, X_j^{(m_j)\mp} \end{pmatrix}.$$

This is shown in the following *Note 1.*

NOTE 1. We must show that

$$\boldsymbol{\sigma} \cdot \boldsymbol{n}\, X_j^{(m_j)\pm} = X_j^{(m_j)\mp}.$$

Since $\boldsymbol{\sigma} \cdot \boldsymbol{n}$ commutes with M_i, $(i = 1, 2, 3,)$ $\boldsymbol{\sigma} \cdot \boldsymbol{n}\, X_j^{(m_j)\pm}$ has the same quantum numbers m_j and j as $X_j^{(m_j)\pm}$. From the fact that $\boldsymbol{\sigma} \cdot \boldsymbol{n}$ anti-commutes with K, it follows that $\boldsymbol{\sigma} \cdot \boldsymbol{n}\, X_j^{(m_j)\pm}$ is an eigenfunction of K belonging to an eigenvalue differing by a minus sign from that for $X_j^{(m_j)\pm}$.

Therefore, one may write

$$\boldsymbol{\sigma} \cdot \boldsymbol{n}\, X_j^{(m_j)\pm} = a_j^{(m_j)\pm}\, X_j^{(m_j)\mp} \,,$$

where $a_j^{(m_j)\pm}$ are constants to be determined.
It is easily seen that $a_j^{(m_j)\pm}$ does not depend on m_j. In fact, from

$$\boldsymbol{\sigma} \cdot \boldsymbol{n}(M_1 - \mathrm{i}\, M_2)\, X_j^{(m_j)\pm} = a_j^{(m_j)\pm}(M_1 - \mathrm{i}\, M_2)\, X_j^{(m_j)\mp} = \\ = a_j^{(m_j)\pm}\, \hbar[(j + m_j)(j - m_j + 1)]^{\frac{1}{2}}\, X_j^{(m_j-1)\mp}$$

and

$$\boldsymbol{\sigma} \cdot \boldsymbol{n}(M_1 - \mathrm{i}\, M_2)\, X_j^{(m_j)\pm} = \\ = \hbar[(j + m_j)(j - m_j + 1)]^{\frac{1}{2}}\, \boldsymbol{\sigma} \cdot \boldsymbol{n}\, X_j^{(m_j-1)\pm} = \\ = \hbar[(j + m_j)(j - m_j + 1)]^{\frac{1}{2}}\, a_j^{(m_j-1)\pm}\, X_j^{(m_j-1)\mp} \,,$$

it follows that

$$a_j^{(m_j)\pm} = a_j^{(m_j-1)\pm} \,.$$

Therefore one may write $a_j^{\pm}$ omitting the eigenvalue of M_3.

From the normalization of $X_j^{(m_j)\pm}$ one must have

$$\int (X_j^{(m_j)\mp})^\dagger\, \boldsymbol{\sigma} \cdot \boldsymbol{n}\, X_j^{(m_j)\pm}\, \mathrm{d}\, \Omega = a_j^{\pm} \,.$$

Hence, since $\boldsymbol{\sigma} \cdot \boldsymbol{n}$ is an Hermitian matrix,

$$a_j^{+} = (a_j^{-})^* \,.$$

On the other hand, since $(\boldsymbol{\sigma} \cdot \boldsymbol{n})^2 = 1$, one has

$$(\boldsymbol{\sigma} \cdot \boldsymbol{n})^2\, X_j^{(m_j)+} = a_j^{+}\, a_j^{-}\, X_j^{(m_j)+} \,, \quad \text{giving} \quad a_j^{+}\, a_j^{-} = 1 \,,$$

and finally

$$|a_j^{+}| = |a_j^{-}| = 1 \,.$$

In order to complete the determination of $a_j^{\pm}$, note that, from

$$\boldsymbol{\sigma} \cdot \boldsymbol{n}\, X_j^{(j)+} = \begin{pmatrix} n_3 & n_1 - \mathrm{i}n_2 \\ n_1 + \mathrm{i}n_2 & -n_3 \end{pmatrix} \begin{pmatrix} c_1^{+}\, Y_{j-\frac{1}{2}}^{(j-\frac{1}{2})} \\ 0 \end{pmatrix} = a_j^{+}\, X_j^{(j)-} = \\ = a_j^{+} \begin{pmatrix} c_1^{-}\, Y_{j+\frac{1}{2}}^{(j-\frac{1}{2})} \\ c_2^{-}\, Y_{j+\frac{1}{2}}^{(j+\frac{1}{2})} \end{pmatrix},$$

it follows that

$$c_1^{+}(n_1 + \mathrm{i}n_2)\, Y_{j-\frac{1}{2}}^{(j-\frac{1}{2})} = a_j^{+}\, c_2^{-}\, Y_{j+\frac{1}{2}}^{(j+\frac{1}{2})}. \tag{6}$$

A general property of the spherical harmonics is that $Y_l^{(l)} = (-1)^l \times \times C_l(n_1 + \mathrm{i}n_2)^l$, where C_l is a positive real constant. Combining this relation with (6) we obtain

$$c_1^+(n_1 + \mathrm{i}\, n_2)\, Y_{j-\frac{1}{2}}^{(j-\frac{1}{2})} = C_{j-\frac{1}{2}}\, c_1^+(-1)^{j-\frac{1}{2}}(n_1 + \mathrm{i}\, n_2)^{j+\frac{1}{2}} =$$
$$= (-1)^{j+\frac{1}{2}}\, c_2^-\, C_{j+\frac{1}{2}}(n_1 + \mathrm{i}\, n_2)^{j+\frac{1}{2}}\, a_j^+ ,$$

and because $c_1^+ > 0$, $c_2^- < 0$, one must have $a_j^+ > 0$.
This determines the values

$$a_j^+ = a_j^- = 1 .$$

2. The hydrogen atom

We shall now determine the eigenfunctions of the hydrogen atom by solving the radial equations for f and g.

Recalling the results of § 1, for a stationary solution of energy E eq. (4) becomes ($A_0 = Ze/4\pi r$)

$$\frac{1}{c}\begin{pmatrix}(E + e A_0 - mc^2) f\; X_j^{(m_j)+\varepsilon} \\ (E + e A_0 + mc^2)\, g\; X_j^{(m_j)-\varepsilon}\end{pmatrix} = -\,\mathrm{i}\,\hbar \begin{pmatrix} g'\; X_j^{(m_j)+\varepsilon} \\ f'\; X_j^{(m_j)-\varepsilon}\end{pmatrix} + \tag{7}$$
$$+ \frac{\mathrm{i}\,\hbar}{r}\begin{pmatrix} -(k+1)\, g\; X_j^{(m_j)+\varepsilon} \\ (k-1) f\; X_j^{(m_j)-\varepsilon}\end{pmatrix},$$

where $\varepsilon = \pm 1$ for $k \lessgtr 0$ respectively.

From eq. (7) one obtains a system of equations for $f(r)$ and $g(r)$

$$(E + e A_0 - mc^2) f = -\,\mathrm{i}\,\hbar c \left(g' + \frac{g}{r} + \frac{k}{r}\, g \right),$$
$$(E + e A_0 + mc^2)\, g = -\,\mathrm{i}\,\hbar c \left(f' + \frac{f}{r} - \frac{k}{r}\, f \right). \tag{8}$$

Normalizing the wave functions to one ($\int \psi^\dagger \psi\, \mathrm{d}^3 x = 1$) one must have, for the radial functions,

$$\int_0^\infty (|f|^2 + |g|^2)\, r^2\, \mathrm{d}r = 1 .$$

It is convenient to introduce the new functions $F = rf$ and $G = \mathrm{i}\, rg$. Eqs. (8) then become

$$(E + e A_0 - mc^2)\, F = -\,\hbar c \left(G' + \frac{k}{r}\, G \right),$$
$$(E + e A_0 + mc^2)\, G = \hbar c \left(F' - \frac{k}{r}\, F \right), \tag{9}$$

and the normalization condition reads

$$\int_0^\infty (|F|^2 + |G|^2)\, \mathrm{d}r = 1\,.$$

In analogy with the procedure followed for particles of spin zero, we introduce the new variable

$$\rho = 2r\sqrt{-A}\,,$$

with

$$A = (E^2 - E_0^2)/\hbar^2c^2\,.$$

Clearly, ρ is real for bound states.

Putting $\varepsilon \equiv E/E_0$, and writing F and G in the form

$$F = (1+\varepsilon)^{\frac{1}{2}}\, \mathrm{e}^{-\varrho/2}\,(u+v)\,,$$
$$G = (1-\varepsilon)^{\frac{1}{2}}\, \mathrm{e}^{-\varrho/2}\,(u-v)\,,$$

with $v \gg u$ for $\rho \to \infty$, the system of equations (9) becomes *

$$\begin{aligned} \mathrm{d}u/\mathrm{d}\rho &= [\rho - Z\alpha\,\varepsilon(1-\varepsilon^2)^{-\frac{1}{2}}](u/\rho) + [k - Z\alpha(1-\varepsilon^2)^{-\frac{1}{2}}](v/\rho)\,,\\ \mathrm{d}v/\mathrm{d}\rho &= Z\alpha\,\varepsilon(1-\varepsilon^2)^{-\frac{1}{2}}(v/\rho) + [k + Z\alpha(1-\varepsilon^2)^{-\frac{1}{2}}](u/\rho)\,. \end{aligned} \tag{10}$$

Inserting in (10) u and v expressed as power series

$$u = \rho^\lambda \sum a_\nu \rho^\nu\,, \quad v = \rho^\lambda \sum b_\nu \rho^\nu\,, \tag{11}$$

and comparing the coefficients of $\rho^{\nu+\lambda-1}$ on both sides, we have

$$\begin{aligned} (\nu+\lambda)a_\nu &= a_{\nu-1} - Z\alpha\,\varepsilon(1-\varepsilon^2)^{-\frac{1}{2}}\, a_\nu + [k - \alpha Z(1-\varepsilon^2)^{-\frac{1}{2}}]\, b_\nu,\\ (\nu+\lambda)b_\nu &= [k + \alpha Z(1-\varepsilon^2)^{-\frac{1}{2}}]\, a_\nu + \alpha Z\,\varepsilon(1-\varepsilon^2)^{-\frac{1}{2}}\, b_\nu\,. \end{aligned} \tag{12}$$

For $\nu = 0$, putting $a_{-1} = 0$, one finds two equations for a_0 and b_0, which have a solution other than $a_0 = b_0 = 0$ if

$$\begin{vmatrix} \lambda + \alpha Z\,\varepsilon(1-\varepsilon^2)^{-\frac{1}{2}} & -k + \alpha Z(1-\varepsilon^2)^{-\frac{1}{2}} \\ -k - \alpha Z(1-\varepsilon^2)^{-\frac{1}{2}} & \lambda - \alpha Z\,\varepsilon(1-\varepsilon^2)^{-\frac{1}{2}} \end{vmatrix} = 0\,,$$

* We note that from eq. (9) it follows, as $\rho \to \infty$, that

$$\frac{\mathrm{d}F}{\mathrm{d}\rho} \sim \tfrac{1}{2}\left(\frac{E_0+E}{E_0-E}\right)^{\frac{1}{2}} G, \qquad \frac{\mathrm{d}G}{\mathrm{d}\rho} \sim \tfrac{1}{2}\left(\frac{E_0-E}{E_0+E}\right)^{\frac{1}{2}} F, \qquad \frac{\mathrm{d}^2F}{\mathrm{d}\rho^2} \sim \tfrac{1}{4}\,F, \qquad \frac{\mathrm{d}^2G}{\mathrm{d}\rho^2} \sim \tfrac{1}{4}\,G,$$

whence

$$F \sim a\mathrm{e}^{-\varrho/2}, \qquad G \sim b\mathrm{e}^{-\varrho/2}, \qquad \text{with } a/b = -[(E_0+E)/(E_0-E)]^{\frac{1}{2}}\,.$$

As $\rho \to \infty$ the solutions behave like decreasing exponentials. (They would be oscillating functions if $E^2 > E_0^2$.)

whence *

$$\lambda = \pm (k^2 - \alpha^2 Z^2)^{\frac{1}{2}}.$$

The positive root must be chosen so that $\int_0^\infty (|f|^2 + |g|^2)\, r^2\, dr < \infty$.

From the second of eqs. (12), putting

$$n_r = -\lambda + \alpha Z\, \varepsilon (1 - \varepsilon^2)^{-\frac{1}{2}} = \alpha Z\, \varepsilon (1 - \varepsilon^2)^{-\frac{1}{2}} - (k^2 - \alpha^2 Z^2)^{\frac{1}{2}}$$

one has

$$\frac{b_\nu}{a_\nu} = -\frac{k + \alpha Z(1 - \varepsilon^2)^{-\frac{1}{2}}}{n_r - \nu}. \tag{13}$$

In the first of eqs. (12), b_ν may be replaced by its expression in terms of a_ν as given by eq. (13). Thus one finds the recurrence formula

$$a_\nu = \frac{-(n_r - \nu)\, a_{\nu-1}}{\nu(2\lambda + \nu)} = (-1)^\nu \frac{(n_r - 1) \dots (n_r - \nu)}{\nu!\,(2\lambda + 1) \dots (2\lambda + \nu)}\, a_0 .$$

Since, as follows from eq. (13),

$$b_\nu / a_\nu = n_r\, b_0 / (n_r - \nu)\, a_0,$$

one has also

$$b_\nu = (-1)^\nu \frac{n_r(n_r - 1) \dots (n_r - \nu + 1)}{\nu!\,(2\lambda + 1) \dots (2\lambda + \nu)}\, b_0 .$$

3. Energy levels

The series (11) must reduce to polynomials for the wave function to be square integrable, and so n_r must be a positive integer **. This gives

$$\varepsilon = E/mc^2 = \left[1 + \left(\frac{\alpha Z}{n_r + \lambda}\right)^2\right]^{-\frac{1}{2}},$$

or also, putting $n = n_r + |k|$, $(|k| = j + \frac{1}{2})$ ***,

$$E = mc^2 \left[1 + \left(\frac{\alpha Z}{n - |k| + (k^2 - \alpha^2 Z^2)^{\frac{1}{2}}}\right)^2\right]^{-\frac{1}{2}}. \tag{14}$$

Thus the energy levels are expressed as functions of n and $j + \frac{1}{2}$, by the same

* Remember that for particles of spin zero $\lambda = \pm [(l + \frac{1}{2})^2 - \alpha^2 Z^2]^{\frac{1}{2}} - \frac{1}{2}$.

** The case $n_r = 0$ is discussed in the article by Bethe and Salpeter, *loc. cit.*, p. 153 (for $n_r = 0$, k may not be negative).

*** For particles of spin zero, we had $n = n_r + l + 1$.

formula, both for positive and negative k. The expansion of (14) in a power series of $\alpha^2 Z^2$, terminated at the second term, coincides with eq. (13), Part I, Chap. VI.

For a given n, two cases are possible

$$\alpha)\ k > 0 \qquad \begin{array}{llll} k = 1, & 2, & \ldots & n, \\ j = \frac{1}{2}, & \frac{3}{2}, & \ldots & n - \frac{1}{2}, \\ l = 0, & 1, & \ldots & n - 1; \end{array}$$

$$\beta)\ k < 0 \qquad \begin{array}{llll} k = -1, & -2, & \ldots & -(n-1), \\ j = \ \frac{1}{2}, & \frac{3}{2}, & \ldots & n - \frac{3}{2}, \\ l = \ 1, & 2, & \ldots & n - 1. \end{array}$$

Note that $k = -n$ would correspond to $n_r = 0$ and $k < 0$, which is not allowed (see footnote ** on p. 204).

Thus an energy level of the non-relativistic theory splits into n sublevels depending on $|k|$. However, whereas in the non-relativistic theory neglecting the spin there are n^2 levels characterized by the same n and by different l and m, in Dirac's theory there are twice as many levels. In fact, for a given j, $2j + 1$ values of m_j are possible. Therefore, for $k = 1, 2, \ldots n$, one has 2, 4, ..., $2n$ states, respectively, giving altogether $n(n + 1)$ states, while for $k < 0$ one has 2, 4, ... $2(n - 1)$ states, giving altogether $n(n - 1)$. Thus, for a given n, there are $2n^2$ states.

The relativistic corrections always lower the levels from their Schroedinger-Rydberg values. For a given n the total splitting, i.e. the distance from the lowest to the highest sublevel, is, in units $mc^2\alpha^2$,

$$\Delta E = \tfrac{1}{2}\, \alpha^2 Z^4 (n - 1)/n^4 .$$

Fig. 7 shows the displacement of the levels of the hydrogen atom as given by eq. (14).

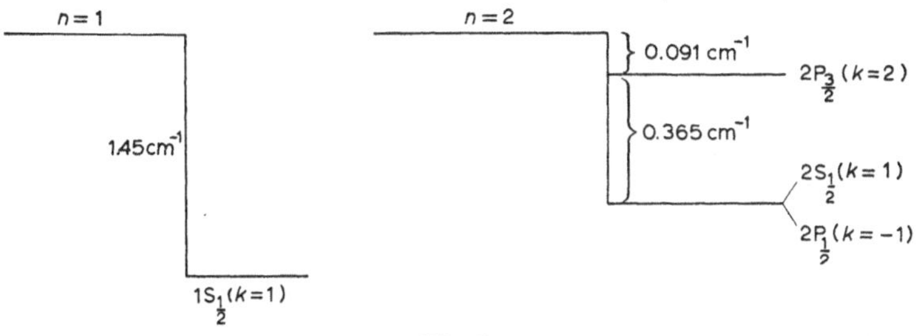

Fig. 7

We see that, according to Dirac's theory, the $2S_{\frac{1}{2}}$ and $2P_{\frac{1}{2}}$ sublevels, belonging to the same value of $|k|$, should be lowered by the same amount 0.456 cm^{-1}. Experiment shows, however, that the $2S_{\frac{1}{2}}$ sublevel is 1057.77 MHz higher than $2P_{\frac{1}{2}}$, the difference being about 10% the separation of $2P_{\frac{1}{2}}$ and $2P_{\frac{3}{2}}$. This effect (Lamb shift) has been explained by taking into account the interaction of the electron with the quantized electromagnetic field.

PART III

COLLISION AND RADIATION PROCESSES

CHAPTER I

TIME-INDEPENDENT SCATTERING OF A SPINLESS PARTICLE

1. Introduction

This chapter deals with the time-independent theory of scattering of a spin zero particle by a fixed centre of force, with which it is assumed to interact through the intermediacy of a static potential.

The non-relativistic and the relativistic theory, based on the use of the Schroedinger and of the Klein-Gordon equation respectively, are almost identical in structure and will be developed side by side.

In §§ 2, 3 and 4 the problem is formulated in the coordinate, momentum and angular momentum representations, respectively. In some particular calculations, of which a wealth of examples may be found in the classic book of Mott and Massey *, one of the representations may prove more convenient than the others.

2. Coordinate representation

The Schroedinger equation

$$-\frac{\hbar^2}{2m}\Delta\,\psi + V(\boldsymbol{x})\,\psi = \hbar\mathrm{i}\frac{\partial}{\partial t}\psi$$

and the Klein-Gordon equation

$$\left[\Delta - \frac{1}{c^2}\left(\frac{\partial}{\partial t} - \frac{\mathrm{i}e}{\hbar}A_0\right)^2 - \kappa^2 + \mathfrak{U}(\boldsymbol{x})\right]\psi = 0$$

($A_0 = A_0(\boldsymbol{x})$, an electrostatic potential, is the fourth component of a four-vector, $\mathfrak{U}(\boldsymbol{x})$ is a static scalar potential), for a stationary solution

$$\psi(\boldsymbol{x}, t) = \psi(\boldsymbol{x})\,\mathrm{e}^{-\mathrm{i}Et/\hbar}$$

reduce to the form

$$(\Delta + k^2)\,\psi(\boldsymbol{x}) = \mathfrak{B}(\boldsymbol{x})\,\psi(\boldsymbol{x}), \tag{1}$$

* N. F. Mott and H. S. W. Massey, *The Theory of Atomic Collisions* (Oxford, 1949).

where (Schroedinger)

$$k^2 = 2mE/\hbar^2 ,$$

$E \equiv$ total energy without rest energy,

$$\mathfrak{B}(x) = 2mV(x)/\hbar^2 ;$$

(Klein-Gordon)

$$k^2 = (E^2 - m^2c^4)/\hbar^2c^2 ,$$

$E \equiv$ total energy including rest energy,

$$\mathfrak{B}(x) = - \mathfrak{U}(x) - (e/\hbar c)^2 (A_0(x))^2 - 2(\hbar c)^{-2} E\, e\, A_0(x) .$$

It should be stressed that, for $eA_0(x) \neq 0$, the "potential" $\mathfrak{B}(x)$ of the Klein-Gordon equation is always energy dependent.

Eq. (1) is a partial differential equation, which may possess solutions representing bound states of the particle. Such solutions are square integrable, i.e. $\int |\psi(x)|^2\, d^3x$ is finite, and exist only for special values of E forming a discrete energy spectrum. However, our sole aim here is the study of solutions describing scattering processes, and these are not square integrable. For large values of $r = |x|$, they must consist of a plane wave, which represents the particle moving freely towards the scattering centre (*impinging* or *incident wave*), and an *outgoing wave* produced by the interaction.

By using the expression of Δ in polar coordinates, it is easy to see that, if $\mathfrak{B}(x)$ falls off at large distances more rapidly than $1/r$ (this excludes the Coulomb potential, which will be treated separately), eq. (1) has a solution with the asymptotic behaviour

$$(2\pi)^{\frac{3}{2}}\, \psi(x) \sim e^{ik \cdot x} + f(\theta)\, \frac{e^{ikr}}{r} , \tag{2}$$

which is precisely what we require.* Here θ is the angle between $\boldsymbol{k}$ and $\boldsymbol{x}$ and $k = |\boldsymbol{k}|$.

Note that the energy has a value that pertains to both the incident and the

* In fact, one has

$$\frac{1}{r}\frac{d^2}{dr^2}(r\psi) + k^2\psi - \left[\frac{1}{r^2}\Lambda(\theta, \phi) + \mathfrak{B}(x)\right]\psi = 0,$$

and, putting $r\psi = \varphi$,

$$\left(\frac{d^2}{dr^2} + k^2\right)\varphi - \left[\frac{1}{r^2}\Lambda(\theta, \phi) + \mathfrak{B}(x)\right]\varphi = 0 .$$

Then, if $\lim_{r\to\infty} r\mathfrak{B}(x) = 0$, asymptotically one has

$$\left(\frac{d^2}{dr^2} + k^2\right)(\varphi - r\, e^{i\, k\cdot x}) = 0 \qquad \text{i.e.} \qquad \varphi \sim r\, e^{i\, k\cdot x} + f(\theta)\, e^{ikr} .$$

outgoing wave. This can be expressed, more briefly, by saying that the energy is conserved at large distances from the region where the interaction is active.

The partial differential equation (1), with the prescribed asymptotic behaviour (2) for solutions, is equivalent to the integral equation

$$\psi(\boldsymbol{x}) = \psi^{(0)}(\boldsymbol{x}) + \int K(\boldsymbol{x}, \boldsymbol{x}')\, \psi(\boldsymbol{x}')\, \mathrm{d}^3 x' \,, \tag{3}$$

where

$$\psi^{(0)}(\boldsymbol{x}) = (2\pi)^{-\frac{3}{2}}\, \mathrm{e}^{\mathrm{i}\boldsymbol{k}\cdot\boldsymbol{x}}$$

and

$$K(\boldsymbol{x}, \boldsymbol{x}') \equiv - \frac{\mathrm{e}^{\mathrm{i}k|\boldsymbol{x}-\boldsymbol{x}'|}}{4\pi\, |\boldsymbol{x} - \boldsymbol{x}'|}\, \mathfrak{V}(\boldsymbol{x}') = -\, \mathfrak{G}^{(0)}(\boldsymbol{x}, \boldsymbol{x}')\, \mathfrak{V}(\boldsymbol{x}') \,. \tag{4}$$

In fact the Green's function $\mathfrak{G}^{(0)}(\boldsymbol{x}, \boldsymbol{x}')$ satisfies the equation

$$(\Delta + k^2)\, \frac{\mathrm{e}^{\mathrm{i}k|\boldsymbol{x}-\boldsymbol{x}'|}}{|\boldsymbol{x} - \boldsymbol{x}'|} = -\, 4\pi\, \delta(\boldsymbol{x} - \boldsymbol{x}') \,, \tag{5}$$

as follows from the Fourier representation of $\mathrm{e}^{\mathrm{i}kr}/r$, given in § 3. Thus, acting on both sides of eq. (3) with the operator $(\Delta + k^2)$, one finds eq. (1), since $(\Delta + k^2)\, \psi^{(0)}(\boldsymbol{x}) = 0$. On the other hand, assuming for simplicity that $\mathfrak{V}(\boldsymbol{x}) = 0$ for $r > r_0$ ($r_0 \equiv$ range of the interaction), eq. (3), with the expression (4) for K, gives *

$$\psi(\boldsymbol{x}) \sim (2\pi)^{-\frac{3}{2}}\, \mathrm{e}^{\mathrm{i}\boldsymbol{k}\cdot\boldsymbol{x}} - \frac{1}{4\pi}\frac{\mathrm{e}^{\mathrm{i}kr}}{r} \int \mathrm{e}^{-\mathrm{i}k\boldsymbol{n}\cdot\boldsymbol{x}'}\, \mathfrak{V}(\boldsymbol{x}')\, \psi(\boldsymbol{x}')\, \mathrm{d}^3 x'$$

(with $\boldsymbol{n} = \boldsymbol{x}/r$), which is of the form (2) with

$$f(\theta) = -\, (\pi/2)^{\frac{1}{2}} \int \mathrm{e}^{-\mathrm{i}k\boldsymbol{n}\cdot\boldsymbol{x}'}\, \mathfrak{V}(\boldsymbol{x}')\, \psi(\boldsymbol{x}')\, \mathrm{d}^3 x' \,.$$

It is worth noting that the replacement of $\mathfrak{G}^{(0)}(\boldsymbol{x}, \boldsymbol{x}')$, in eqs. (4) and (3), by its complex conjugate $\mathfrak{G}^{(0)*}(\boldsymbol{x}, \boldsymbol{x}')$, would amount to replacing $\mathrm{e}^{\mathrm{i}kr}$ by $\mathrm{e}^{-\mathrm{i}kr}$ in the second term of eq. (2), i.e. the outgoing wave by an incident wave. Thus $\mathfrak{G}^{(0)}$ and $\mathfrak{G}^{(0)*}$ are the Green's functions of two closely related problems.

3. Momentum representation

The wave function

$$\varphi(\boldsymbol{k}) = (2\pi)^{-\frac{3}{2}} \int \psi(\boldsymbol{x})\, \mathrm{e}^{-\mathrm{i}\boldsymbol{k}\cdot\boldsymbol{x}}\, \mathrm{d}^3 x$$

* In fact, expanding $|\boldsymbol{x} - \boldsymbol{x}'|/r$ in a power series of r'/r, one has

$$|\boldsymbol{x} - \boldsymbol{x}'| = r\left(1 - \frac{2\boldsymbol{x}\cdot\boldsymbol{x}'}{r^2} + \frac{\boldsymbol{x}'^2}{r^2}\right)^{\frac{1}{2}} = r\left[1 - \frac{\boldsymbol{n}\cdot\boldsymbol{x}'}{r} + O\left(\frac{r'^2}{r^2}\right) + \dots\right].$$

Then, as $r \to \infty$,

$$\mathrm{e}^{\mathrm{i}k|\boldsymbol{x}-\boldsymbol{x}'|}/|\boldsymbol{x} - \boldsymbol{x}'| \sim \mathrm{e}^{\mathrm{i}k(r - \boldsymbol{n}\cdot\boldsymbol{x}')}/r \,.$$

obeys the equation

$$(k^2 - k'^2)\,\varphi(\boldsymbol{k}') = \int (\boldsymbol{k}'\,|\mathfrak{V}|\,\boldsymbol{k}'')\,\varphi(\boldsymbol{k}'')\,\mathrm{d}^3\mathrm{k}'' , \tag{6}$$

where

$$(\boldsymbol{k}'\,|\mathfrak{V}|\,\boldsymbol{k}'') = (2\pi)^{-3} \int \mathrm{e}^{\mathrm{i}\,(\boldsymbol{k}''-\boldsymbol{k}')\cdot\boldsymbol{x}}\,\mathfrak{V}(\boldsymbol{x})\,\mathrm{d}^3\mathrm{x} ,$$

as can be seen by inserting

$$\psi(\boldsymbol{x}) = (2\pi)^{-\frac{3}{2}} \int \varphi(\boldsymbol{k})\,\mathrm{e}^{\mathrm{i}\boldsymbol{k}\cdot\boldsymbol{x}}\,\mathrm{d}^3\mathrm{k}$$

in eq. (1).

The integral equation (6), which is merely the Fourier transform of eq. (1), contains no indication as to the asymptotic behaviour which the solutions would exhibit, were they transformed to the coordinate representation.

In the momentum representation, a solution having the physical character described by (2), obeys the equation

$$\varphi(\boldsymbol{k}') = \delta(\boldsymbol{k}' - \boldsymbol{k}) + \int K(\boldsymbol{k}', \boldsymbol{k}'')\,\varphi(\boldsymbol{k}'')\,\mathrm{d}^3\mathrm{k}'' , \tag{7}$$

with

$$K(\boldsymbol{k}', \boldsymbol{k}'') = \int (\boldsymbol{k}'|\boldsymbol{x})\,K(\boldsymbol{x}, \boldsymbol{x}')(\boldsymbol{x}'|\boldsymbol{k}'')\,\mathrm{d}^3\mathrm{x}\,\mathrm{d}^3\mathrm{x}' , \tag{8}$$

$$(\boldsymbol{k}'|\boldsymbol{x}) = (\boldsymbol{x}|\boldsymbol{k}')^* = (2\pi)^{-\frac{3}{2}}\,\mathrm{e}^{-\mathrm{i}\boldsymbol{k}'\cdot\boldsymbol{x}} ,$$

as follows from eq. (3) by simple manipulations.

NOTE 1. It is possible to derive an explicit formula for $K(\boldsymbol{k}', \boldsymbol{k}'')$. For this purpose the equation

$$\mathrm{e}^{\mathrm{i}kr}/4\pi r = \lim_{\varepsilon\to 0^+} (2\pi)^{-3} \int \frac{\mathrm{e}^{\mathrm{i}\boldsymbol{k}'\cdot\boldsymbol{x}}}{k'^2 - k^2 - \mathrm{i}\,\varepsilon}\,\mathrm{d}^3\mathrm{k}' \tag{9}$$

must be used. This may be verified as follows. In the first place, performing the angular integration, one obtains

$$\lim_{\varepsilon\to 0^+} (2\pi)^{-2} \int_0^\infty \frac{k'^2\,\mathrm{d}k'}{k'^2 - k^2 - \mathrm{i}\,\varepsilon} \int_0^\pi \mathrm{e}^{\mathrm{i}k'r\cos\theta} \sin\theta\,\mathrm{d}\theta =$$

$$= \lim_{\varepsilon\to 0^+} (2\pi)^{-2} \int_0^\infty \frac{k'^2\,\mathrm{d}k'}{k'^2 - k^2 - \mathrm{i}\,\varepsilon} \cdot \frac{(\mathrm{e}^{\mathrm{i}k'r} - \mathrm{e}^{-\mathrm{i}k'r})}{\mathrm{i}\,k'r} \tag{α}$$

$$= \lim_{\varepsilon\to 0^+} \frac{1}{4\pi^2\,\mathrm{i}\,r} \int_{-\infty}^{+\infty} \frac{k'\,\mathrm{e}^{\mathrm{i}k'r}}{k'^2 - k^2 - \mathrm{i}\,\varepsilon}\,\mathrm{d}k' .$$

The integral over k' is easily calculated by transforming it into an integral along a closed contour in the complex k plane. This is done by adding to the integration path along the real axis a semi-circle of infinite radius in the upper half-plane. In fact the contribution from such a semi-circle vanishes, since

$$e^{ik'r} = e^{i(k_1 + i k_2)r} = e^{ik_1 r} e^{-k_2 r}$$

tends to zero as $k_2 \to \infty$. Then the integral is equal to the sum of the residues for all the poles of the integrand which are within the contour, i.e. in the upper half-plane. In our case, the poles of the integrand are at the points $k' = \pm (k^2 + i\,\varepsilon)^{\frac{1}{2}} = \pm (k + i\,\eta)$, with η real and infinitely small. Only the first of these is above the real axis, therefore the integral (α) reduces to the integral along a circle of infinitely small radius ρ with centre at $k' = k + i\,\eta$, which gives

$$\frac{1}{4\pi^2\, i\, r} \lim_{\eta \to 0^+} \int_{-\infty}^{+\infty} \frac{e^{ik'r}\, k'\, dk'}{(k' - k - i\,\eta)(k' + k + i\,\eta)} = \frac{e^{ikr}}{8\pi^2\, i\, r} \oint \frac{dk'}{k' - k} .$$

Putting $k' = \rho\, e^{i\theta} + k$, the right-hand side becomes

$$\frac{e^{ikr}}{8\pi^2\, i\, r} \int_0^{2\pi} \frac{i\rho\, e^{i\theta}\, d\theta}{\rho\, e^{i\theta}} = \frac{e^{ikr}}{4\pi r} ,$$

which proves eq. (9).

We could dispense with the ε in eq. (9) by following the prescription that the integration be performed along the path shown in Fig. 8a.

In analogy with (9), we have also

$$\frac{e^{-ikr}}{4\pi r} = \lim_{\varepsilon \to 0^+} (2\pi)^{-3} \int \frac{e^{ik' \cdot x}}{k'^2 - k^2 + i\,\varepsilon}\, d^3k' , \qquad (\beta)$$

where ε can be dropped if the integration is performed along the path shown in Fig. 8b.

The factors multiplying the exponentials within the integrals (9) and

Fig. 8

(β) are the Fourier transforms of the Green's functions $\mathfrak{G}^{(0)}(\boldsymbol{x}, \boldsymbol{x}')$ and $\mathfrak{G}^{(0)*}(\boldsymbol{x}, \boldsymbol{x}')$. These can also be written in the form

$$\frac{1}{k'^2 - k^2 \mp \mathrm{i}\,\varepsilon} = \pm\, 2\pi\, \mathrm{i}\, \delta_\pm(k'^2 - k^2)\,,$$

where

$$\delta_+(x) = \lim_{\varepsilon \to 0^+} \frac{1}{2\pi} \int_0^\infty \mathrm{e}^{-\mathrm{i}tx - \varepsilon t}\, \mathrm{d}t = \lim_{\varepsilon \to 0^+} \frac{-\,\mathrm{i}}{2\pi(x - \mathrm{i}\,\varepsilon)}\,, \tag{10}$$

$$\delta_-(x) = \lim_{\varepsilon \to 0^+} \frac{1}{2\pi} \int_{-\infty}^0 \mathrm{e}^{-\mathrm{i}tx + \varepsilon t}\, \mathrm{d}t = \lim_{\varepsilon \to 0^+} \frac{\mathrm{i}}{2\pi(x + \mathrm{i}\,\varepsilon)}\,. \tag{10'}$$

Note that

$$\delta_+(x) + \delta_-(x) = \frac{1}{2\pi} \int_{-\infty}^{+\infty} \mathrm{e}^{-\mathrm{i}tx}\, \mathrm{d}t = \lim_{\varepsilon \to 0^+} \frac{\varepsilon}{\pi(x^2 + \varepsilon^2)} = \delta(x)\,. \tag{11}$$

An alternative form for the transforms of $\mathfrak{G}^{(0)}(\boldsymbol{x}, \boldsymbol{x}')$ and $\mathfrak{G}^{(0)*}(\boldsymbol{x}, \boldsymbol{x}')$ is

$$\frac{1}{k'^2 - k^2 \mp \mathrm{i}\,\varepsilon} = P\left(\frac{1}{k'^2 - k^2}\right) \pm \mathrm{i}\,\pi\, \delta(k'^2 - k^2)\,, \tag{12}$$

where P denotes the principal value as defined by Cauchy,

$$\int_{-\infty}^{+\infty} P\left(\frac{1}{k'^2 - k^2}\right) \mathrm{d}k' \equiv P \int_{-\infty}^{+\infty} \frac{\mathrm{d}k'}{k'^2 - k^2} =$$

$$= \lim_{\varepsilon \to 0^+} \left(\int_{-\infty}^{-k-\varepsilon} + \int_{-k+\varepsilon}^{k-\varepsilon} + \int_{k+\varepsilon}^{+\infty} \right) \frac{\mathrm{d}k'}{k'^2 - k^2}\,.$$

The proof of (12) is based on the identity

$$\frac{1}{x \mp \mathrm{i}\,\varepsilon} = \frac{x \pm \mathrm{i}\,\varepsilon}{x^2 + \varepsilon^2} = \frac{x}{x^2 + \varepsilon^2} \pm \mathrm{i}\, \frac{\varepsilon}{x^2 + \varepsilon^2}\,.$$

In the limit $\varepsilon \to 0$ the real and imaginary parts possess the properties of a principal value and a δ function, respectively,

$$\lim_{\varepsilon\to 0^+}\int_{-a}^{+b}\frac{x}{x^2+\varepsilon^2}f(x)\,\mathrm{d}x = P\int_{-a}^{+b}\frac{f(x)}{x}\,\mathrm{d}x\,,$$

$$\lim_{\varepsilon\to 0^+}\frac{1}{\pi}\int_{-a}^{+b}\frac{\varepsilon}{\varepsilon^2+x^2}f(x)\,\mathrm{d}x = f(0)\,.$$

Using eqs. (4), (8) and (9) we find

$$K(\boldsymbol{k}',\boldsymbol{k}'') = -\frac{(\boldsymbol{k}'\,|\mathfrak{V}|\,\boldsymbol{k}'')}{k'^2-k^2-\mathrm{i}\,\varepsilon} = -\,2\pi\,\mathrm{i}\,\delta_+(k'^2-k^2)(\boldsymbol{k}'\,|\mathfrak{V}|\,\boldsymbol{k}''). \quad (13)$$

For reasons which will later become clear, it is customary to introduce in eq. (7) the notation

$$(\boldsymbol{k}'\,|U|\,\boldsymbol{k}) = \int(\boldsymbol{k}'\,|\mathfrak{V}|\,\boldsymbol{k}'')\,\varphi(\boldsymbol{k}'')\,\mathrm{d}^3\mathrm{k}'' = (2\pi)^{-\frac{3}{2}}\int \mathrm{e}^{-\mathrm{i}\boldsymbol{k}'\cdot\boldsymbol{x}}\,\mathfrak{V}(\boldsymbol{x})\,\psi(\boldsymbol{x})\,\mathrm{d}^3\mathrm{x}\,, \quad (14)$$

where the argument to the right of the matrix $(\boldsymbol{k}'\,|U|\,\boldsymbol{k})$ serves to remind us that we are concerned with solutions containing an incident wave with wavevector $\boldsymbol{k}$. Thus (7) reads *

$$\varphi(\boldsymbol{k}') = \delta(\boldsymbol{k}'-\boldsymbol{k}) - 2\pi\,\mathrm{i}\,\delta_+(k'^2-k^2)\,(\boldsymbol{k}'\,|U|\,\boldsymbol{k})\,. \quad (15)$$

An equation for the matrix $(\boldsymbol{k}'\,|U|\,\boldsymbol{k})$ can very easily be found by substituting (15) into (7), and by remembering that $x\delta(x) = 0$ and

$$2\pi\,\mathrm{i}\,x\,\delta_+(x) = x\left[P\left(\frac{1}{x}\right) + \mathrm{i}\,\pi\,\delta(x)\right] = 1\,.$$

Thus we obtain

$$(\boldsymbol{k}'\,|U|\,\boldsymbol{k}) = (\boldsymbol{k}'\,|\mathfrak{V}|\,\boldsymbol{k}) - 2\pi\,\mathrm{i}\int(\boldsymbol{k}'\,|\mathfrak{V}|\,\boldsymbol{k}'')\,\delta_+(k''^2-k^2)(\boldsymbol{k}''\,|U|\,\boldsymbol{k})\,\mathrm{d}^3\mathrm{k}''. \quad (16)$$

---

* Eq. (15) may also be put into the form

$$\varphi(\boldsymbol{k}') = \delta(\boldsymbol{k}'-\boldsymbol{k}) - 2\pi\mathrm{i}\delta_+(E'-E)(\boldsymbol{k}'\,|\,T\,|\,\boldsymbol{k})\,,$$

with

$$(\boldsymbol{k}'\,|\,T\,|\,\boldsymbol{k}) = \hbar^2(\boldsymbol{k}'\,|\,U\,|\,\boldsymbol{k})/2m \qquad \text{(Schroedinger)},$$

and

$$(\boldsymbol{k}'\,|\,T\,|\,\boldsymbol{k}) = \frac{\hbar^2c^2}{E'+E}(\boldsymbol{k}'\,|\,U\,|\,\boldsymbol{k}) \qquad \text{(Klein-Gordon)}\,.$$

This is useful for comparison with the general theory of collisions (see B. A. Lippmann and J. Schwinger, Phys. Rev. **79** (1950), 469).

4. Angular momentum representation (central potentials)

In this section we shall confine ourselves to central potentials by assuming that $\mathfrak{B}$ depends only on the distance from the scattering centre. We shall therefore write

$$\mathfrak{B} = \mathfrak{B}(r) \, .$$

The use of the angular momentum representation is especially convenient in this case.

The transformation function from the coordinate to the angular momentum representation is

$$(\boldsymbol{x}'|rlm) = \frac{1}{r}\, \delta(r - |\boldsymbol{x}'|)\, Y_l^{(m)}(\theta', \phi') \, ,$$

where θ' and ϕ' are the polar coordinates of $\boldsymbol{x}'$, and $Y_l^{(m)}$ is a spherical harmonic.

We now transform eq. (3) by multiplication with the transformation function $(rlm|\boldsymbol{x})$ and integration over $\boldsymbol{x}$. Here we shall use the shorter notation

$$\Phi_{lm}(r) = (2\pi)^{\frac{3}{2}} \int (rlm|\boldsymbol{x})\, \psi(\boldsymbol{x})\, \mathrm{d}^3\mathrm{x} \tag{17}$$

for the wave function in the new representation, multiplied by $(2\pi)^{\frac{3}{2}}$, and conversely

$$\psi(\boldsymbol{x}) = (2\pi)^{-\frac{3}{2}} \sum_{l,m} \int_0^\infty (\boldsymbol{x}|rlm)\, \Phi_{lm}(r)\, \mathrm{d}r \, , \tag{18}$$

having employed

$$\int_0^\infty \mathrm{d}r \sum_{l,m} (\boldsymbol{x}'|rlm)(rlm|\boldsymbol{x}'') = \delta(\boldsymbol{x}' - \boldsymbol{x}'') \, .$$

Thus we have

$$\Phi_{lm}(r) = \int (rlm|\boldsymbol{x})\, \mathrm{e}^{\mathrm{i}\boldsymbol{k}\cdot\boldsymbol{x}}\, \mathrm{d}^3\mathrm{x} - \sum_{l',m'} \int_0^\infty (rlm|\,\mathfrak{G}^{(0)}|r'l'm')\, \mathfrak{B}(r')\, \Phi_{l'm'}(r')\, \mathrm{d}r' \, , \tag{19}$$

where

$$(rlm|\,\mathfrak{G}^{(0)}|r'l'm') = \int (rlm|\boldsymbol{x}'')\, \mathfrak{G}^{(0)}(\boldsymbol{x}'', \boldsymbol{x}''')(\boldsymbol{x}'''|r'l'm')\, \mathrm{d}^3\mathrm{x}''\, \mathrm{d}^3\mathrm{x}''' \, .$$

A notable simplification is achieved by using the formula *

* Use the expansion of a spherical wave in a series of Legendre polynomials (see e.g. A. Sommerfeld, *Partial Differential Equations in Physics*, (Acad. Press, New York, 1949) p. 145):

$$\mathfrak{G}^{(0)}(\boldsymbol{x}, \boldsymbol{x}') = \tfrac{1}{4} \sum_{l=0}^{\infty} (l + \tfrac{1}{2})\, P_l(\cos\Theta)\, \frac{1}{\sqrt{rr'}}\, J_{l+\frac{1}{2}}(kr)\left[(-1)^l J_{-l-\frac{1}{2}}(kr') + \mathrm{i}\, J_{l+\frac{1}{2}}(kr')\right],$$

$$r' > r \, ,$$

where Θ is the angle between $\boldsymbol{x}$ and $\boldsymbol{x}'$, and employ the addition theorem of spherical

$$(rlm|\, \mathfrak{G}^{(0)}|r'l'm') = \delta_{ll'}\, \delta_{mm'}\, \mathfrak{G}_l^{(0)}(r, r')\,,$$

where

$$\mathfrak{G}_l^{(0)}(r,r') = \tfrac{1}{2}\pi\sqrt{rr'}\, J_{l+\frac{1}{2}}(kr)\left[(-1)^l J_{-l-\frac{1}{2}}(kr') + \mathrm{i}\, J_{l+\frac{1}{2}}(kr')\right], \quad (r' > r)\,. \tag{20}$$

For $r' < r$, $\mathfrak{G}_l{}^{(0)}(r, r')$ is defined as the right-hand side of this equation with r and r' interchanged. The functions $J_{l+\frac{1}{2}}$ and $J_{-l-\frac{1}{2}}$ are Bessel functions.

Moreover, if the momentum of the impinging wave is in the direction of the z axis, we have

$$\int (rlm|\boldsymbol{x})\, \mathrm{e}^{\mathrm{i}k\cdot x}\, \mathrm{d}^3\mathrm{x} = (2\pi^2(2l+1)r/k)^{\frac{1}{2}}\, \mathrm{i}^l\, J_{l+\frac{1}{2}}(kr)\, \delta_{m,0}\,,$$

and (19) becomes

$$\Phi_{lm}(r) = \left[2\pi^2(2l+1)r/k\right]^{\frac{1}{2}}\, \mathrm{i}^l\, J_{l+\frac{1}{2}}(kr)\, \delta_{m,0} - \\ - \int_0^\infty \mathfrak{G}_l^{(0)}(r, r')\, \mathfrak{V}(r')\, \Phi_{lm}(r')\, \mathrm{d}r'\,. \tag{21}$$

This is an inhomogeneous equation for $\Phi_{l0}(r)$ and a homogeneous one for $\Phi_{lm}(r) (m > 0)$. It is clear from eq. (17) and from the cylindrical symmetry around the z axis, that all $\Phi_{lm}(r)$ vanish except $\Phi_{l0}(r)$. The latter is conveniently replaced by

$$\Phi_l(r) = k\, \Phi_{l0}(r)/\mathrm{i}^l \left[4\pi(2l+1)\right]^{\frac{1}{2}}\,,$$

which obeys the equation

$$\Phi_l(r) = (\pi kr/2)^{\frac{1}{2}}\, J_{l+\frac{1}{2}}(kr) - \int_0^\infty \mathfrak{G}_l^{(0)}(r, r')\, \mathfrak{V}(r')\, \Phi_l(r')\, \mathrm{d}r'\,. \tag{22}$$

This integral equation for $\Phi_l(r)$ is the analogue of eqs. (3) and (15), in so far as it implies the asymptotic behaviour (2) at large distances in the coordinate representation.

A differential equation for $\Phi_l(r)$ can be obtained from eq. (22). In fact, applying the differential operator

$$\left(\frac{\mathrm{d}^2}{\mathrm{d}r^2} + k^2 - \frac{l(l+1)}{r^2}\right) \tag{23}$$

---

harmonics

$$P_l(\cos\Theta) = \sum_{m=-l}^{+l} \frac{4\pi}{2l+1}\, Y_l{}^{(m)*}(\theta, \phi)\, Y_l{}^{(m)}(\theta', \phi')\,,$$

where Θ is the angle between the directions fixed by the polar angles θ, ϕ and θ', ϕ'.

to both members, the first term on the right-hand side gives zero, while the second term, by virtue of the equation *

$$\left(\frac{d^2}{dr^2} + k^2 - \frac{l(l+1)}{r^2}\right) \mathfrak{G}_l^{(0)}(r, r') = -\,\delta(r - r')\,, \tag{24}$$

yields $\mathfrak{V}\,\Phi_l$. Thus we obtain

$$\Phi_l'' + \left(k^2 - \frac{l(l+1)}{r^2} - \mathfrak{V}\right)\Phi_l = 0\,.$$

This equation clearly contains less than (22), as the asymptotic behaviour of $\Phi_l(r)$ for large r must be suitably assigned in order to reproduce (2). What this asymptotic behaviour must be is seen from eq. (22), which, for large r, reduces to

$$\Phi_l(r) \sim (1 + \mathrm{i}\,I)\sin(kr - \tfrac{1}{2}l\pi) + I\cos(kr - \tfrac{1}{2}l\pi)\,,$$

with

$$I = -\int_0^\infty (\pi r'/2k)^{\frac{1}{2}}\, J_{l+\frac{1}{2}}(kr')\, \mathfrak{V}(r')\, \Phi_l(r')\, dr'\,.$$

This is more conveniently written in the two alternative forms

$$\Phi_l(r) \sim e^{\mathrm{i}\eta_l}\sin(kr - \tfrac{1}{2}l\pi + \eta_l)\,,$$

$$\Phi_l(r) \sim \sin(kr - \tfrac{1}{2}l\pi) + \frac{S_l - 1}{2\mathrm{i}}\, e^{\mathrm{i}kr - \frac{1}{2}\mathrm{i}l\pi}\,, \tag{25}$$

where η_l and S_l are defined by

$$e^{\mathrm{i}\eta_l}\sin\eta_l = I\,, \qquad S_l = e^{2\mathrm{i}\eta_l}\,. \tag{26}$$

The constants η_l are known as the *phase shifts* (or, simply, *phases*), and S_l are the eigenvalues of the S-matrix (see § 8).

* In fact, for $r \neq r'$ the operator (23) gives zero when applied to (20), since

$$\left[\frac{d^2}{dr^2} + k^2 - \frac{l(l+1)}{r^2}\right]\begin{bmatrix}\sqrt{r}\,J_{l+\frac{1}{2}}\\ \sqrt{r}\,J_{-l-\frac{1}{2}}\end{bmatrix} = 0\,.$$

In order to exhibit the δ-like behaviour for $r = r'$, it is convenient to multiply eq. (24) by a regular function $f(r')$ and integrate with respect to r' from $r - \varepsilon$ to $r + \varepsilon$, letting ε tend to zero at the end.

Using the identity

$$\sqrt{x}\,J_{-l-\frac{1}{2}}(x)\frac{d}{dx}[\sqrt{x}\,J_{l+\frac{1}{2}}(x)] - \sqrt{x}\,J_{l+\frac{1}{2}}(x)\frac{d}{dx}[\sqrt{x}\,J_{-l-\frac{1}{2}}(x)] = 2(-1)^l/\pi$$

for the Bessel functions, which is a consequence of $\sqrt{x}\,J_{l+\frac{1}{2}}(x)$ and $\sqrt{x}\,J_{-l-\frac{1}{2}}(x)$ obeying the same differential equation, the integral on the left-hand side gives $-f(r)$.

Now we shall verify that the asymptotic behaviour (25) for $\Phi_l(r)$ actually corresponds to the asymptotic behaviour (2) for $\psi(\boldsymbol{x})$. From eq. (18) we have

$$(2\pi)^{\frac{3}{2}}\,\psi(\boldsymbol{x}) = (1/kr)\sum_{l=0}^{\infty}(2l+1)\,\mathrm{i}^l\,P_l(\cos\theta)\,\Phi_l(r)\,.$$

If now, for large r, $\Phi_l(r)$ behaves according to (25), we have

$$(2\pi)^{\frac{3}{2}}\,\psi(\boldsymbol{x}) \sim \sum_{l=0}^{\infty}(2l+1)\,P_l(\cos\theta)\,\frac{(-1)^{l+1}\,\mathrm{e}^{-\mathrm{i}kr}+\mathrm{e}^{\mathrm{i}kr+2\mathrm{i}\eta_l}}{2\,\mathrm{i}\,kr} =$$

$$= \sum_{l=0}^{\infty}(2l+1)\,P_l(\cos\theta)\,\frac{\mathrm{e}^{\mathrm{i}kr}-(-1)^l\,\mathrm{e}^{-\mathrm{i}kr}}{2\,\mathrm{i}\,kr} +$$

$$+\frac{\mathrm{e}^{\mathrm{i}kr}}{2\mathrm{i}kr}\sum_{l=0}^{\infty}(2l+1)(\mathrm{e}^{2\mathrm{i}\eta_l}-1)\,P_l(\cos\theta)\,.$$

While the first term on the right can easily be recognized as the asymptotic expression for large r of the expansion of $\mathrm{e}^{\mathrm{i}kr\cos\theta}$ in a series of Legendre polynomials, corresponding to the first term in eq. (2), the second term corresponds to the second term in eq. (2) with

$$f(\theta) = \frac{1}{2\mathrm{i}k}\sum_{l=0}^{\infty}(2l+1)(\mathrm{e}^{2\mathrm{i}\eta_l}-1)\,P_l(\cos\theta)\,. \tag{27}$$

This formula establishes the relationship between the descriptions of scattering processes in the coordinate and in the angular momentum representation.

NOTE 2. Instead of the function $\Phi_l(r)$, one often encounters in literature the function $\varphi_l(r)$, which obeys the integral equation

$$\varphi_l(r) = (\pi kr/2)^{\frac{1}{2}}\,J_{l+\frac{1}{2}}(kr) - \int_0^{\infty} g_l(r,r')\,\mathfrak{V}(r')\,\varphi_l(r')\,\mathrm{d}r'\,, \tag{28}$$

where now

$$g_l(r,r') = \tfrac{1}{2}(-1)^l\pi\sqrt{rr'}\,J_{l+\frac{1}{2}}(kr)\,J_{-l-\frac{1}{2}}(kr')\,,\quad (r'>r)\,.$$

(For $r'<r$, r and r' are interchanged on the right-hand side.)

For $l=0$, one has

$$g_0(r, r') = \frac{1}{k} \sin(kr) \cos(kr') , \qquad (r' > r) .$$

Like $\mathfrak{G}_l^{(0)}(r, r')$, $g_l(r, r')$ obeys the equation

$$\left(\frac{\mathrm{d}^2}{\mathrm{d}r^2} + k^2 - \frac{l(l+1)}{r^2}\right) g_l(r, r') = -\,\delta(r - r') ,$$

so that $\varphi_l(r)$ satisfies the differential equation

$$\varphi_l'' + \left(k^2 - \frac{l(l+1)}{r^2} - \mathfrak{V}\right) \varphi_l = 0 .$$

The asymptotic behaviour for large r is, however, different from (25), as can be seen from eq. (28) which yields,

$$\varphi_l(r) \sim \frac{1}{\cos \eta_l} \sin\left(kr - \frac{l\pi}{2} + \eta_l\right) =$$

$$= \sin\left(kr - \frac{l\pi}{2}\right) + \mathrm{tg}\, \eta_l \cos\left(kr - \frac{l\pi}{2}\right) ,$$

with

$$k \,\mathrm{tg}\, \eta_l = - \int_0^\infty (\pi k r'/2)^{\frac{1}{2}} J_{l+\frac{1}{2}}(kr') \mathfrak{V}(r') \varphi_l(r') \,\mathrm{d}r' . \qquad (29)$$

The phase shifts η_l in these formulae coincide with those defined by eq. (26), as can be seen from the fact that $\varphi_l(r)$ and $\Phi_l(r)$ are solutions of the same differential equation both vanishing for $r = 0$.

Thus we have the relation

$$\Phi_l(r) = \mathrm{e}^{\mathrm{i}\eta_l} \cos \eta_l \quad \varphi_l(r)$$

between these functions, which have different values of the derivative at $r = 0$.

5. Scattering amplitudes and cross sections in the coordinate representation

So far we have merely collected a set of equations for wave functions all corresponding to the asymptotic behaviour (2) at large distances from the scattering centre.

We shall now regard the problem from a more physical viewpoint, basing our discussion on the coordinate representation.

We consider the current density

$$\boldsymbol{j} = (\hbar/2m\mathrm{i})(\psi^* \nabla_x \psi - (\nabla_x \psi^*)\,\psi)\,,$$

constructed from the wave function $\psi(\boldsymbol{x})$. It obeys the continuity equation

$$\operatorname{div} \boldsymbol{j} = 0\,,$$

as can be seen from eq. (1) (for $\mathfrak{V}(\boldsymbol{x})$ real). The operations performed by the experimental physicist consist essentially in measuring the flux of $\boldsymbol{j}$ through a surface element dS, normal to $\boldsymbol{x}$, at a large distance from the scattering centre. This is given by $j_r\, r^2\, \mathrm{d}\Omega$, where

$$j_r = \frac{\hbar}{2m\mathrm{i}}\left(\psi^* \frac{\partial \psi}{\partial r} - \frac{\partial \psi^*}{\partial r}\,\psi\right), \tag{30}$$

and dΩ is the solid angle spanned by dS.

Since j_r is needed only for large r, we can insert the asymptotic expression (2) in (30), which gives

$$j_r \sim (2\pi)^{-3}\frac{\hbar k}{m} \times \\ \times \left\{\cos\theta + \frac{1}{r^2}\,|f(\theta)|^2 + \frac{1}{2r}(1+\cos\theta)\left[f(\theta)\,\mathrm{e}^{\mathrm{i}kr(1-\cos\theta)} + \text{c.c.}\right]\right\} \tag{31}$$

The right-hand member consists only of three terms. The first originates from the impinging wave $(2\pi)^{-\frac{3}{2}}\,\mathrm{e}^{\mathrm{i}\boldsymbol{k}\cdot\boldsymbol{x}}$; its total flux through a sphere is equal to zero. The second term is what is loosely called the scattered current, and comes from $(2\pi)^{-\frac{3}{2}} f(\theta)\,\mathrm{e}^{\mathrm{i}kr}/r$. The third term, which arises from interference of the incident and scattered waves, is a rapidly oscillating function of r for $r \gg 1/k$, except when θ is very small. Thus, for sufficiently large angles, only the first two terms are of importance, as the third term averages out to zero over the measuring apparatus.

The quantity

$$(2\pi)^{-3}\frac{\hbar k}{m}\,|f(\theta)|^2\,\mathrm{d}\Omega$$

represents the additional flux, through the surface element $\mathrm{d}S = r^2\,\mathrm{d}\Omega$, resulting from the scattering process. Divided by the current density $(2\pi)^{-3}\hbar k/m$ of the impinging wave in the forward direction, it defines the *differential cross section*

$$\mathrm{d}\sigma(\theta) = |f(\theta)|^2\,\mathrm{d}\Omega\,,$$

which has the physical dimensions of the square of a length.

The quantity

$$\sigma = 2\pi \int_0^{\pi} |f(\theta)|^2 \sin\theta \, \mathrm{d}\theta$$

is known as the *total cross section*.

As the continuity equation must be satisfied, the flux of $\boldsymbol{j}$ through a sphere must be zero. But the flux of the first term on the right in eq. (31) has been seen to vanish. Therefore, the fluxes of the second and third terms must cancel out, so that the total cross section can be expressed in the form

$$\sigma = \int |f(\theta)|^2 \, \mathrm{d}\Omega =$$
$$= -\tfrac{1}{2}r \left[\int_{-1}^{1} \int_0^{2\pi} (1+x) f(\cos^{-1} x)\, \mathrm{e}^{\mathrm{i}kr(1-x)} \, \mathrm{d}x \, \mathrm{d}\phi + \text{complex conjugate} \right], \quad (32)$$

with $x = \cos\theta$. If r is large enough, the x integration over any domain which does not include the value $x = 1$ ($\theta = 0$) vanishes, as can be seen by partial integration with respect to x (see below). This is why, for large r and for sufficiently large angles, the interference terms could be neglected, and the current assumed to consist of the current of the impinging wave and that of the outgoing wave. This is not true for small angles. The interference terms at small angles are indispensable in order to satisfy eq. (32).

We now carry out the partial integration

$$-\frac{r}{2} \int_{-1}^{1} \int_0^{2\pi} (1+x) f \, \mathrm{e}^{\mathrm{i}kr(1-x)} \, \mathrm{d}x \, \mathrm{d}\phi =$$
$$= -\int_0^{2\pi} \mathrm{d}\phi \left\{ \left[\frac{\mathrm{i}(1+x)}{2k} f \mathrm{e}^{\mathrm{i}kr(1-x)} \right]_{-1}^{+1} - \int_{-1}^{+1} \frac{\mathrm{i}}{2k} \mathrm{e}^{\mathrm{i}kr(1-x)} \frac{\mathrm{d}}{\mathrm{d}x} [(1+x) f] \, \mathrm{d}x \right\}.$$

The second term on the right is of the order of $1/r$. The first term gives $2\pi f(0)/\mathrm{i}\,k$.

Thus finally we find

$$\sigma = \int |f(\theta)|^2 \, \mathrm{d}\Omega = (2\pi/\mathrm{i}\,k)[f(0) - f^*(0)] = (4\pi/k) \,\mathrm{Im} f(0). \quad (33)$$

This important relation, connecting the total cross section with the scattering amplitude in the forward direction, is called the *optical theorem*.

NOTE 3. Eq. (33) would not hold if the potential $\mathfrak{V}(\boldsymbol{x})$ were not real. Complex potentials are used to describe the action of centres of force which, besides scattering particles, are also capable of absorbing or emitting them.

For a complex potential the continuity equation holds in the form

$$\operatorname{div} \boldsymbol{j} = \tilde{\rho} = (\hbar/2mi)\, \psi^* (\mathfrak{B} - \mathfrak{B}^*)\, \psi\,,$$

where $\tilde{\rho}$ describes a source or a sink.

When inelastic processes are possible, eq. (33) becomes

$$\sigma = \sigma_{\text{el}} + \sigma_{\text{inel}} = (4\pi/k)\, \operatorname{Im} f_{\text{el}}(0)\,,$$

where σ_{el} and σ_{inel} are, respectively, the elastic and the inelastic total cross section, and $f_{\text{el}}(0)$ is the elastic scattering amplitude in the forward direction. We prove this for the case of absorption.

The absorption cross section may be defined as the flux of the current $\boldsymbol{j}$ through a sphere with centre at the scatterer, divided by the current density for the incident wave. Introducing a minus sign since the flux is ingoing, one has

$$\sigma_{\text{abs}} = (\mathrm{i}/2k)\,(2\pi)^{-3} \lim_{r\to\infty} \int_0^r \int \psi^* (\mathfrak{B} - \mathfrak{B}^*)\, \psi\, r'^2\, \mathrm{d}r'\, \mathrm{d}\Omega'\,.$$

In this case the imaginary part of $\mathfrak{B}$ must be chosen so that σ_{abs} turns out to be positive.

Using the continuity equation in its new form, one has

$$\sigma_{\text{el}} + \sigma_{\text{abs}} + \lim_{r\to\infty} \left[\frac{r}{2} \int_{-1}^{+1} \int_0^{2\pi} (1 + x) f\, \mathrm{e}^{\mathrm{i}kr(1-x)}\, \mathrm{d}x\, \mathrm{d}\phi + \text{c.c.} \right] = 0\,,$$

and so

$$\sigma_{\text{el}} + \sigma_{\text{abs}} = (4\,\pi/k)\, \operatorname{Im} f_{\text{el}}\,(0).$$

On the other hand, one can see that when, and only when, (27) obeys the indentity (33) the phase shifts are real.

In fact one has

$$\sigma = \int |f(\theta)|^2\, \mathrm{d}\Omega =$$

$$= \frac{\pi}{2k^2} \sum_{l=0}^{\infty} \sum_{l'=0}^{\infty} (2l + 1)(2l' + 1)(\mathrm{e}^{2\mathrm{i}\eta_l} - 1)(\mathrm{e}^{-2\mathrm{i}\eta_{l'}^*} - 1) \int_0^{\pi} P_l P_{l'} \sin\theta\, \mathrm{d}\theta$$

and this, according to (33), must be equal to

$$2\pi\, k^{-2} \operatorname{Im} \left[\frac{1}{\mathrm{i}} \sum_{l=0}^{\infty} (2l + 1)(\mathrm{e}^{2\mathrm{i}\eta_l} - 1)\, P_l(1) \right].$$

Since

$$P_l(1) = 1\,, \qquad \int_0^\pi P_l P_{l'} \sin\theta\, \mathrm{d}\theta = 2(2l+1)^{-1}\,\delta_{ll'}\,,$$

this equation gives

$$2 - \mathrm{e}^{2\mathrm{i}\eta_l} - \mathrm{e}^{-2\mathrm{i}\eta_l^*} = \mathrm{e}^{2\mathrm{i}(\eta_l-\eta_l^*)} - \mathrm{e}^{-2\mathrm{i}\eta_l^*} - \mathrm{e}^{2\mathrm{i}\eta_l} + 1\,,$$

namely

$$\eta_l = \eta_l^*\,,$$

so that the phase shifts are real. On the other hand, in the case of a complex potential, where (33) is not valid, the phase shifts are complex.

6. Scattering amplitudes and cross sections in the momentum and angular momentum representations

The cross section can easily be expressed in the momentum representation by finding a relation between $f(\theta)$ and the scattering matrix $(\boldsymbol{k}'\,|U|\,\boldsymbol{k})$. We proceed by retransforming eq. (15) into the coordinate representation. The transform of the first term is the incident wave, i.e. the first term of (2). We shall expand the transform of the second term in powers of $1/r$, and derive an expression for $f(\theta)$ by identifying the first term of the expansion with $(2\pi)^{-\frac{3}{2}} f(\theta)\,\mathrm{e}^{\mathrm{i}kr}/r$.

The quantity to be expanded is

$$-(2\pi)^{-\frac{1}{2}}\,\mathrm{i} \int \mathrm{e}^{\mathrm{i}\boldsymbol{k}'\cdot\boldsymbol{x}}\,(\boldsymbol{k}'\,|U|\,\boldsymbol{k})\;\delta_+(\boldsymbol{k}'^2 - \boldsymbol{k}^2)\,\mathrm{d}^3\mathrm{k}'\,.$$

Taking polar coordinates around the $\boldsymbol{x}$ direction, $\boldsymbol{k}'$ and $\boldsymbol{k}$ will be represented by (k', ω', ϕ') and (k, ω, ϕ) respectively. Therefore, we can write the above integral in the form

$$-(2\pi)^{-\frac{3}{2}} \int \frac{\mathrm{e}^{\mathrm{i}k'r\cos\omega'}}{k'^2 - k^2 - \mathrm{i}\,\varepsilon} \times$$

$$\times\; U(k'\sin\omega'\cos\phi', k'\sin\omega'\sin\phi', k'\cos\omega'; k\sin\omega\cos\phi, k\sin\omega\sin\phi, k\cos\omega)\times$$
$$\times\; k'^2\,\mathrm{d}k'\,\sin\omega'\,\mathrm{d}\omega'\,\mathrm{d}\phi'\,.$$

By partial integration with respect to ω' one finds

$$\int_0^\pi \mathrm{e}^{\mathrm{i}k'r\cos\omega'}\,U(k'\sin\omega'\cos\phi', k'\sin\omega'\sin\phi', k'\cos\omega'; \boldsymbol{k})\sin\omega'\,\mathrm{d}\omega' =$$
$$= (1/\mathrm{i}\,k'r)\left[\mathrm{e}^{\mathrm{i}k'r}\,U(0,0,k';\boldsymbol{k}) - \mathrm{e}^{-\mathrm{i}k'r}\,U(0,0,-k';\boldsymbol{k})\right] + O(1/r^2)\,.$$

On dropping the $1/r^2$ terms, the integration over ϕ' can be performed in the original integral, and gives

$$(2\pi)^{-\frac{1}{2}}\,\mathrm{i}\,r^{-1}\int_{-\infty}^{+\infty}\frac{k'\,\mathrm{e}^{\mathrm{i}k'r}}{k'^2-k^2-\mathrm{i}\,\varepsilon}\,U(0,0,k';\boldsymbol{k})\,\mathrm{d}k'\,.$$

Now the integration with respect to k' can be performed in the same way as that for the Fourier transform of $\mathfrak{G}^{(0)}(\boldsymbol{x},\boldsymbol{x}')$ in § 3. One obtains

$$-\,(2\pi)^{\frac{1}{2}}\,(2r)^{-1}\,\mathrm{e}^{\mathrm{i}kr}\,U(0,0,k;\boldsymbol{k})\,. \tag{34}$$

This shows that only that part of $(\boldsymbol{k}'\,|U|\,\boldsymbol{k})$ is needed, for which $k'^2=k^2$, i.e. the elements of the matrix which are diagonal with respect to the energy, or, as is often said, *on the energy shell*. Such matrix elements will be denoted by $(\boldsymbol{k}'\,|\underline{U}|\,\boldsymbol{k})$.

From eqs. (2) and (34) we see that

$$f(\theta)=-\,2\pi^2(\boldsymbol{k}'\,|\underline{U}|\,\boldsymbol{k})\,, \tag{35}$$

where the matrix element is taken for $k'^2=k^2$ (as indicated by underlining U) and for a direction of $\boldsymbol{k}'$ such that $\cos\theta=\boldsymbol{k}\cdot\boldsymbol{k}'/k^2$.

The differential cross section can be written in the two equivalent forms

$$\mathrm{d}\sigma(\theta)=|f(\theta)|^2\,\mathrm{d}\Omega=4\pi^4\,|(\boldsymbol{k}'\,|\underline{U}|\,\boldsymbol{k})|^2\,\mathrm{d}\Omega\,, \tag{36}$$

and, using eq. (27) (central potentials), can also be expressed in the angular momentum representation as

$$\mathrm{d}\sigma(\theta)=(2k)^{-2}\,\Big|\sum_{l=0}^{\infty}(2l+1)(\mathrm{e}^{2\mathrm{i}\eta_l}-1)\,P_l(\cos\theta)\Big|^2\,\mathrm{d}\Omega\,. \tag{37}$$

Finally, note the formula

$$\sigma=4\pi\,k^{-2}\sum_{l=0}^{\infty}(2l+1)\sin^2\eta_l$$

for the total cross section, which results immediately from eq. (37) and from the orthonormality properties of the Legendre polynomials.

7. Calculation of phase shifts from experimental cross sections

In the discussion of experimental results, the suitability of a certain potential $\mathfrak{V}(r)$ may be tested by comparing the theoretical cross section (36) directly with the experimental one. On the other hand, a direct comparison of (37) with the experimental cross section is not very convenient. It is often more useful * to find a set of phase shifts which reproduces the experimental angular distribution, and compare these empirical phase shifts with the theoretical ones, evaluated by means of eqs. (26), (29).

In other words, let us assume that the experimental $I(\theta) = |f_{\exp}(\theta)|^2$ is given as an expansion

$$I(\theta) = \sum_{s=0}^{\infty} c_s P_s(\cos\theta) ,$$

with

$$c_s = (s + \tfrac{1}{2}) \int_0^{\pi} I(\theta) P_s(\cos\theta) \sin\theta \, d\theta ,$$

and that $f_{\exp}(\theta)$ be of the form (27). The experimental phase shifts are then found by equating $I(\theta)\, d\Omega$ to (37), which gives

$$\sum_{s=0}^{\infty} c_s P_s(\cos\theta) = \tag{38}$$

$$= k^{-2} \sum_{m,n=0}^{\infty} (2m+1)(2n+1) \sin\eta_m \sin\eta_n \cos(\eta_m - \eta_n) P_m(\cos\theta) P_n(\cos\theta) .$$

Using the formula ($m \leq n$)

$$P_m(z) P_n(z) = \sum_{r=0}^{m} \frac{A_{m-r} A_r A_{n-r}}{A_{m+n-r}} \cdot \frac{2m+2n-4r+1}{2m+2n-2r+1} \cdot P_{m+n-2r}(z) ,$$

with

$$A_m = 1.3.5. \dots (2m-1)/m! ,$$

and putting

$$a_{mn} = a_{nm} = k^{-2}(2m+1)(2n+1) \sin\eta_m \sin\eta_n \cos(\eta_m - \eta_n) ,$$

$$\alpha_{mn,r} = \alpha_{nm,r} = \frac{A_{m-r} A_r A_{n-r}}{A_{m+n-r}} \cdot \frac{2m+2n-4r+1}{2m+2n-2r+1} ,$$

we see that the right-hand member of eq. (38) can be written as

$$\sum_{m,n=0}^{\infty} a_{mn} \sum_{r=0}^{\min[m,n]} \alpha_{mn,r} P_{m+n-2r}(\cos\theta) ,$$

* Especially when the phase shifts decrease rapidly with l, $\eta_0 \gg \eta_1 \gg \eta_2 \dots$ This is the case in low-energy collisions.

and, by comparison with the left-hand side, that it must be

$$c_s = \sum_{r=0}^{\infty} \sum_{m=r}^{r+s} a_{m,2r+s-m}\, \alpha_{m,2r+s-m,r}\,.$$

This is a system of algebraic equations

$$\begin{aligned} k^2 c_0 &= \sin^2 \eta_0 + 3 \sin^2 \eta_1 + 5 \sin^2 \eta_2 + \dots, \\ k^2 c_1 &= 6 \sin \eta_0 \sin \eta_1 \cos(\eta_0 - \eta_1) + 12 \sin \eta_1 \sin \eta_2 \cos(\eta_1 - \eta_2) + \dots, \\ k^2 c_2 &= 6 \sin^2 \eta_1 + 10 \sin \eta_0 \sin \eta_2 \cos(\eta_0 - \eta_2) + \dots, \\ &\dots\dots\dots\dots \end{aligned} \tag{39}$$

Note first that, if $\eta_0, \eta_1, \eta_2 \dots$ is a solution of this system of equations, $-\eta_0, -\eta_1, -\eta_2 \dots$ is also one. Therefore the experimental phase shifts are not determined uniquely by the experimental cross section.

If all the phase shifts are small and, furthermore, $\eta_0 \gg \eta_1 \gg \eta_2 \dots$, eqs. (39) can be solved by a perturbation method. In the first approximation

$$\sin \eta_0 = k \sqrt{c_0}\,.$$

Putting $\cos(\eta_0 - \eta_1) = 1$, we get from the second equation

$$\sin \eta_1 = k c_1 / 6 \sqrt{c_0}\,.$$

Then $\sin \eta_2$ is obtained from the third equation, and so on. The second approximation for $\sin \eta_0$ is calculated from the first equation by replacing $\sin \eta_1$ by its first approximation and neglecting higher order terms,

$$\sin^2 \eta_0 = k^2 \left(c_0 - \frac{c_1^2}{12\, c_0} \right),$$

etc.

8. The *S*-matrix

Most important in the general theory of collisions is the *S*-matrix. For the problem of scattering of one particle by a fixed centre, it is defined, in the momentum representation, by

$$(\boldsymbol{k}' |S| \boldsymbol{k}) = \delta(\boldsymbol{k}' - \boldsymbol{k}) - 2\pi \mathrm{i}\, \delta(k'^2 - k^2)(\boldsymbol{k}' |\underline{U}| \boldsymbol{k})\,.$$

The right-hand side consists of two terms, of which the first is the unit matrix and the second vanishes outside the energy shell.

By means of the transformation function

$$(\boldsymbol{k}' | klm) = k^{-1}\, \delta(|\boldsymbol{k}'| - k)\, Y_l^{(m)}(\boldsymbol{k}')\,,$$

we now go over to the (k, l, m) representation. Thus, employing the formula

$$P_l(\cos \widehat{\boldsymbol{kk}'}) = 4\pi \sum_{m=-l}^{l} (2l+1)^{-1}\, Y_l^{(m)*}(\boldsymbol{k})\, Y_l^{(m)}(\boldsymbol{k}')\,,$$

we obtain (central potentials)

$$(k'l'm'\,|S|\,klm) = \delta(k' - k)\,\delta_{l'l}\,\delta_{m'm}\,S_l\,,$$

with $S_l = e^{2i\eta_l}$.

This shows that the S-matrix is diagonal in the (k, l, m) representation, and also, how the phase shifts are related to its eigenvalues.

From the form of the eigenvalues S_l and the fact that the phase shifts are real for real potentials, it follows that the S-matrix must be unitary. Conversely the reality of the phase shifts follows from the S-matrix being unitary. In the momentum representation this property is expressed by

$$(\boldsymbol{k}\,|SS^\dagger|\,\boldsymbol{k}') = (\boldsymbol{k}\,|S^\dagger S|\,\boldsymbol{k}') = \delta(\boldsymbol{k} - \boldsymbol{k}')\,.$$

Therefore, we can write

$$\begin{aligned}\delta(\boldsymbol{k} - \boldsymbol{k}') &= \int (\boldsymbol{k}\,|S|\,\boldsymbol{k}'')(\boldsymbol{k}''\,|S^\dagger|\,\boldsymbol{k}')\,d^3k'' = \\ &= \delta(\boldsymbol{k} - \boldsymbol{k}') + 2\pi i\delta(k^2 - k'^2)\big[(\boldsymbol{k}\,|\underline{U}^\dagger|\,\boldsymbol{k}') - (\boldsymbol{k}\,|\underline{U}|\,\boldsymbol{k}')\big] + \\ &\quad + 2\pi^2 k\delta(k^2 - k'^2)\int (\boldsymbol{k}\,|\underline{U}|\,\boldsymbol{k}'')(\boldsymbol{k}''\,|\underline{U}^\dagger|\,\boldsymbol{k}')\,d\Omega''\,,\end{aligned}$$

where $d\Omega''$ is an element of solid angle in the direction of $\boldsymbol{k}''$. Introducing the scattering amplitude $f(\theta)$ by eq. (35), and letting $\boldsymbol{k}'$ coincide with $\boldsymbol{k}$ (scattering in the forward direction!), we obtain the optical theorem

$$4\pi\,\mathrm{Im} f(0) = k \int |f(\theta)|^2\, d\Omega\,,$$

from which the reality of the phase shifts for real central potentials, and therefore the unitarity of the S-matrix, has been deduced in § 5.

9. Exact theory for non-relativistic Coulomb scattering

In this section we treat the scattering of a particle of charge $Z_1 e$ by a fixed centre of charge $Z_2 e$, according to the Schroedinger equation (for the scattering of an electron by a nucleus, $Z_1 = -1$ and $Z_2 = Z$).

In this case the Schroedinger equation can be solved exactly. The solutions cannot have the asymptotic behaviour (2), since $V(\boldsymbol{x})$ does not fall off sufficiently rapidly at large distances.

9.1. Coordinate representation

The Schroedinger equation for

$$V(\boldsymbol{x}) = Z_1 Z_2\, e^2/4\pi r\,,$$

reads

$$\left(\Delta + k^2 - \frac{2m}{\hbar^2}\frac{Z_1 Z_2\, e^2}{4\pi r}\right)\psi(\boldsymbol{x}) = 0\,,$$

and has the exact solution

$$(2\pi)^{\frac{3}{2}}\,\psi(\boldsymbol{x}) = \mathrm{e}^{-\frac{1}{2}\pi n}\,\Gamma(1 + \mathrm{i}\, n)\,\mathrm{e}^{\mathrm{i}\boldsymbol{k}\cdot\boldsymbol{x}}\,F(-\,\mathrm{i}\, n, 1, \mathrm{i}\,\rho)\,, \tag{40}$$

where $\rho = kr - \boldsymbol{k}\cdot\boldsymbol{x}$, $n = Z_1 Z_2\, e^2\, m/4\pi\hbar^2 k$.

The function F in this formula is of the hypergeometric type. It can be expressed either as a series

$$F(-\,\mathrm{i}\, n, 1, \mathrm{i}\,\rho) = 1 + (1\,!)^{-2}\, n\rho + (2\,!)^{-2}\, n(n+\mathrm{i})\,\rho^2 + (3\,!)^{-2}\, n(n+\mathrm{i})(n+2\mathrm{i})\,\rho^3 + \ldots,$$

or as the sum of two functions

$$F(-\,\mathrm{i}\, n, 1, \mathrm{i}\,\rho) = W_1(-\,\mathrm{i}\, n, 1, \mathrm{i}\,\rho) + W_2(-\,\mathrm{i}\, n, 1, \mathrm{i}\,\rho)\,, \tag{41}$$

having the integral representation

$$W_s(-\,\mathrm{i}\, n, 1, \mathrm{i}\,\rho) = \frac{1}{2\pi\mathrm{i}}\int_{\gamma_s}\left(1 - \frac{\mathrm{i}\,\rho}{t}\right)^{\mathrm{i}n}\frac{\mathrm{e}^t}{t}\,\mathrm{d}t\,, \qquad (s = 1, 2)\,. \tag{42}$$

The paths of integration γ_1, γ_2 in the complex t plane start at, and return to $-\,\infty$, and encircle the points $t = 0$ and $t = \mathrm{i}\,\rho$ as indicated in Fig. 9.

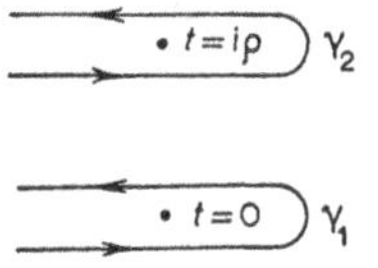

Fig. 9

For large ρ these functions have the asymptotic behaviour

$$W_1 \sim \frac{1}{\Gamma(1+\mathrm{i}\, n)}(-\,\mathrm{i}\,\rho)^{\mathrm{i}n}\left[1 + \frac{n^2}{1!\,\mathrm{i}\,\rho} + \frac{n^2(n+\mathrm{i})^2}{2!(\mathrm{i}\,\rho)^2} + \ldots\right],$$

$$W_2 \sim \frac{1}{\Gamma(-\,\mathrm{i}\, n)}\,\mathrm{e}^{\mathrm{i}\rho}(\mathrm{i}\,\rho)^{-\mathrm{i}n-1}\left[1 + \frac{(1+\mathrm{i}\, n)^2}{1!\,\mathrm{i}\,\rho} + \frac{(1+\mathrm{i}\, n)^2(2+\mathrm{i}\, n)^2}{2!\,(\mathrm{i}\rho)^2} + \ldots\right].$$

This shows that the asymptotic behaviour of $\psi(\boldsymbol{x})$ for large r is

$$(2\pi)^{\frac{3}{2}}\,\psi(\boldsymbol{x}) \sim \mathrm{e}^{\mathrm{i}[\boldsymbol{k}\cdot\boldsymbol{x} + n\log(kr - \boldsymbol{k}\cdot\boldsymbol{x})]}\left(1 + \frac{n^2}{\mathrm{i}(kr - \boldsymbol{k}\cdot\boldsymbol{x})}\right) + f_{\mathrm{c}}(\theta)\,\frac{\mathrm{e}^{\mathrm{i}(kr - n\log kr)}}{r}\,,$$

with

$$f_c(\theta) = (n/2k)\,\mathrm{cosec}^2\,(\tfrac{1}{2}\theta)\,\exp\left[-\mathrm{i}\,n\log(1-\cos\theta) + \mathrm{i}\,\pi + 2\mathrm{i}\eta_0^{(c)}\right],$$
$$\exp 2\mathrm{i}\eta_0^{(c)} = \Gamma(1+\mathrm{i}\,n)/\Gamma(1-\mathrm{i}\,n)\,. \tag{43}$$

It differs considerably from (2), as the first term is a distorted plane wave and the second term also presents a distortion, though only in the form of a phase factor.

The differential cross section $\mathrm{d}\sigma(\theta)$ may be evaluated by following the procedure of § 5:

$$\mathrm{d}\sigma(\theta) = |f_c(\theta)|^2\,\mathrm{d}\Omega = (n/2k)^2\,\mathrm{cosec}^4(\tfrac{1}{2}\theta)\,\mathrm{d}\Omega\,. \tag{44}$$

Note that the total cross section is infinite, as can be shown by performing the angular integration, or by having recourse to (33).

9.2. *Momentum representation*

It is of interest to go over to the momentum representation and derive the wave function $\varphi(\boldsymbol{k}') = (2\pi)^{-\frac{3}{2}} \int \psi(\boldsymbol{x})\,\mathrm{e}^{-\mathrm{i}\boldsymbol{k}'\cdot\boldsymbol{x}}\,\mathrm{d}^3\mathrm{x}$.

The momentum wave function $\varphi(\boldsymbol{k})$ for Coulomb scattering has been given explicitly by Guth and Mullin *, and can be obtained following a method used by Sommerfeld ** in a somewhat different context.

We must evaluate the integral

$$X = \int \mathrm{e}^{-\mathrm{i}\boldsymbol{k}'\cdot\boldsymbol{x}}\,\mathrm{e}^{\mathrm{i}\boldsymbol{k}\cdot\boldsymbol{x}}\,F(-\mathrm{i}\,n, 1, \mathrm{i}\,kr - \mathrm{i}\,\boldsymbol{k}\cdot\boldsymbol{x})\,\mathrm{d}^3\mathrm{x}\,.$$

Starting with the representation of the F function

$$F(-\mathrm{i}\,n, 1, \mathrm{i}\,\rho) = \frac{1}{2\pi\mathrm{i}}\oint\left(1-\frac{\mathrm{i}\,\rho}{t}\right)^{\mathrm{i}n}\frac{\mathrm{e}^t}{t}\,\mathrm{d}t\,,$$

which is obtained from (41) and (42) by combining the integration paths γ_1 and γ_2 to a single closed path encircling the points $t = 0$ and $t = \mathrm{i}\,\rho$, a simple change of the integration variable leads to the more convenient form

$$F(-\mathrm{i}\,n, 1, \mathrm{i}\,\rho) = \frac{1}{2\pi\mathrm{i}}\oint\frac{(v+1)^{\mathrm{i}n}}{v^{\mathrm{i}n+1}}\,\mathrm{e}^{-\mathrm{i}v\varrho}\,\mathrm{d}v\,, \tag{45}$$

where, now, the integration path encircles the points $v = 0$ and $v = -1$.

Thus we have

$$X = \frac{1}{2\pi\mathrm{i}}\oint\frac{(v+1)^{\mathrm{i}n}}{v^{\mathrm{i}n+1}}\,\mathrm{d}v\int\mathrm{e}^{-\mathrm{i}krv}\,\mathrm{e}^{\mathrm{i}(\boldsymbol{k}-\boldsymbol{k}')\cdot\boldsymbol{x}+\mathrm{i}v\boldsymbol{k}\cdot\boldsymbol{x}}\,\mathrm{d}^3\mathrm{x}\,.$$

---

* E. Guth and C. J. Mullin, Phys. Rev. (L) **83** (1951), 667.

** A. Sommerfeld, *Atombau und Spektrallinien*, Band II (Braunschweig, 1951), p. 456.

In order to make the second integral convergent, it is necessary to replace $ikrv$ by $ikrv + \varepsilon r$ ($\varepsilon > 0$), which amounts to replacing k by $k - \mathrm{i}(\varepsilon/v)$. As the v integration path can be made to coincide with the negative real axis except at the points $v = 0$ and $v = -1$, this is essentially equivalent to giving k a small positive imaginary part.

Thus

$$X = -(2\pi)^{-1}\,\mathrm{i} \oint (v+1)^{\mathrm{i}n}\, v^{-(\mathrm{i}n+1)}\, \mathrm{d}v \int \mathrm{e}^{-K_0 r}\, \mathrm{e}^{\mathrm{i}\boldsymbol{K}\cdot\boldsymbol{x}}\, \mathrm{d}^3\mathrm{x}\,,$$

which can be written more conveniently

$$X = (2\pi)^{-1}\,\mathrm{i} \oint (v+1)^{\mathrm{i}n}\, v^{-\mathrm{i}n-1}\, \mathrm{d}v \lim_{\varepsilon\to 0} \frac{\partial}{\partial\varepsilon} \int \frac{1}{r}\, \mathrm{e}^{\mathrm{i}\boldsymbol{K}\cdot\boldsymbol{x} - K_0 r}\, \mathrm{d}^3\mathrm{x} =$$

$$= 2\mathrm{i} \oint (v+1)^{\mathrm{i}n}\, v^{-\mathrm{i}n-1}\, \mathrm{d}v \lim_{\varepsilon\to 0} \frac{\partial}{\partial\varepsilon} (K_0^2 + \boldsymbol{K}^2)^{-1}\,,$$

with $\boldsymbol{K} = (\boldsymbol{k} - \boldsymbol{k}') + v\boldsymbol{k}$, $K_0 = \mathrm{i}\,kv + \varepsilon$.

Putting now $K_0^2 + \boldsymbol{K}^2 = U(v - v_0)$ with $U = 2\boldsymbol{k}\cdot(\boldsymbol{k} - \boldsymbol{k}') + 2\,\mathrm{i}\,\varepsilon\, k$, $U v_0 = -(\boldsymbol{k} - \boldsymbol{k}')^2 - \varepsilon^2$, one has

$$X = 2\mathrm{i} \lim_{\varepsilon\to 0} \frac{\partial}{\partial\varepsilon} \oint \frac{(v+1)^{\mathrm{i}n}}{v^{\mathrm{i}n+1}} \frac{\mathrm{d}v}{U(v - v_0)}\,.$$

The integral with respect to v can be reduced to the residue around the pole v_0 which lies outside the integration path. This is due to the fact that the integrand decreases at infinity as $1/v^2$, so that the original integration path encircling the points $v = 0$ and $v = -1$ can be completed with a circle of infinite radius, together with which it forms a path encircling v_0.

Thus finally we have

$$X = 4\pi \lim_{\varepsilon\to 0} \frac{\partial}{\partial\varepsilon} \left[(Uv_0 + U)^{\mathrm{i}n}\,(Uv_0)^{-\mathrm{i}n-1}\right],$$

so that

$$\varphi(\boldsymbol{k}') =$$

$$= -\frac{1}{2\pi^2}\, \mathrm{e}^{-\frac{1}{2}\pi n}\, \Gamma(1 + \mathrm{i}n) \lim_{\varepsilon\to 0} \frac{\partial}{\partial\varepsilon} \left\{[k'^2 + (\varepsilon - \mathrm{i}\,k)^2]^{\mathrm{i}n} / [\varepsilon^2 + (\boldsymbol{k}' - \boldsymbol{k})^2]^{\mathrm{i}n+1}\right\}.$$

To compare this with eq. (15), we note that two terms result from differentiation with respect to ε. The one which arises from the differentiation of the denominator corresponds to $\delta(\boldsymbol{k} - \boldsymbol{k}')$ of eq. (15), while the derivative of the numerator gives the analogue of the second term.

Employing the formula

$$\delta(\boldsymbol{x}) = \frac{1}{\pi^2} \lim_{\varepsilon\to 0} \frac{\varepsilon}{(\varepsilon^2 + \boldsymbol{x}^2)^2},$$

and eqs. (10), (11), we obtain

$$\varphi(\boldsymbol{k}') = \left\{ \Gamma(2 + \mathrm{i}\, n)\, \mathrm{e}^{-\frac{1}{2}\pi n} \left[\frac{k'^2 - k^2}{(\boldsymbol{k}' - \boldsymbol{k})^2} \right]^{\mathrm{i}n} \right\} \delta(\boldsymbol{k}' - \boldsymbol{k})$$

$$- 2\pi\mathrm{i}\, (kn/\pi^2) \left\{ \Gamma(1 + \mathrm{i}\, n)\, \mathrm{e}^{-\frac{1}{2}\pi n} \frac{(k'^2 - k^2)^{\mathrm{i}n}}{(\boldsymbol{k}' - \boldsymbol{k})^{2(\mathrm{i}n+1)}} \right\} \delta_+(k'^2 - k^2)\,.$$

Of special interest is the limiting case for small n. The first Born approximation consists in expanding $\varphi(\boldsymbol{k}')$ in powers of n and neglecting all terms of order higher than n,

$$\varphi(\boldsymbol{k}') = \delta(\boldsymbol{k}' - \boldsymbol{k}) - 2\pi\,\mathrm{i} \left[\frac{kn}{\pi^2(\boldsymbol{k}' - \boldsymbol{k})^2} \right] \delta_+(k'^2 - k^2) + \ldots$$

This is of the form (15) with

$$(\boldsymbol{k}'\,|U|\,\boldsymbol{k}) = \frac{kn}{\pi^2(\boldsymbol{k}' - \boldsymbol{k})^2}\,.$$

The scattering cross section, as given by eq. (36), is

$$\mathrm{d}\sigma_{\mathrm{Born}}(\theta) = (n/2k)^2 \operatorname{cosec}^4(\tfrac{1}{2}\theta)\, \mathrm{d}\Omega\,,$$

and coincides with (44). The fact that the first Born approximation yields the exact value for the cross section is a peculiarity of the Coulomb potential scattering.

9.3. *Angular momentum representation*

Finally, a few words about the angular momentum representation may be useful.

From eqs. (40) and (45), with the help of the well known expansion of a plane wave, we have

$$(2\pi)^{\frac{3}{2}}\, \psi(\boldsymbol{x}) = \mathrm{e}^{-\frac{1}{2}\pi n}\, \Gamma(1 + \mathrm{i}\, n) \frac{1}{2\pi\mathrm{i}} (\pi/2kr)^{\frac{1}{2}} \sum_{l=0}^{\infty} \mathrm{i}^l (2l + 1) P_l(\cos\theta) \times$$

$$\times \oint (v + 1)^{\mathrm{i}n - \frac{1}{2}}\, v^{-(\mathrm{i}n+1)}\, \mathrm{e}^{-\mathrm{i}vkr} J_{l+\frac{1}{2}}(kr(1 + v))\, \mathrm{d}v\,. \qquad (46)$$

Substituting the relation

$$J_{l+\frac{1}{2}}(kr) = (2kr/\pi)^{\frac{1}{2}} \frac{\Gamma(l + 1)}{(2l + 1)!} (2kr)^l\, \mathrm{e}^{\mathrm{i}kr} F(l + 1, 2l + 2, -2\mathrm{i}\, kr)$$

into (46), we have

$$(2\pi)^{\frac{3}{2}}\,\psi(\boldsymbol{x}) = \mathrm{e}^{-\frac{1}{2}\pi n}\Gamma\,(1+\mathrm{i}\,n)\,\frac{1}{2\pi\mathrm{i}}\sum_{l=0}^{\infty}\mathrm{i}^l\,\frac{l!}{(2l)!}\,(2kr)^l\,P_l(\cos\theta)\,\mathrm{e}^{\mathrm{i}kr}\times$$

$$\times\oint(v+1)^{\mathrm{i}n+l}\,v^{-\mathrm{i}n-1}\,F(l+1,2l+2,-2\mathrm{i}\,kr(1+v))\,\mathrm{d}v\,.$$

Now, considering the general term of the expansion of $F(l+1, 2l+2, -2\mathrm{i}\,kr(1+v))$ and using Euler's integral of the first kind

$$\Gamma(\beta)/\Gamma(1-\alpha)\,\Gamma(\alpha+\beta) = \frac{1}{2\pi\,\mathrm{i}}\oint v^{\alpha-1}(1+v)^{\beta-1}\,\mathrm{d}v\,,\qquad(\mathrm{Re}\,\beta>0)\,,$$

it is easy to prove the formula

$$F(\mathrm{i}\,n+l+1,2l+2,-2\mathrm{i}\,kr) =$$

$$=\frac{\Gamma(1+\mathrm{i}\,n)\,\Gamma(l+1)}{2\pi\mathrm{i}\,\Gamma(\mathrm{i}\,n+l+1)}\oint F(l+1,\,2l+2,\,-2\mathrm{i}\,kr(1+v))\,\frac{(1+v)^{\mathrm{i}n+l}}{v^{\mathrm{i}n+1}}\,\mathrm{d}v\,.$$

Employing this, we obtain

$$(2\pi)^{\frac{3}{2}}\,\psi(\boldsymbol{x}) = \mathrm{e}^{-\frac{1}{2}\pi n}\,\mathrm{e}^{\mathrm{i}kr}\sum_{l=0}^{\infty}\frac{\mathrm{i}^l}{(2l)!}\times$$

$$\times\,(2kr)^l\,\Gamma(\mathrm{i}\,n+l+1)\,P_l(\cos\theta)\,F(\mathrm{i}\,n+l+1,2l+2,-2\mathrm{i}\,kr)\,,$$

which can be written as

$$(2\pi)^{\frac{3}{2}}\,\psi(\boldsymbol{x}) = \sum_{l=0}^{\infty}(2l+1)\,\mathrm{i}^l\,\mathrm{e}^{\mathrm{i}\eta_l{}^{(c)}}\,L_l(r)\,P_l(\cos\theta)\,,$$

with $\eta_l^{(c)} = \arg\Gamma(l+1+\mathrm{i}\,n)$ and

$$L_l(r) = \mathrm{e}^{-\frac{1}{2}\pi n}\,\frac{|\Gamma(l+1+\mathrm{i}\,n)|}{(2l+1)!}\,(2kr)^l\,\mathrm{e}^{\mathrm{i}kr}\;F(\mathrm{i}\,n+l+1,2l+2,-2\mathrm{i}\,kr)\,.$$

The function $L_l(r)$ has the asymptotic behaviour

$$L_l(r)\sim\frac{1}{kr}\sin(kr-\tfrac{1}{2}l\pi+\eta_l^{(c)}-n\log 2kr)\,,$$

which differs by the logarithmic term from that of the radial functions for scattering by short range forces. Expressed in terms of the phase shifts, $f_c(\theta)$ has the form

$$f_c(\theta) = (1/2ik) \sum_{l=0}^{\infty} (2l+1)\,(e^{2i\eta_l^{(c)}} - 1)\, P_l(\cos\theta)\,.$$

This formula, though merely a complicated way of writing a function which is already known in the simpler form (43), is often useful.

In cases where A_0 deviates from the Coulomb form at short distances (for instance, for electrostatic potentials of nuclei) and/or a short range potential is present, the scattering wave function has the asymptotic behaviour

$$(2\pi)^{\frac{3}{2}}\, \psi(\boldsymbol{x}) \sim \exp\left[i(\boldsymbol{k}\cdot\boldsymbol{x} + n\log(kr - \boldsymbol{k}\cdot\boldsymbol{x}))\right]\left(1 + \frac{n^2}{i(kr - \boldsymbol{k}\cdot\boldsymbol{x})}\right) +$$

$$+ f(\theta)\,\frac{1}{r}\,e^{i(kr - n\log kr)}\,.$$

This is also the asymptotic behaviour of the scattering solution of the Klein-Gordon equation of § 2, with $A_0 = Z_1 Z_2 e^2/4\pi r$ and $\mathfrak{U}(\boldsymbol{x}) = 0$.

10. Born expansions

From now on we shall write $g\mathfrak{V}$ instead of $\mathfrak{V}$. The constant g, which determines the strength of the potential, will be assumed small.

The Born expansion consists in expressing the scattering amplitude as a power series in g. This is achieved by a process of iteration of one of the integral equations (3), (15), (22), according to the representation in which one is working.

10.1. Coordinate and momentum representations

By iterating (3) we obtain

$$(2\pi)^{\frac{3}{2}}\, \psi(\boldsymbol{x}) = e^{i\boldsymbol{k}\cdot\boldsymbol{x}} + \sum_{n=1}^{\infty} g^n \int K_n(\boldsymbol{x}, \boldsymbol{x}')\, e^{i\boldsymbol{k}\cdot\boldsymbol{x}'}\, d^3x'\,, \tag{47}$$

with

$$K_1(\boldsymbol{x}, \boldsymbol{x}') = K(\boldsymbol{x}, \boldsymbol{x}')\,, \qquad K_n(\boldsymbol{x}, \boldsymbol{x}') = \int K(\boldsymbol{x}, \boldsymbol{x}'')\, K_{n-1}(\boldsymbol{x}'', \boldsymbol{x}')\, d^3x''\,.$$

The expansion of $f(\theta)$ is readily obtained from (47) and (4),

$$f(\theta) = -\frac{1}{4\pi} \sum_{n=1}^{\infty} g^n \int e^{-i\boldsymbol{k}'\cdot\boldsymbol{x}}\, \mathfrak{V}(\boldsymbol{x})\, K_{n-1}(\boldsymbol{x}, \boldsymbol{x}')\, e^{i\boldsymbol{k}\cdot\boldsymbol{x}'}\, d^3x\, d^3x'\,, \tag{48}$$

where $K_0(\boldsymbol{x}, \boldsymbol{x}') = \delta(\boldsymbol{x} - \boldsymbol{x}')$. Here θ denotes the angle between $\boldsymbol{k}'$ and $\boldsymbol{k}$, and $k'^2 = k^2$.

On the other hand, iterating eq. (16) we obtain

$$(\boldsymbol{k}' |U| \boldsymbol{k}) = g(\boldsymbol{k}' |\mathfrak{V}| \boldsymbol{k}) + \sum_{n=1}^{\infty} g^{n+1} \int K_n(\boldsymbol{k}', \boldsymbol{k}'')(\boldsymbol{k}'' |\mathfrak{V}| \boldsymbol{k})\, \mathrm{d}^3\mathrm{k}'' , \quad (49)$$

where

$$K_1(\boldsymbol{k}', \boldsymbol{k}'') = -2\pi \mathrm{i} (\boldsymbol{k}' |\mathfrak{V}| \boldsymbol{k}'')\, \delta_+(k''^2 - k^2) ,$$
$$K_n(\boldsymbol{k}', \boldsymbol{k}'') = \int K_1(\boldsymbol{k}', \boldsymbol{k}''')\, K_{n-1}(\boldsymbol{k}''', \boldsymbol{k}'')\, \mathrm{d}^3\mathrm{k}''' .$$

According to eq. (35), $f(\theta)$ is simply proportional to $(\boldsymbol{k}' |\underline{U}| \boldsymbol{k})$, and its iterated expansion is readily obtained from eq. (49).

It will depend on the particular type of potential as to whether it will be convenient to make use of either (48) or (49) in an actual calculation.

We rewrite (48), and the equivalent expression obtained from (35) and (49), up to terms which are quadratic in the potential strength:

$$f(\theta) = -(g/4\pi) \int \mathrm{e}^{-\mathrm{i}\boldsymbol{k}' \cdot \boldsymbol{x}}\, \mathfrak{V}(\boldsymbol{x})\, \mathrm{e}^{\mathrm{i}\boldsymbol{k} \cdot \boldsymbol{x}}\, \mathrm{d}^3\mathrm{x} +$$
$$+ (g/4\pi)^2 \int \mathrm{e}^{-\mathrm{i}\boldsymbol{k}' \cdot \boldsymbol{x}}\, \mathfrak{V}(\boldsymbol{x}) \frac{\mathrm{e}^{\mathrm{i}k|\boldsymbol{x}-\boldsymbol{x}'|}}{|\boldsymbol{x} - \boldsymbol{x}'|} \mathfrak{V}(\boldsymbol{x}')\, \mathrm{e}^{\mathrm{i}\boldsymbol{k} \cdot \boldsymbol{x}'}\, \mathrm{d}^3\mathrm{x}\, \mathrm{d}^3\mathrm{x}' + \ldots , \quad (50)$$

$$f(\theta) = -2\pi^2 g (\boldsymbol{k}' |\mathfrak{V}| \boldsymbol{k}) + 2\pi^2 g^2 \int \frac{(\boldsymbol{k}' |\mathfrak{V}| \boldsymbol{k}'')(\boldsymbol{k}'' |\mathfrak{V}| \boldsymbol{k})}{k''^2 - k^2 - \mathrm{i}\,\varepsilon}\, \mathrm{d}^3\mathrm{k}'' + \ldots ,$$

with $k'^2 = k^2$.

In the current terminology, the integration over $\boldsymbol{k}''$ in the second line of the above formula is often said to represent a *virtual state* of the system produced by the matrix element $(\boldsymbol{k}'' |\mathfrak{V}| \boldsymbol{k})$ from the initial state $\boldsymbol{k}$, and converted by $(\boldsymbol{k}' |\mathfrak{V}| \boldsymbol{k}'')$ into the final state of momentum $\boldsymbol{k}'$. The denominator $(k''^2 - k^2 - \mathrm{i}\,\varepsilon)$ (and any such denominator occurring in the higher approximations) is referred to as an *energy denominator*, being essentially the difference between the energy of the virtual state and that of the initial and final states. Note that there is no energy conservation in the virtual states.

10.2. Computation of terms in the Born expansion

The evaluation of a term in the Born expansion becomes more difficult, the higher the order of the term. In the case of central potentials the first order term can often be quite easily evaluated. In fact, the threefold integral is readily reduced to a single integration by taking the z axis parallel to the vector $\boldsymbol{k} - \boldsymbol{k}'$,

$$\int e^{-i\boldsymbol{k}'\cdot\boldsymbol{x}}\,\mathfrak{V}(r)\,e^{i\boldsymbol{k}\cdot\boldsymbol{x}}\,d^3x = 4\pi\,|\boldsymbol{k} - \boldsymbol{k}'|^{-1}\int_0^\infty \sin(|\boldsymbol{k} - \boldsymbol{k}'|\,r)\,\mathfrak{V}(r)\,r\,dr\,.$$

Therefore, the first Born approximation gives

$$f_1(\theta) = -\,g\,|\boldsymbol{k} - \boldsymbol{k}'|^{-1}\int_0^\infty \sin(|\boldsymbol{k} - \boldsymbol{k}'|\,r)\,\mathfrak{V}(r)\,r\,dr\,. \tag{51}$$

The calculation of the second term of (50), the second Born approximation, is a much more difficult problem, as it involves the evaluation of a sixfold integral. There is no general way of proceeding, though some artifices are known to be of help in special cases (see the case of the Yukawa potential below).

10.3. Yukawa potential

By this we mean a potential of the type

$$\mathfrak{V}(\boldsymbol{x}) = \mathfrak{V}(r) = -\,g\,e^{-\mu r}/r\,, \qquad \mu > 0\,.$$

The scattering amplitude, in the *first Born approximation*, is *

$$f_1(\theta) = \frac{g}{\mu^2 + (\boldsymbol{k} - \boldsymbol{k}')^2}\,,$$

and the differential cross section

$$\frac{d\sigma}{d\Omega} = \frac{g^2}{[(\boldsymbol{k} - \boldsymbol{k}')^2 + \mu^2]^2} = \frac{g^2}{[2k^2(1 - \cos\theta) + \mu^2]^2}\,, \tag{52}$$

where θ is the scattering angle.

It is interesting to note that the total cross section (first Born approximation) is convergent, since the denominator in (52) cannot vanish for $\mu \neq 0$. This is in contrast to the case of the cross section (44) for Coulomb scattering, which may be obtained from (52) by the replacement

$$g \to -\,Z_1 Z_2 e^2 m/2\pi\hbar^2\,, \qquad \mu \to 0\,.$$

---

* Remember that

$$e^{-\mu r}/4\pi r = (2\pi)^{-3}\int e^{i\boldsymbol{k}\cdot\boldsymbol{x}}/(k^2 + \mu^2)\,d^3k\,,$$

as may be easily verified by evaluating the integral on the right by the method of contour integration (see § 3, where a similar calculation is performed; here the poles of the integrand are at $\pm\,i\mu$).

Let us now consider the *second Born approximation*. Using the expression for $f_1(\theta)$ given above, we see that the second term may be written as

$$f_2(\theta) = \frac{g^2}{2\pi^2}\int \frac{1}{\mu^2 + (\boldsymbol{k}' - \boldsymbol{k}'')^2}\,\frac{1}{k''^2 - k^2 - i\varepsilon}\,\frac{1}{\mu^2 + (\boldsymbol{k}'' - \boldsymbol{k})^2}\,d^3k'' .$$

A trick, used by Feynman to deal with similar integrals in quantum electrodynamics, can be applied here advantageously. It consists in employing the formula

$$\frac{1}{abc} = 2\int_0^1 dx_1 \int_0^{x_1} \left[a + (b-a)\,x_1 + (c-b)\,x_2\right]^{-3} dx_2\,, \tag{53}$$

or

$$1/abc = 2\int (ax_1 + bx_2 + cx_3)^{-3}\,\delta(x_1 + x_2 + x_3 - 1)\,dx_1\,dx_2\,dx_3\,. \tag{53'}$$

NOTE 3. Eq. (53) is a special case of the general formula

$$J(a_0, \ldots a_n) = \prod_{i=0}^{n} \frac{1}{a_i}\,, \tag{α}$$

where

$$\begin{aligned} J(a_0, \ldots a_n) = n!\int_0^1 dx_1 \int_0^{x_1} dx_2 \ldots \int_0^{x_{n-1}} dx_n \times \\ \times \left[a_0 + (a_1 - a_0)\,x_1 + \ldots (a_n - a_{n-1})\,x_n\right]^{-(n+1)}. \end{aligned} \tag{β}$$

This formula can easily be proved by induction. In fact, it is trivial for $n = 1$. Assume now that it holds for $n - 1$, i.e. that

$$J(a_0, \ldots a_{n-1}) = \prod_{i=0}^{n-1} \frac{1}{a_i}\,. \tag{γ}$$

Since

$$\begin{aligned} \int_0^{x_{n-1}} \left[a_0 + (a_1 - a_0)\,x_1 + \ldots (a_n - a_{n-1})\,x_n\right]^{-n-1} dx_n = \\ = \left[n(a_n - a_{n-1})\right]^{-1}\left\{\left[a_0 + (a_1 - a_0)\,x_1 + \ldots + (a_{n-1} - a_{n-2})\,x_{n-1}\right]^{-n} - \right. \\ \left. - \left[a_0 + (a_1 - a_0)\,x_1 + \ldots + (a_n - a_{n-2})\,x_{n-1}\right]^{-n}\right\}, \end{aligned} \tag{δ}$$

using (γ) and (δ) in (β), we have

$$J(a_0, \ldots a_n) = \frac{1}{a_n - a_{n-1}}\left[\prod_{i=0}^{n-1} \frac{1}{a_i} - \frac{1}{a_n}\prod_{i=0}^{n-2} \frac{1}{a_i}\right] = \prod_{i=0}^{n} \frac{1}{a_i}\,.$$

An alternative way of writing (α) is obtained by making the change of variables

$$\xi_0 = 1 - x_1\,, \qquad \xi_1 = x_1 - x_2, \ldots \qquad \xi_{n-1} = x_{n-1} - x_n\,,$$

so that

$$x_n = 1 - \xi_0 \ldots - \xi_{n-1}.$$

Then

$$\prod_{i=0}^{n} \frac{1}{a_i} = n! \int_0^1 \mathrm{d}\xi_0 \int_0^{1-\xi_0} \mathrm{d}\xi_1 \ldots \int_0^{1-\xi_0 \ldots - \xi_{n-2}} \mathrm{d}\xi_{n-1} \quad \times$$
$$\times \; [a_0\xi_0 + a_1\xi_1 + \ldots + a_{n-1}\xi_{n-1} + a_n x_n]^{-n-1},$$

or also

$$\prod_{i=0}^{n} \frac{1}{a_i} = n! \int_0^1 \mathrm{d}\xi_0 \int_0^{1-\xi_0} \mathrm{d}\xi_1 \ldots \int_0^{1-\xi_0 \ldots - \xi_{n-2}} \mathrm{d}\xi_{n-1} \int \mathrm{d}\xi_n \times$$
$$\times [a_0\xi_0 + \ldots + a_n\xi_n]^{-n-1} \times \delta(\xi_0 + \ldots + \xi_n - 1)$$
$$= n! \int [a_0 x_0 + \ldots + a_n x_n]^{-n-1}\, \delta(x_0 + \ldots + x_n - 1)\, \mathrm{d}x_0 \ldots \mathrm{d}x_n$$

the range of the variables being confined to $x_i \geqslant 0$, $(i = 0, \ldots n)$.

Using (53′) we can write

$$f_2(\theta) = g^2\pi^{-2} \int \big[x_0(\mu^2 + (\boldsymbol{k}' - \boldsymbol{k}'')^2) + x_1(k''^2 - k^2 - \mathrm{i}\,\varepsilon) +$$
$$+ x_2(\mu^2 + (\boldsymbol{k}'' - \boldsymbol{k})^2)\big]^{-3}\, \delta(x_0 + x_1 + x_2 - 1)\, \mathrm{d}^3\mathrm{k}''\, \mathrm{d}x_0\, \mathrm{d}x_1\, \mathrm{d}x_2\,.$$

Introducing the new variable $\boldsymbol{K} = \boldsymbol{k}'' - x_0\boldsymbol{k}' - x_2\boldsymbol{k}$, we obtain the integral

$$f_2(\theta) = g^2\pi^{-2} \int \big[\boldsymbol{K}^2 + A\big]^{-3}\, \delta(x_0 + x_1 + x_2 - 1)\, \mathrm{d}x_0\, \mathrm{d}x_1\, \mathrm{d}x_2\, \mathrm{d}^3\mathrm{K}$$

with

$$A = k^2\big[x_0(1 - x_0) + x_2(1 - x_2) - x_1 - 2\xi x_0 x_2\big] + \mu^2(x_0 + x_2) - \mathrm{i}\varepsilon x_1$$

where we have used the properties of the δ function, the relation $k^2 = k'^2$, and the notation $\xi = \cos\theta$. As

$$\int (\boldsymbol{K}^2 + A)^{-3}\, \mathrm{d}^3\mathrm{K} = 4\pi \int_0^\infty (K^2 + A)^{-3}\, K^2\, \mathrm{d}K = (\pi/2)^2 A^{-\frac{3}{2}}\,,$$

introducing the new variables $u = x_0 + x_2$, $v = x_0 - x_2$, we get

$$f_2(\theta) =$$

$$= (g^2/8) \int_0^1 du \int_{-u}^{+u} \left[\mu^2 u + k^2 (2u - \tfrac{1}{2}(u^2 + v^2) - 1 + \tfrac{1}{2}\xi(v^2 - u^2)) + \right.$$

$$\left. + \mathrm{i}\,\varepsilon(u-1)\right]^{-\frac{3}{2}} dv = (g^2/4) \int_0^1 \frac{u\, du}{b(u)\,[b(u) - au^2]^{\frac{1}{2}}},$$

with $a = \frac{1}{2}k^2(1-\xi)$ and $b(u) = \mu^2 u + k^2(2u - 1 - \frac{1}{2}u^2(1+\xi)) + \mathrm{i}\,\varepsilon(u-1)$.

Performing this last integration, the result *

$$\mathrm{Re} f_2(\theta) = g^2 [2k^2(1-\xi)(2k^4[1-\xi] + 4k^2\mu^2 + \mu^4)]^{-\frac{1}{2}} \times$$

$$\times \tan^{-1} \left[\frac{\mu^2 k^2 (1-\xi)}{4k^4(1-\xi) + 8k^2\mu^2 + 2\mu^4} \right]^{\frac{1}{2}}$$

is obtained in the limit $\varepsilon = 0$. The imaginary part of $f_2(\theta)$ is not necessary for our purpose. In fact, since $f_1(\theta)$ is real, the differential cross section, up to terms of the third order in g, is

$$d\sigma(\theta) = |f_1(\theta) + f_2(\theta) + \dots|^2 d\Omega = (|f_1(\theta)|^2 + 2f_1(\theta)\,\mathrm{Re} f_2(\theta) + \dots)\, d\Omega,$$

and is immediately obtained in explicit form by inserting $f_1(\theta)$ and $\mathrm{Re} f_2(\theta)$.

NOTE 4. Expressing both members of $\sigma = (4\pi/k)\,\mathrm{Im} f(0)$ in the form of a power series in g, we find

$$\sigma_2 = \int |f_1(\theta)|^2\, d\Omega = (4\pi/k)\,\mathrm{Im} f_2(0),$$
$$\sigma_3 = 2 \int f_1(\theta)\,\mathrm{Re} f_2(\theta)\, d\Omega = (4\pi/k)\,\mathrm{Im} f_3(0),$$

. . . .

Therefore, the fact that $f_1(\theta)$ is real is connected with the leading term of the expansion of σ being of the second order in $g(\sigma_1 = 0)$. In general, the imaginary part of the scattering amplitude $f(\theta)$ must be known to an accuracy of g^n to yield $\sigma_n = 4\pi k^{-1}\,\mathrm{Im} f_n(0)$. On the other hand, σ_n could be computed by angular integration, once the terms $f_1(\theta), f_2(\theta) \dots f_{n-1}(\theta)$ were known. In other words, the advantage derived from dispensing with the angular integration is only apparent, being outweighed by the necessity of knowing $f(\theta)$ to a higher order of accuracy.

10.4. Angular momentum representation

Eqs. (21), (26) or (28), (29) can be used in an iterated form in order to

---

* R. Jost and A. Pais, Phys. Rev. **82** (1951), 840.

derive an expansion for the phase shifts η_l. Here we prefer to work with the latter two equations. Iterating (28) we obtain

$$\varphi_l(r) = (\pi kr/2)^{\frac{1}{2}} J_{l+\frac{1}{2}}(kr) + \sum_{n=1}^{\infty} (-1)^n g^n \int_0^{\infty} g_l(r, r_1)\,\mathfrak{V}(r_1)\, g_l(r_1, r_2)\,\mathfrak{V}(r_2) \ldots$$
$$\ldots g_l(r_{n-1}, r_n)\,\mathfrak{V}(r_n)\,(\pi kr_n/2)^{\frac{1}{2}} J_{l+\frac{1}{2}}(kr_n)\,\mathrm{d}r_1 \ldots \mathrm{d}r_n\,,$$

and, inserting this into (29), we get

$$k \tan \eta_l = -g \int_0^{\infty} \tfrac{1}{2}\pi kr\,(J_{l+\frac{1}{2}}(kr))^2\,\mathfrak{V}(r)\,\mathrm{d}r +$$
$$+ \sum_{n=1}^{\infty} (-1)^{n+1} g^{n+1} \int_0^{\infty} (\pi kr/2)^{\frac{1}{2}} J_{l+\frac{1}{2}}(kr)\;\mathfrak{V}(r) g_l(r, r_1)\,\mathfrak{V}(r_1) \ldots \tag{54}$$
$$\ldots g_l(r_{n-1}, r_n)\,\mathfrak{V}(r_n)(\pi kr_n/2)^{\frac{1}{2}} J_{l+\frac{1}{2}}(kr_n)\,\mathrm{d}r\,\mathrm{d}r_1 \ldots \mathrm{d}r_n\,.$$

Note the special case

$$-k \tan \eta_0 = g \int_0^{\infty} \sin^2(kr)\,\mathfrak{V}(r)\,\mathrm{d}r +$$
$$+ \sum_{n=1}^{\infty} (-1)^n g^{n+1} \int_0^{\infty} \sin(kr)\,\mathfrak{V}(r)\, g_0(r, r_1)\,\mathfrak{V}(r_1) \times$$
$$\times \ldots g_0(r_{n-1}, r_n)\,\mathfrak{V}(r_n)\sin(kr_n)\,\mathrm{d}r\,\mathrm{d}r_1 \ldots \mathrm{d}r_n\,,$$

with $g_0(r, r')$ defined as in *Note 2*.

From the above equations it is easy to derive an expansion for η_l in a power series in g. The first term of this series is

$$\eta_l \simeq -\tfrac{1}{2} g\pi \int_0^{\infty} (J_{l+\frac{1}{2}}(kr))^2\,\mathfrak{V}(r)\, r\,\mathrm{d}r\,. \tag{55}$$

The agreement between the Born approximations in the various representations can be checked by inserting (55) into (27) with $\mathrm{e}^{2\mathrm{i}\eta_l} - 1 \cong 2\mathrm{i}\,\eta_l$. This gives

$$f_1(\theta) = -(g\pi/2k) \int_0^{\infty} \mathfrak{V}(r) \sum_{l=0}^{\infty} (2l+1)\,(J_{l+\frac{1}{2}}(kr))^2\, P_l(\cos\theta)\, r\,\mathrm{d}r$$
$$= -g \int_0^{\infty} \frac{\sin\left[r(2k^2(1-\cos\theta))^{\frac{1}{2}}\right]}{r\,(2k^2(1-\cos\theta))^{\frac{1}{2}}}\,\mathfrak{V}(r)\, r^2\,\mathrm{d}r\,,$$

which coincides with (51). The calculation of the successive terms of the expansion (54) is merely the well-defined problem of performing certain integrations.

Chapter II

NON-RELATIVISTIC STEADY-STATE SCATTERING OF A PARTICLE OF SPIN ONE-HALF

1. Scattering amplitude and cross sections

In analogy with the case of spin zero particles, one may write the time-independent Schroedinger equation

$$(\Delta + k^2)\,\psi = \mathfrak{V}(\boldsymbol{x}, \boldsymbol{\sigma})\,\psi\,, \tag{1}$$

for the stationary two-component wave function

$$\psi(X) = \psi(\boldsymbol{x})\,\mathrm{e}^{-\mathrm{i}Et/\hbar}\,.$$

Here $\mathfrak{V}(\boldsymbol{x}, \boldsymbol{\sigma})$ is an Hermitian operator involving the σ matrices, which may also contain the differential operator $-\,\mathrm{i}\hbar\,\nabla_x$, e.g. in the form of a spin-orbit coupling $\boldsymbol{\sigma}\cdot\boldsymbol{L}$. Potentials of this type are used in nuclear physics to describe the interaction of a nucleon with a nucleus *. For example, the interaction in eq. (5), Part II, Chap. VIII, is of this type.

The wave function may be written as a linear combination of the eigenfunctions of σ_3,

$$\psi(\boldsymbol{x}) = \chi_+\,\psi_+(\boldsymbol{x}) + \chi_-\,\psi_-(\boldsymbol{x})\,.$$

In the formulation of a collision problem, the wave function

$$\psi^{(0)}(\boldsymbol{x}) = (2\pi)^{-\frac{3}{2}}\,\mathrm{e}^{\mathrm{i}\boldsymbol{k}\cdot\boldsymbol{x}}(a\chi_+ + b\chi_-) = (2\pi)^{-\frac{3}{2}}\,\mathrm{e}^{\mathrm{i}\boldsymbol{k}\cdot\boldsymbol{x}}\begin{pmatrix} a \\ b \end{pmatrix} \tag{2}$$

describes the incident wave.

The integral equation

$$\psi(\boldsymbol{x}) = \psi^{(0)}(\boldsymbol{x}) + \int K(\boldsymbol{x}, \boldsymbol{x}')\,\psi(\boldsymbol{x}')\,\mathrm{d}^3x'\,,$$

differs from the corresponding eq. (3) of the preceding chapter only in that K, though still defined by eq. (4), Chap. I, is now a 2×2 matrix, since the potential involves the σ's.

---

* It is easy to recognize that the theory developed below is also suitable for the description of scattering of particles of spin zero by a scattering centre of spin one-half (e.g., scattering of pions by nucleons).

At large distances from the scattering centre, its solution is the superposition of the free wave (2) and of an outgoing scattered wave

$$(2\pi)^{-\frac{3}{2}} f(\theta, \phi) \frac{e^{ikr}}{r} = (2\pi)^{-\frac{3}{2}} \begin{pmatrix} f_+(\theta, \phi) \\ f_-(\theta, \phi) \end{pmatrix} \frac{e^{ikr}}{r} .$$

There are now two scattering amplitudes

$$f_\pm(\theta, \phi) = -(\pi/2)^{\frac{1}{2}} \int e^{-ik\boldsymbol{n}\cdot\boldsymbol{x}'} \chi_\pm^\dagger \mathfrak{V}(\boldsymbol{x}', \boldsymbol{\sigma}) \psi(\boldsymbol{x}') \, d^3x' ,$$

describing processes in which the particle is left in the state with $\sigma_3 = +1$, and in that with $\sigma_3 = -1$, respectively.

If the incident wave function (2) has the coefficients $a = 1$ and $b = 0$, namely if the particle is initially in the state with $\sigma_3 = +1$, f_+ and f_- are referred to as the amplitudes for scattering without and with *spin flip*, respectively. Of course, they assume the rôles of amplitudes with and without spin flip if the initial state of the particle is χ_-.

Retracing the proof of the optical theorem given in Chap. I, § 5, it can be shown that the current

$$\boldsymbol{j} = (\hbar/2mi)[\psi^\dagger \nabla_x \psi - (\nabla_x \psi^\dagger) \psi] ,$$

with

$$\psi(\boldsymbol{x}) = \begin{pmatrix} \psi_+ \\ \psi_- \end{pmatrix} , \quad \text{and} \quad \psi^\dagger(\boldsymbol{x}) = (\psi_+^* \ \ \psi_-^*) ,$$

obeys the continuity equation

$$\operatorname{div} \boldsymbol{j} = 0 ,$$

if ψ is a solution of eq. (1) with Hermitian potential.

The current may be written explicitly in the form

$$\boldsymbol{j} = (\hbar/2mi) [\psi_+^* \nabla_x \psi_+ - (\nabla_x \psi_+^*) \psi_+ + \psi_-^* \nabla_x \psi_- - (\nabla_x \psi_-^*) \psi_-] .$$

Assuming that the incident wave function has $\sigma_3 = +1$, we have, for large r,

$$(2\pi)^{\frac{3}{2}} \psi_+ \sim e^{i\boldsymbol{k}\cdot\boldsymbol{x}} + f_+(\theta, \phi) \, e^{ikr}/r ,$$
$$(2\pi)^{\frac{3}{2}} \psi_- \sim f_-(\theta, \phi) \, e^{ikr}/r .$$

In this case the analogue of eq. (31), § 5 of the previous chapter, is

$$j_r \sim (2\pi)^{-3} \frac{\hbar k}{m} \left\{ \cos\theta + \frac{1}{r^2} |f_+|^2 + \frac{1+\cos\theta}{2r} [f_+ e^{ikr(1-\cos\theta)} + \text{c.c.}] + \frac{1}{r^2} |f_-|^2 \right\} .$$

In this expression, ψ_- does not interfere with the incident wave. The requirement that the flux of the current through a sphere must be zero, leads to the optical theorem

$$\sigma_{\text{tot}} = \int (|f_+|^2 + |f_-|^2)\, \mathrm{d}\Omega = 4\pi\, k^{-1}\, \mathrm{Im}\, f_+(0)$$

(for $\mathfrak{V} = \mathfrak{V}^\dagger$ and for the incident wave $\chi_+\, \mathrm{e}^{\mathrm{i}k \cdot x}$).

The total cross section (combining spin flip and non-spin-flip scattering) is proportional to the imaginary part of the non-spin-flip (i.e. fully elastic) scattering amplitude in the forward direction.

The differential cross section

$$\frac{\mathrm{d}\sigma}{\mathrm{d}\Omega} = \frac{\mathrm{d}\sigma_+}{\mathrm{d}\Omega} + \frac{\mathrm{d}\sigma_-}{\mathrm{d}\Omega}$$

is the sum of

$$\frac{\mathrm{d}\sigma_+}{\mathrm{d}\Omega} = |f_+|^2$$

for a process leaving the particle with the spin $\sigma_3 = +1$, and

$$\frac{\mathrm{d}\sigma_-}{\mathrm{d}\Omega} = |f_-|^2$$

for that corresponding to the final state $\sigma_3 = -1$.

NOTE 1. If the potential involves only the third component of the spin, σ_3 is a constant of the motion and no spin flip can occur. Then, eq. (1) separates into two uncoupled equations for scattering in the states $\sigma_3 = +1$ and $\sigma_3 = -1$, for which the potentials are different.

Denoting by $\mathrm{d}\sigma(+)/\mathrm{d}\Omega$ and $\mathrm{d}\sigma(-)/\mathrm{d}\Omega$ the differential cross sections for scattering in the pure states χ_+ and χ_-, respectively, that for scattering from the initial mixed state $a\chi_+ + b\chi_-$ to the final state χ_+ is

$$|a|^2 \frac{\mathrm{d}\sigma(+)}{\mathrm{d}\Omega}.$$

That to the final state χ_- is

$$|b|^2 \frac{\mathrm{d}\sigma(-)}{\mathrm{d}\Omega},$$

and that to either state

$$|a|^2 \frac{\mathrm{d}\sigma(+)}{\mathrm{d}\Omega} + |b|^2 \frac{\mathrm{d}\sigma(-)}{\mathrm{d}\Omega}. \qquad (\alpha)$$

For a spin independent potential,

$$\frac{d\sigma(+)}{d\Omega} = \frac{d\sigma(-)}{d\Omega} \Rightarrow \frac{d\sigma}{d\Omega}.$$

Then (α) reduces to

$$(|a|^2 + |b|^2)\frac{d\sigma}{d\Omega} = \frac{d\sigma}{d\Omega},$$

i.e. the cross section for scattering of a spinless particle.

2. Spin-orbit representation

For certain spin dependent potentials, eq. (1) is best solved by going over from the representation using the variables (r, l, m_l, s_3) to the new representation (r, j, m_j, l, s), comprising the magnitude of the total angular momentum $\boldsymbol{M} = \hbar \boldsymbol{j} = \boldsymbol{L} + \boldsymbol{s} = \hbar(\boldsymbol{l} + \frac{1}{2}\boldsymbol{\sigma})$, its third component $M_3 = \hbar m_j$ and the magnitudes $\hbar l$ and s ($s = \frac{1}{2}\hbar$) of the orbital angular momentum and of the spin, respectively.

The new representation is especially convenient when the interaction is of the form

$$\mathfrak{V} = \mathfrak{V}_0(r) + \mathfrak{V}_1(r)(\boldsymbol{l} \cdot \boldsymbol{\sigma}), \tag{3}$$

with $\mathfrak{V}_0$ and $\mathfrak{V}_1$ spherically symmetric.

In fact, since

$$\boldsymbol{M}^2 = \boldsymbol{L}^2 + \boldsymbol{s}^2 + 2\boldsymbol{L} \cdot \boldsymbol{s} = \hbar^2(\boldsymbol{l}^2 + \tfrac{1}{4}\boldsymbol{\sigma}^2 + \boldsymbol{l} \cdot \boldsymbol{\sigma}), \tag{4}$$

the spin-orbit coupling $\mathfrak{V}_1(\boldsymbol{l} \cdot \boldsymbol{\sigma})$, and therefore the whole interaction, is diagonal.

The operator $\boldsymbol{l} \cdot \boldsymbol{\sigma}$ can be used to construct the projection operators Λ_l^+ and Λ_l^- for the states of total angular momentum $j = l + \frac{1}{2}$ and $j = l - \frac{1}{2}$, corresponding to the two ways in which the spin can be added to the orbital angular momentum. In fact, for parallel addition $\boldsymbol{j}^2 = (l + \frac{1}{2})(l + \frac{3}{2})$, and for antiparallel addition $\boldsymbol{j}^2 = (l - \frac{1}{2})(l + \frac{1}{2})$.

Since $\boldsymbol{j}^2$ is also given by eq. (4), we have

$$\begin{aligned} \boldsymbol{l} \cdot \boldsymbol{\sigma} &= l & &\text{if } j = l + \tfrac{1}{2}, \\ \boldsymbol{l} \cdot \boldsymbol{\sigma} &= -(l + 1) & &\text{if } j = l - \tfrac{1}{2}. \end{aligned}$$

The operators

$$\Lambda_l^+ = \frac{l + 1 + \boldsymbol{l} \cdot \boldsymbol{\sigma}}{2l + 1}, \quad \Lambda_l^- = \frac{l - \boldsymbol{l} \cdot \boldsymbol{\sigma}}{2l + 1}, \tag{5}$$

are therefore projection operators * with the properties

$$\Lambda_l^+ \psi_{j=l+\frac{1}{2}} = \psi_{j=l+\frac{1}{2}}, \quad \Lambda_l^+ \psi_{j=l-\frac{1}{2}} = 0,$$
$$\Lambda_l^- \psi_{j=l+\frac{1}{2}} = 0, \quad \Lambda_l^- \psi_{j=l-\frac{1}{2}} = \psi_{j=l-\frac{1}{2}},$$
$$\Lambda_l^+ + \Lambda_l^- = I, \quad (\Lambda_l^\pm)^2 = \Lambda_l^\pm, \quad \Lambda_l^+\Lambda_l^- = \Lambda_l^-\Lambda_l^+ = 0.$$

Here $\psi_{j=l\pm\frac{1}{2}}$ denotes eigenfunctions of $\boldsymbol{j}^2$ with $j = l \pm \frac{1}{2}$.

In the angular momentum representation employing the two-component wave functions

$$\psi = \begin{pmatrix} \psi_{+,lm_l}(r, \theta, \phi) \\ \psi_{-,lm_l}(r, \theta, \phi) \end{pmatrix} = Y_l^{(m_l)}(\theta, \phi) \begin{pmatrix} \psi_{+,lm_l}(r) \\ \psi_{-,lm_l}(r) \end{pmatrix},$$

the projection operators are given by eqs. (5), where $\boldsymbol{l}$ must be identified with the differential operator defined in eq. (2), Part. I, Chap. IV, and the components of $\boldsymbol{\sigma}$ with the Pauli matrices.

With the help of $\Lambda_l^\pm$, a two-component eigenfunction of $(\boldsymbol{l}^2, m_l, \sigma_3)$, say $\chi_+ Y_l^{(m_l)}$, can be represented as the sum of two terms

$$\chi_+ Y_l^{(m_l)} = \Lambda_l^+(\chi_+ Y_l^{(m_l)}) + \Lambda_l^-(\chi_+ Y_l^{(m_l)}).$$

The first term is an eigenfunction with the quantum numbers $j = l + \frac{1}{2}$, $m_j = m_l + \frac{1}{2}$, l, and the second is one with $j = l - \frac{1}{2}$, $m_j = m_l + \frac{1}{2}$, l.

In order to obtain an explicit expression for the two terms, we notice that $\boldsymbol{l} \cdot \boldsymbol{\sigma}$ may be written as the sum of three terms which obviously conserve $m_j = m_l + \frac{1}{2}\sigma_3$, namely

$$\boldsymbol{l} \cdot \boldsymbol{\sigma} = \sigma_+ l_- + \sigma_- l_+ + \sigma_3 l_3,$$

with

$$\sigma_\pm = \tfrac{1}{2}(\sigma_1 \pm \mathrm{i}\,\sigma_2), \quad l_\pm = l_1 \pm \mathrm{i}\, l_2.$$

In fact, the matrices $\sigma_+(\sigma_-)$ flip the spin, i.e. increase (decrease) the value of the third component by one unit (see Part II, Chap. I, § 1), while $l_\pm$ act in a

* If A is an operator with the eigenvalues A_i, and ψ_{A_i} is the eigenfunction belonging to the eigenvalue A_i, the projection operator

$$\Lambda^i = \prod_{j \neq i} \frac{A - A_j}{A_i - A_j},$$

has the property

$$\Lambda^i \psi_{A_j} = \delta_{ij} \psi_{A_i}.$$

From a mixture of states $\psi = \sum_j a_j \psi_{A_j}$, it selects the component which is parallel to ψ_{A_i}, $\Lambda^i \sum_j a_j \psi_{A_j} = a_i \psi_{A_i}$.

similar way on $Y_l^{(m_l)}$, since

$$l_\pm\, Y_l^{(m_l)} = \left[(l \mp m_l)(l \pm m_l + 1)\right]^{\frac{1}{2}}\, Y_l^{(m_l \pm 1)} .$$

Therefore

$$\boldsymbol{l} \cdot \boldsymbol{\sigma}\, \chi_\pm\, Y_l^{(m_l)} = \left[(l \mp m_l)(l \pm m_l + 1)\right]^{\frac{1}{2}} \chi_\mp\, Y_l^{(m_l \pm 1)} \pm m_l\, \chi_\pm\, Y_l^{(m_l)} ,$$

and, in particular,

$$\boldsymbol{l} \cdot \boldsymbol{\sigma}\, \chi_\pm\, Y_l^{(0)} = \left[l(l+1)\right]^{\frac{1}{2}} \chi_\mp\, Y_l^{(\pm 1)} .$$

Using these formulae we find, for instance,

$$\Lambda_l^\pm\, \chi_+\, Y_l^{(0)}\, F_l(kr) = \left[(l+\tfrac{1}{2}) \pm \tfrac{1}{2}\right]^{\frac{1}{2}} F^{\frac{1}{2}}_{l,l\pm\frac{1}{2}} .$$

Here $F_l(kr)$ is a solution of the equation

$$F_l'' + \left(k^2 - \frac{l(l+1)}{r^2}\right) F_l = 0 ,$$

having the asymptotic behaviour for large r

$$F_l(kr) \sim \sin(kr - \tfrac{1}{2}\, l\pi) ,$$

and $F^{m_j}_{l,j}$ are eigenfunctions with the quantum numbers j, l, m_j.

For $m_j = \frac{1}{2}$ one has

$$\begin{aligned} F^{\frac{1}{2}}_{l,l+\frac{1}{2}} &= \frac{1}{2l+1}\left[(l+1)^{\frac{1}{2}}\, \chi_+\, Y_l^{(0)} + l^{\frac{1}{2}}\, \chi_-\, Y_l^{(1)}\right] F_l(kr) = \\ &= \frac{1}{2l+1}\begin{pmatrix} (l+1)^{\frac{1}{2}}\, Y_l^{(0)} \\ l^{\frac{1}{2}}\, Y_l^{(1)} \end{pmatrix} F_l(kr) , \\ F^{\frac{1}{2}}_{l,l-\frac{1}{2}} &= \frac{1}{2l+1}\left[l^{\frac{1}{2}}\, \chi_+\, Y_l^{(0)} - (l+1)^{\frac{1}{2}}\, \chi_-\, Y_l^{(1)}\right] F_l(kr) = \\ &= \frac{1}{2l+1}\begin{pmatrix} l^{\frac{1}{2}}\, Y_l^{(0)} \\ -(l+1)^{\frac{1}{2}}\, Y_l^{(1)} \end{pmatrix} F_l(kr) . \end{aligned} \tag{6}$$

Therefore

$$F_l(kr)\, Y_l^{(0)}(\theta)\, \chi_+ = (l+1)^{\frac{1}{2}}\, F^{\frac{1}{2}}_{l,l+\frac{1}{2}} + l^{\frac{1}{2}}\, F^{\frac{1}{2}}_{l,l-\frac{1}{2}} . \tag{7}$$

Similar formulae are easily established for $F^{-\frac{1}{2}}_{l,l\pm\frac{1}{2}}$.

In the next section this last expression will be used to represent a plane wave $\chi_+\, e^{ikz}$ as a series of eigenfunctions of $\boldsymbol{j}^2$, m_j and $\boldsymbol{l}^2$.

The eigenfunctions $F^{m_j}_{l,l+\frac{1}{2}}$ and $F^{m_j}_{l,l-\frac{1}{2}}$ have the same parity, since they are both linear combinations of spherical harmonics of order l, and are therefore multiplied by $(-1)^l$ as the result of a space inversion. On the other hand, for a given j the functions $F_{j-\frac{1}{2},j}$ and $F_{j+\frac{1}{2},j}$ of orbital angular momentum $l = j - \frac{1}{2}$ and $l = j + \frac{1}{2}$ respectively, have opposite parity, since the first is multiplied by $(-1)^{j-\frac{1}{2}}$ and the second by $(-1)^{j+\frac{1}{2}}$ under an inversion. For this reason the set of quantum numbers (j, m_j, parity) is, in the present instance, equivalent to the set (j, m_j, l).

3. Phase shift analysis

With the help of eq. (7), the plane wave

$$\chi_+ \, \mathrm{e}^{\mathrm{i}kz} = \chi_+ \sum_{l=0}^{\infty} [4\pi(2l+1)]^{\frac{1}{2}} \, \mathrm{i}^l \, \frac{F_l(kr)}{kr} \, Y_l^{(0)}(\theta) \, ,$$

representing a free particle moving in the direction of the z axis with $\sigma_3 = +1$, may be written in the form

$$\chi_+ \, \mathrm{e}^{\mathrm{i}kz} = \sum_{l=0}^{\infty} [4\pi(2l+1)]^{\frac{1}{2}} \, \mathrm{i}^l \, \frac{1}{kr} \left[(l+1)^{\frac{1}{2}} \, F^{\frac{1}{2}}_{l,l+\frac{1}{2}} + l^{\frac{1}{2}} \, F^{\frac{1}{2}}_{l,l-\frac{1}{2}}\right] .$$

Similarly

$$\chi_- \, \mathrm{e}^{\mathrm{i}kz} = \sum_{l=0}^{\infty} [4\pi(2l+1)]^{\frac{1}{2}} \, \mathrm{i}^l \, \frac{1}{kr} \left[(l+1)^{\frac{1}{2}} \, F^{-\frac{1}{2}}_{l,l+\frac{1}{2}} + l^{\frac{1}{2}} \, F^{-\frac{1}{2}}_{l,l-\frac{1}{2}}\right] .$$

An interaction such as (3), which is diagonal in the (r, j, m_j, l) representation, will scatter each of these waves by shifting the phase of the outgoing part of $F_{l,l+\frac{1}{2}}$ by the amount $2\eta_{l,l+\frac{1}{2}}$, and that of $F_{l,l-\frac{1}{2}}$ by $2\eta_{l,l-\frac{1}{2}}$. The phase shifts $\eta_{l,l\pm\frac{1}{2}}$ depend on j and l, since the interaction discriminates between waves of different total and orbital angular momentum *, but not on m_j, since the interaction is invariant under rotations.

Denoting by

$$f^{(+)}(\theta, \phi) \, \frac{\mathrm{e}^{\mathrm{i}kr}}{r} \, , \quad (r \to \infty) \, ,$$

the scattered wave corresponding to the incident wave $\chi_+ \, \mathrm{e}^{\mathrm{i}kz}$, we have

$$f^{(+)}(\theta, \phi) = \frac{\sqrt{\pi}}{\mathrm{i}k} \sum_{l=0}^{\infty} \frac{1}{(2l+1)^{\frac{1}{2}}} \times$$

$$\times \left\{\left[(l+1)\,\Delta_{l,l+\frac{1}{2}} + l\,\Delta_{l,l-\frac{1}{2}}\right] \chi_+ \, Y_l^{(0)} + [l(l+1)]^{\frac{1}{2}} (\Delta_{l,l+\frac{1}{2}} - \Delta_{l,l-\frac{1}{2}}) \, \chi_- \, Y_l^{(1)}\right\} ,$$

* Thus $\eta_{0,\frac{1}{2}}$, $\eta_{1,\frac{1}{2}}$, $\eta_{1,\frac{3}{2}}$ ... account for scattering in the states $s_{\frac{1}{2}}(l = 0, j = \frac{1}{2})$, $p_{\frac{1}{2}}(l = 1, j = \frac{1}{2})$, $p_{\frac{3}{2}}(l = 1, j = \frac{3}{2})$ etc.

where

$$\Delta_{l,l\pm\frac{1}{2}} = e^{2i\eta_{l,l\pm\frac{1}{2}}} - 1 .$$

Similarly, in the scattered wave

$$f^{(-)}(\theta, \phi)\, e^{ikr}/r , \quad (r \to \infty) ,$$

corresponding to the incident wave $\chi_-\, e^{ikz}$,

$$f^{(-)}(\theta, \phi) = \frac{\sqrt{\pi}}{ik} \sum_{l=0}^{\infty} (2l+1)^{-\frac{1}{2}} \{[(l+1)\, \Delta_{l,l+\frac{1}{2}} + l\, \Delta_{l,l-\frac{1}{2}}]\, \chi_-\, Y_l^{(0)} + \\ + [l(l+1)]^{\frac{1}{2}} (\Delta_{l,l+\frac{1}{2}} - \Delta_{l,l-\frac{1}{2}})\, \chi_+\, Y_l^{(-1)}\} .$$

Another useful formula for the scattering amplitudes is

$$f^{(\pm)}(\theta,\phi) = \frac{\sqrt{\pi}}{ik} \sum_{l=0}^{\infty} (2l+1)^{\frac{1}{2}} (\Delta_{l,l+\frac{1}{2}}\, \Lambda_l^+ + \Delta_{l,l-\frac{1}{2}}\, \Lambda_l^-)\, \chi_\pm\, Y_l^{(0)}$$

$$= \frac{\sqrt{\pi}}{ik} \sum_{l=0}^{\infty} (2l+1)^{-\frac{1}{2}} \{[(l+1)\Delta_{l,l+\frac{1}{2}} + l\Delta_{l,l-\frac{1}{2}}] + (\Delta_{l,l+\frac{1}{2}} - \Delta_{l,l-\frac{1}{2}})\, \boldsymbol{l}\cdot\boldsymbol{\sigma}\}\chi_\pm\, Y_l^{(0)} ,$$

which shows that $f^{(+)}$and $f^{(-)}$ are the results of operating on the angular part of the incident wave function with the same differential-and-spin operator.

4. Cross sections

We distinguish the four cross sections

$$\frac{d\sigma^{(++)}}{d\Omega} = |\chi_+^\dagger f^{(+)}|^2 = \pi k^{-2} |\sum_{l=0}^{\infty} (2l+1)^{-\frac{1}{2}} [(l+1)\, \Delta_{l,l+\frac{1}{2}} + l\Delta_{l,l-\frac{1}{2}}]\, Y_l^{(0)}\, |^2 ,$$

$$\frac{d\sigma^{(--)}}{d\Omega} = |\chi_-^\dagger f^{(-)}|^2 = \frac{d\sigma^{(++)}}{d\Omega} ,$$

$$\frac{d\sigma^{(-+)}}{d\Omega} = |\chi_-^\dagger f^{(+)}|^2 = \pi k^{-2} |\sum_{l=1}^{\infty} [l(l+1)/(2l+1)]^{\frac{1}{2}} (\Delta_{l,l+\frac{1}{2}} - \Delta_{l,l-\frac{1}{2}})\, Y_l^{(1)}|^2 ,$$

$$\frac{d\sigma^{(+-)}}{d\Omega} = |\chi_+^\dagger f^{(-)}|^2 = \frac{d\sigma^{(-+)}}{d\Omega} .$$

The former two describe processes in which the spin has the same orientation in the initial and final state, the latter ($\sigma^{(-+)}$ and $\sigma^{(+-)}$) processes with spin flip. These differential cross sections do not depend on ϕ *. This results from the choice of the axis of quantization for the spin along the direction of propagation of the incident wave, in both the initial and the final state.

If the spin in the initial state is unpolarized, while the measuring device selects only one of the final states χ_+ and χ_-, then

$$\frac{d\sigma_{unp}}{d\Omega} = \tfrac{1}{2}\left(\frac{d\sigma^{(++)}}{d\Omega} + \frac{d\sigma^{(+-)}}{d\Omega}\right) = \tfrac{1}{2}\left(\frac{d\sigma^{(-+)}}{d\Omega} + \frac{d\sigma^{(--)}}{d\Omega}\right) =$$

$$= \tfrac{1}{2}\,\pi k^{-2}\{|\sum_{l=0}^{\infty}(2l+1)^{-\frac{1}{2}}[(l+1)\,\Delta_{l,l+\frac{1}{2}} + l\,\Delta_{l,l-\frac{1}{2}}]\,Y_l^{(0)}|^2 +$$

$$+ |\sum_{l=1}^{\infty}[l(l+1)/(2l+1)]^{\frac{1}{2}}(\Delta_{l,l+\frac{1}{2}} - \Delta_{l,l-\frac{1}{2}})\,Y_l^{(1)}|^2\}$$

is the cross section for scattering of a beam of unpolarized particles. On the other hand, if the orientation of the spin in the final state is also not observed, the cross section is $2d\sigma_{unp}/d\Omega$.

Using the orthonormality of the spherical harmonics, the total cross section is found to be

$$\sigma_{tot} = \pi k^{-2}\sum_{l=0}^{\infty}(2l+1)^{-1}\left[|(l+1)\,\Delta_{l,l+\frac{1}{2}} + l\,\Delta_{l,l-\frac{1}{2}}|^2 + l(l+1)\,|\Delta_{l,l+\frac{1}{2}} - \Delta_{l,l-\frac{1}{2}}|^2\right].$$

For a given l, there is interference between the $j = l + \frac{1}{2}$ and the $j = l - \frac{1}{2}$ partial waves.

The total cross section for $s_{\frac{1}{2}}$-scattering is

$$\sigma(s_{\frac{1}{2}}) = \pi k^{-2}\,|\Delta_{0,\frac{1}{2}}|^2 = 4\pi\,k^{-2}\sin^2\eta_{0,\frac{1}{2}}\,,$$

while for pure $p_{\frac{1}{2}}$—scattering

$$\sigma(p_{\frac{1}{2}}) = \tfrac{1}{3}\pi\,k^{-2}(|\Delta_{1,\frac{1}{2}}|^2 + 2\,|\Delta_{1,\frac{1}{2}}|^2) = 4\pi\,k^{-2}\sin^2\eta_{1,\frac{1}{2}}\,,$$

and for pure $p_{\frac{3}{2}}$—scattering

$$\sigma(p_{\frac{3}{2}}) = \tfrac{1}{3}\pi\,k^{-2}(4\,|\Delta_{1,\frac{3}{2}}|^2 + 2\,|\Delta_{1,\frac{3}{2}}|^2) = 8\pi\,k^{-2}\sin^2\eta_{1,\frac{3}{2}}\,.$$

* Since $Y_l^{(1)} \sim e^{i\phi}$, $|\sum a_l Y_l^{(1)}|^2 = |\sum a_l Y_l^{(1)}\,e^{-i\phi}|^2$ is independent of ϕ.

CHAPTER III

TIME-INDEPENDENT SCATTERING OF DIRAC PARTICLES

1. The continuity equation

The wave function is assumed to obey a Dirac equation of the form

$$\left(\gamma_\mu \frac{\partial}{\partial x_\mu} + \kappa + \frac{1}{\hbar c}\,\mathfrak{B}\right)\psi = 0\,, \tag{1}$$

where $\mathfrak{B}$ is an interaction, in general involving the γ's.

The adjoint $\bar{\psi}$ obeys the equation

$$\frac{\partial \bar{\psi}}{\partial x_\mu}\gamma_\mu - \kappa\bar{\psi} - \frac{1}{\hbar c}\,\bar{\psi}\,\beta\,\mathfrak{B}^\dagger\,\beta = 0\,, \tag{2}$$

which follows from eq. (2) on taking its Hermitian conjugate and employing the Hermiticity of the γ's and their anticommutation relations.

Eq. (1), multiplied on the left by $\hbar c\gamma_4$, becomes

$$(H^{(0)} + H^{(1)})\,\psi = \hbar \mathrm{i}\,\frac{\partial}{\partial t}\,\psi\,,$$

with

$$H^{(0)} = -\,\mathrm{i}\,\hbar c\,\boldsymbol{\alpha}\cdot\boldsymbol{\nabla}_x + \beta\, mc^2\,,\quad H^{(1)} = \beta\,\mathfrak{B}\,.$$

If $\mathfrak{B}$ is a (Lorentz) scalar, eqs. (1) and (2) are relativistically invariant.

The sum of eq. (1), multiplied on the left by $\bar{\psi}$, and of eq. (2), multiplied on the right by ψ, gives

$$\frac{\partial j_\mu}{\partial x_\mu} = \frac{\mathrm{i}}{\hbar}\,\bar{\psi}(\beta\,\mathfrak{B}^\dagger\,\beta - \mathfrak{B})\,\psi\,,$$

where $j_\mu = \mathrm{i}\,c\,\bar{\psi}\,\gamma_\mu\,\psi$ are the components of the four-current.

Hence the four-current obeys the continuity equation

$$\frac{\partial j_\mu}{\partial x_\mu} = 0\,,$$

if

$$\mathfrak{V} = \beta\, \mathfrak{V}^{\dagger}\, \beta\,.$$

Examples of potentials satisfying this condition are

$$\mathfrak{V} = \mathfrak{V}(x)\,, \quad \mathfrak{V} = \mathrm{i}\, \gamma_5\, \mathfrak{V}(x)\,,$$
$$\mathfrak{V} = \mathrm{i}\, \gamma_5\, \gamma_\mu\, \partial \mathfrak{V}(x)/\partial x_\mu\,, \quad \mathfrak{V} = \mathrm{i}\, A_\mu\, \gamma_\mu\,,$$

where $\mathfrak{V}(x)$ is a real function, and A_μ is a four-vector with components $\boldsymbol{A}$ and A_0 all real.

If the interaction $\mathfrak{V}$ is static, eq. (1) admits steady solutions of the type

$$\psi = \psi(\boldsymbol{x})\, \mathrm{e}^{-\mathrm{i}Et/\hbar}\,,$$

the function $\psi(\boldsymbol{x})$ being a solution of the time-independent Dirac equation

$$(-\,\mathrm{i}\,\hbar c\, \boldsymbol{\alpha} \cdot \nabla_x + \beta\, mc^2 + \beta\, \mathfrak{V}(\boldsymbol{x}))\, \psi(\boldsymbol{x}) = E\, \psi(\boldsymbol{x})\,.$$

2. Steady-state scattering

Proceeding as in § 2 of Chap. I, we can write an integral equation for the scattering solutions of eq. (1):

$$\psi(\boldsymbol{x}) = \psi^{(0)}(\boldsymbol{x}) + \int K_{\mathrm{D}}(\boldsymbol{x}, \boldsymbol{x}')\, \psi(\boldsymbol{x}')\, \mathrm{d}^3\mathrm{x}'\,. \tag{3}$$

Here $\psi^{(0)}(\boldsymbol{x})$ is a positive energy plane wave, and the Dirac kernel $K_{\mathrm{D}}(\boldsymbol{x}, \boldsymbol{x}')$ is a solution of the equation

$$(H^{(0)} - E)\, K_{\mathrm{D}}(\boldsymbol{x}, \boldsymbol{x}') = -\,\delta(\boldsymbol{x} - \boldsymbol{x}')\, H^{(1)}\,.$$

It is convenient to write

$$K_{\mathrm{D}}(\boldsymbol{x}, \boldsymbol{x}') = G_{\mathrm{D}}(\boldsymbol{x}, \boldsymbol{x}')\, H^{(1)}(\boldsymbol{x}')\,.$$

The steady-state Green's function is defined by

$$(H^{(0)} - E)\, G_{\mathrm{D}}(\boldsymbol{x}, \boldsymbol{x}') = -\,\delta(\boldsymbol{x} - \boldsymbol{x}')\, I\,,$$

or, explicitly,

$$\left[-\,\mathrm{i}\,\boldsymbol{\alpha} \cdot \nabla_x + \beta\, \kappa - (k^2 + \kappa^2)^{\frac{1}{2}}\right] G_{\mathrm{D}}(\boldsymbol{x}, \boldsymbol{x}') = -\,\delta(\boldsymbol{x} - \boldsymbol{x}')\, I/\hbar c\,.$$

Using the anticommutation relations for the Dirac matrices, it is easy to show that

$$G_{\mathrm{D}}(\boldsymbol{x}, \boldsymbol{x}') = \frac{1}{\hbar c}\left[\mathrm{i}\,\boldsymbol{\alpha} \cdot \nabla_x - \beta\, \kappa - (k^2 + \kappa^2)^{\frac{1}{2}}\right] \mathfrak{G}^{(0)}(\boldsymbol{x}, \boldsymbol{x}')\,, \tag{4}$$

where $\mathfrak{G}^{(0)}(\boldsymbol{x}, \boldsymbol{x}')$ is the Green's function satisfying eq. (5) of Chap. I, § 2.

Substituting in (4) the Fourier representation of $\mathfrak{G}^{(0)}$ given by eq. (9) of Chap. I, § 3, we obtain that of G_{D},

$$G_{\mathrm{D}}(\boldsymbol{x}, \boldsymbol{x}') = (2\pi)^{-3} \frac{1}{\hbar c} \lim_{\varepsilon \to 0^+} \int \frac{\boldsymbol{\alpha} \cdot \boldsymbol{k}' + \beta \kappa + (k^2 + \kappa^2)^{\frac{1}{2}}}{k^2 - k'^2 + \mathrm{i}\, \varepsilon} \mathrm{e}^{\mathrm{i}\boldsymbol{k}' \cdot (\boldsymbol{x} - \boldsymbol{x}')} \mathrm{d}^3 \mathrm{k}' .$$

The asymptotic behaviour of the integral on the right of (3) is *

$$\sim (2\pi)^{-\frac{3}{2}} f(\theta, \phi) \frac{\mathrm{e}^{\mathrm{i}kr}}{r} . \tag{5}$$

The scattering amplitude $f(\theta, \phi)$ is now a four-component spinor.

3. Cross section

The optical theorem can easily be proved for Dirac particles. The component $\mathrm{i}c\, \bar{\psi}\, \gamma \cdot \boldsymbol{n}\, \psi$ of the current of a steady solution behaving asymptotically as the sum of a positive energy solution $\psi^{(0)}(\boldsymbol{x}) = (2\pi)^{-\frac{3}{2}} u(\boldsymbol{k}, \lambda)\, \mathrm{e}^{\mathrm{i}\boldsymbol{k} \cdot \boldsymbol{x}}$ of the free-particle equation (see Part II, Chap. V) and an outgoing wave (5), for large r is proportional to

$$\bar{u}(\boldsymbol{k}, \lambda)\, \gamma \cdot \boldsymbol{n}\, u(\boldsymbol{k}, \lambda) +$$
$$+ \frac{1}{r^2} \bar{f} \gamma \cdot \boldsymbol{n} f + \frac{1}{r} [\bar{f} \gamma \cdot \boldsymbol{n}\, u(\boldsymbol{k}, \lambda)\, \mathrm{e}^{\mathrm{i}(\boldsymbol{k} \cdot \boldsymbol{x} - kr)} + \bar{u}(\boldsymbol{k}, \lambda)\, \gamma \cdot \boldsymbol{n} f\, \mathrm{e}^{\mathrm{i}(kr - \boldsymbol{k} \cdot \boldsymbol{x})}] .$$

Upon integrating with respect to the direction of $\boldsymbol{n}$, the first term, which differs from $\boldsymbol{k} \cdot \boldsymbol{n}$ by a constant, gives zero. Such a term comes from the incident wave. In the second term, $\bar{f} \gamma \cdot \boldsymbol{n} f$ may be replaced ** by

$$\frac{1}{\mathrm{i}\sqrt{E^2 - E_0^2}} \bar{f} (E\, \gamma_4 - E_0) f$$

($\boldsymbol{n} \equiv \boldsymbol{x}/r, E_0 = mc^2$). This, divided by the magnitude of the current of the incident wave, gives the differential cross section ***

$$\frac{\mathrm{d}\sigma}{\mathrm{d}\Omega} = \frac{E}{E^2 - E_0^2} \bar{f} (E\, \gamma_4 - E_0) f =$$

* If $\mathfrak{B}$ has a short range.

** This can be shown without difficulty using the fact that $f\,\mathrm{e}^{\mathrm{i}kr}/r$ is asymptotically a solution of the free Dirac equation.

*** With the aid of eq. (20) of Part II, Chap. V, the cross section may be expressed in terms

$$= \frac{E}{E + E_0}(f_1^* f_1 + f_2^* f_2) + \frac{E}{E - E_0}(f_3^* f_3 + f_4^* f_4) .$$

It is easy to show that

$$\sigma = \int_{4\pi} \frac{\mathrm{d}\sigma}{\mathrm{d}\Omega}\,\mathrm{d}\Omega =$$

$$= \frac{-\mathrm{i}\,E}{\sqrt{E^2 - E_0^2}} \lim_{r\to\infty} r \int \left[\bar{f}\gamma \cdot \boldsymbol{n}\, u(\boldsymbol{k},\lambda)\, \mathrm{e}^{-\mathrm{i}kr(1-\cos\theta)} + \bar{u}(\boldsymbol{k},\lambda)\, \gamma \cdot \boldsymbol{n} f \mathrm{e}^{\mathrm{i}kr(1-\cos\theta)}\right] \mathrm{d}\Omega$$

$$= \frac{2\pi}{k} \frac{E}{\sqrt{E^2 - E_0^2}} \left[\bar{u}(\boldsymbol{k},\lambda)\, \gamma \cdot \boldsymbol{n} f - \bar{f}\gamma \cdot \boldsymbol{n}\, u(\boldsymbol{k},\lambda)\right]_{\boldsymbol{n} = \boldsymbol{k}/|\boldsymbol{k}|} =$$

$$= \frac{4\pi}{k} \operatorname{Im} A\left(\frac{\boldsymbol{k}}{k}\right) .$$

Here $A(\boldsymbol{k}/k)$ is the elastic scattering amplitude

$$A(\boldsymbol{k}/k) = A(\boldsymbol{n}_0) = \frac{\mathrm{i}E}{\sqrt{E^2 - E_0^2}}\bar{u}(\boldsymbol{k},\lambda)\, \gamma \cdot \boldsymbol{n}_0 f = u^\dagger(\boldsymbol{k},\lambda) f$$

in the forward direction. It is proportional to the probability amplitude for a process in which the particle is scattered into a positive energy state $u(\boldsymbol{k},\lambda)$ identical with the incident state.

4. Phase shift analysis

Here we consider the scattering, by a short range potential, of a positive frequency plane wave propagating in the direction of the third axis, with $\sigma_3 = +1$,

$$u(\boldsymbol{k},\uparrow)\, \mathrm{e}^{\mathrm{i}kz} = \tfrac{1}{2}\left[k_0(k_0 + k)\right]^{-\frac{1}{2}} (\alpha_3 k + k_0 + \beta\kappa)\, \mathrm{e}^{\mathrm{i}kz} \begin{pmatrix} \chi_+ \\ \chi_+ \end{pmatrix} . \tag{6}$$

(Pauli representation!).

---

of the large components as follows

$$\frac{\mathrm{d}\sigma}{\mathrm{d}\Omega} = \frac{2E}{E + E_0}(f_1^* f_1 + f_2^* f_2),$$

(Pauli representation!).
Since

$$U_1^\dagger U_1 = u_1^* u_1 + u_2^* u_2 = \frac{E + E_0}{2E},$$

from eqs. (20) and (8) of Part II, Chap. V, the cross section may also be written in the form

$$\frac{\mathrm{d}\sigma}{\mathrm{d}\Omega} = \frac{f_1^* f_1 + f_2^* f_2}{u_1^* u_1 + u_2^* u_2},$$

which involves only the large components (u_1, u_2) of the incident wave and (f_1, f_2) of the scattered wave.

If the interaction is rotationally invariant, the eigenvalues of $K = \rho_3(\boldsymbol{\sigma} \cdot \boldsymbol{L} + \hbar)$, $\boldsymbol{M}^2$ and M_3 are good quantum numbers.

It is therefore possible to represent the incident wave function (6) as a superposition of eigenfunctions of K, $\boldsymbol{M}^2$ and M_3 ($m_j = \frac{1}{2}$). To do this, we first write

$$\mathrm{e}^{\mathrm{i}kz}\begin{pmatrix}\chi_+\\ \chi_+\end{pmatrix} = (kr)^{-1}\sum_j [4\pi(2j+1)]^{\frac{1}{2}}\left[\mathrm{i}^{j-\frac{1}{2}}j^{\frac{1}{2}}\mathfrak{F}^{\frac{1}{2}}_{j-\frac{1}{2},j} + \mathrm{i}^{j+\frac{1}{2}}(j+1)^{\frac{1}{2}}\mathfrak{F}^{\frac{1}{2}}_{j+\frac{1}{2},j}\right],$$

where

$$\mathfrak{F}^{m_j}_{l,j} = \begin{pmatrix}F^{m_j}_{l,j}\\ F^{m_j}_{l,j}\end{pmatrix}$$

is a four-element column matrix, and $F^{m_j}_{l,j}$ is the two-component wave function defined by eqs. (6) of the previous chapter.

Using this we have

$$u(\boldsymbol{k}, \uparrow)\,\mathrm{e}^{\mathrm{i}kz} = \tfrac{1}{2}\left[k_0(k_0+k)\right]^{-\frac{1}{2}}\frac{1}{kr}\sum_j [4\pi(2j+1)]^{\frac{1}{2}} \times$$
$$\times\,(k_0 + \kappa\rho_3 + k\rho_1)\left[\mathrm{i}^{j-\frac{1}{2}}j^{\frac{1}{2}}\mathfrak{F}^{\frac{1}{2}}_{j-\frac{1}{2},j} + \mathrm{i}^{j+\frac{1}{2}}(j+1)^{\frac{1}{2}}\mathfrak{F}^{\frac{1}{2}}_{j+\frac{1}{2},j}\right].$$

Each term of the sum is an eigenfunction of $\boldsymbol{M}^2$ belonging to the eigenvalue $\hbar^2 j(j+1)$, and it can be divided into the sum of two eigenfunctions of K belonging to the eigenvalues $\hbar(j+\frac{1}{2})$ and $-\hbar(j+\frac{1}{2})$ respectively. This is done by means of the projection operators

$$\Lambda^{\pm}_j = \frac{j+\frac{1}{2} \pm \hbar^{-1}K}{2j+1},$$

and noticing that

$$\Lambda^{\pm}_j\,\mathfrak{F}^{\frac{1}{2}}_{j\mp\frac{1}{2},j} = \begin{pmatrix}F^{\frac{1}{2}}_{j\mp\frac{1}{2},j}\\ 0\end{pmatrix}, \quad \Lambda^{\pm}_j\,\mathfrak{F}^{\frac{1}{2}}_{j\pm\frac{1}{2},j} = \begin{pmatrix}0\\ F^{\frac{1}{2}}_{j\pm\frac{1}{2},j}\end{pmatrix}.$$

Thus one finds

$$u(\boldsymbol{k}, \uparrow)\,\mathrm{e}^{\mathrm{i}kz} = \frac{1}{2kr}\left[k_0(k_0+k)\right]^{-\frac{1}{2}}\sum_j [4\pi(2j+1)]^{\frac{1}{2}} \times$$
$$\times\left\{\begin{pmatrix}(k_0+k+\kappa)\,\mathrm{i}^{j-\frac{1}{2}}j^{\frac{1}{2}}F^{\frac{1}{2}}_{j-\frac{1}{2},j}\\ (k_0+k-\kappa)\,\mathrm{i}^{j+\frac{1}{2}}(j+1)^{\frac{1}{2}}F^{\frac{1}{2}}_{j+\frac{1}{2},j}\end{pmatrix} + \begin{pmatrix}(k_0+k+\kappa)\,\mathrm{i}^{j+\frac{1}{2}}(j+1)^{\frac{1}{2}}F^{\frac{1}{2}}_{j+\frac{1}{2},j}\\ (k_0+k-\kappa)\,\mathrm{i}^{j-\frac{1}{2}}j^{\frac{1}{2}}F^{\frac{1}{2}}_{j-\frac{1}{2},j}\end{pmatrix}\right\}.$$

For the first term in braces, $K = \hbar(j+\frac{1}{2})$, whereas, for the second, $K = -\hbar(j+\frac{1}{2})$.

The first term is given explicitly by

$$\psi_0^+ = \begin{bmatrix} \frac{1}{2}\left(\dfrac{j+\frac{1}{2}}{j}\right)^{\frac{1}{2}} F\, Y^{(0)}_{j-\frac{1}{2}} \\[2ex] \frac{1}{2}\left(\dfrac{j-\frac{1}{2}}{j}\right)^{\frac{1}{2}} F\, Y^{(0)}_{j-\frac{1}{2}} \\[2ex] -\frac{1}{2}\mathrm{i}\left(\dfrac{j+\frac{1}{2}}{j+1}\right)^{\frac{1}{2}} G\, Y^{(0)}_{j+\frac{1}{2}} \\[2ex] \frac{1}{2}\mathrm{i}\left(\dfrac{j+\frac{3}{2}}{j+1}\right)^{\frac{1}{2}} G\, Y^{(1)}_{j+\frac{1}{2}} \end{bmatrix}, \tag{7}$$

with

$$F = F_0^{(j,+)} = (k_0 + k + \kappa)\, \mathrm{i}^{j-\frac{1}{2}}\, F_{j-\frac{1}{2}}(kr)\,,$$
$$G = G_0^{(j,+)} = (k_0 + k - \kappa)\, \mathrm{i}^{j+\frac{3}{2}}\, F_{j+\frac{1}{2}}(kr)\,.$$

It is easy to see that these functions obey the system of differential equations

$$(k_0 - \kappa)\, F_0^{(j,+)} = -\left(G_0^{(j,+)\prime} + \frac{j+\frac{1}{2}}{r}\, G_0^{(j,+)}\right),$$

$$(k_0 + \kappa)\, G_0^{(j,+)} = F_0^{(j,+)\prime} - \frac{j+\frac{1}{2}}{r}\, F_0^{(j,+)}\,,$$

and have the asymptotic behaviour for large r

$$F_0^{(j,+)} \sim (k_0 + k + \kappa)\, \mathrm{i}^{j-\frac{1}{2}} \sin\left[kr - \tfrac{1}{2}\left(j - \tfrac{1}{2}\right)\pi\right],$$
$$G_0^{(j,+)} \sim (k_0 + k - \kappa)\, \mathrm{i}^{j+\frac{3}{2}} \sin\left[kr - \tfrac{1}{2}\left(j + \tfrac{1}{2}\right)\pi\right].$$

Similarly, the second term is

$$\psi_0^- = \begin{bmatrix} \frac{1}{2}\left(\dfrac{j+\frac{1}{2}}{j+1}\right)^{\frac{1}{2}} F\, Y^{(0)}_{j+\frac{1}{2}} \\[2ex] -\frac{1}{2}\left(\dfrac{j+\frac{3}{2}}{j+1}\right)^{\frac{1}{2}} F\, Y^{(1)}_{j+\frac{1}{2}} \\[2ex] -\frac{1}{2}\mathrm{i}\left(\dfrac{j+\frac{1}{2}}{j}\right)^{\frac{1}{2}} G\, Y^{(0)}_{j-\frac{1}{2}} \\[2ex] -\frac{1}{2}\mathrm{i}\left(\dfrac{j-\frac{1}{2}}{j}\right)^{\frac{1}{2}} G\, Y^{(1)}_{j-\frac{1}{2}} \end{bmatrix}, \tag{7'}$$

with

$$F = F_0^{(j,-)} = (k_0 + k + \kappa)\, \mathrm{i}^{j+\frac{1}{2}} F_{j+\frac{1}{2}}(kr)\,,$$
$$G = G_0^{(j,-)} = (k_0 + k - \kappa)\, \mathrm{i}^{j+\frac{1}{2}} F_{j-\frac{1}{2}}(kr)\,,$$

satisfying the equations

$$(k_0 - \kappa)\, F_0^{(j,-)} = -\left(G_0^{(j,-)\prime} - \frac{j+\frac{1}{2}}{r} G_0^{(j,-)}\right),$$

$$(k_0 + \kappa)\, G_0^{(j,-)} = F_0^{(j,-)\prime} + \frac{j+\frac{1}{2}}{r} F_0^{(j,-)}\,,$$

and having the asymptotic behaviour

$$F_0^{(j,-)} \sim (k_0 + k + \kappa)\, \mathrm{i}^{j+\frac{1}{2}} \sin\left[kr - \tfrac{1}{2}(j+\tfrac{1}{2})\,\pi\right],$$
$$G_0^{(j,-)} \sim (k_0 + k - \kappa)\, \mathrm{i}^{j+\frac{1}{2}} \sin\left[kr - \tfrac{1}{2}(j-\tfrac{1}{2})\,\pi\right].$$

A solution of the Dirac equation for a particle under the action of a centre of force, for example of an electron in a central electrostatic potential A_0 of short range, which is asymptotically the sum of the incident wave (6) and of outgoing waves, may be written in the form

$$(2\pi)^{\frac{3}{2}}\, \psi = \frac{1}{2kr}\left[k_0(k_0+k)\right]^{-\frac{1}{2}} \sum_j \left[4\pi(2j+1)\right]^{\frac{1}{2}} (\psi^+ + \psi^-)\,.$$

Here $\psi^\pm$ are defined by eqs. (7) and (7′) with $F = F^{(j,\pm)}$ and $G = G^{(j,\pm)}$ satisfying the equations

$$(k_0 - \kappa + e\,A_0)\, F^{(j,\pm)} = -\left(G^{(j,\pm)\prime} \pm \frac{j+\frac{1}{2}}{r} G^{(j,\pm)}\right),$$

$$(k_0 + \kappa + e\,A_0)\, G^{(j,\pm)} = F^{(j,\pm)\prime} \mp \frac{j+\frac{1}{2}}{r} F^{(j,\pm)}\,,$$

and having the asymptotic behaviour

$$F^{(j,\pm)} \sim \mathrm{e}^{\mathrm{i}\eta_j^\pm} (k_0 + k + \kappa)\, \mathrm{i}^{j\mp\frac{1}{2}} \sin\left[kr - \tfrac{1}{2}(j\mp\tfrac{1}{2})\,\pi + \eta_j^\pm\right],$$
$$G^{(j,\pm)} \sim \mathrm{e}^{\mathrm{i}\eta_j^\pm} (k_0 + k - \kappa)\, \mathrm{i}^{j+1\pm\frac{1}{2}} \sin\left[kr - \tfrac{1}{2}(j\pm\tfrac{1}{2})\,\pi + \eta_j^\pm\right].$$

The phase shifts $\eta_j^\pm$, corresponding to scattering in the states of angular momentum j and $K = \pm\hbar(j+\frac{1}{2})$ respectively, do not depend on m_j because of the rotation invariance of the interaction.

By some manipulations, it can be shown that, for large r,

$$(2\pi)^{\frac{3}{2}}\, \psi \sim u(\boldsymbol{k}, \uparrow)\, \mathrm{e}^{\mathrm{i}kz} + f\,\frac{\mathrm{e}^{\mathrm{i}kr}}{r}\,,$$

where

$$f = \frac{1}{2ik}\left(\frac{\pi}{2k_0(k_0+k)}\right)^{\frac{1}{2}}(k_0+k+\kappa\rho_3)\sum_j (2j+1)^{\frac{1}{2}}(\Delta_j^+ + \Delta_j^-\rho_1)\times$$

$$\times\begin{bmatrix} (2j)^{-\frac{1}{2}}\begin{pmatrix}(j+\frac{1}{2})^{\frac{1}{2}}\, Y_{j-\frac{1}{2}}^{(0)}\\ (j-\frac{1}{2})^{\frac{1}{2}}\, Y_{j-\frac{1}{2}}^{(1)}\end{pmatrix} \\ [2(j+1)]^{-\frac{1}{2}}\begin{pmatrix}(j+\frac{1}{2})^{\frac{1}{2}}\, Y_{j+\frac{1}{2}}^{(0)}\\ -(j+\frac{3}{2})^{\frac{1}{2}}\, Y_{j+\frac{1}{2}}^{(1)}\end{pmatrix}\end{bmatrix},$$

with

$$\Delta_j^{\pm} = e^{2i\eta_j^{\pm}} - 1\,,$$

which can be recast into the form

$$f = \frac{1}{2ik}\left(\frac{\pi}{k_0(k_0+k)}\right)^{\frac{1}{2}}\times$$

$$\times\sum_j (2j+1)^{\frac{1}{2}}\,(\Delta_j^+\Lambda_j^+ + \Delta_j^-\Lambda_j^-)\,(k_0+k+\kappa\rho_3)\left[j^{\frac{1}{2}}\,\Xi_{j-\frac{1}{2},j}^{\frac{1}{2}} + (j+1)^{\frac{1}{2}}\,\Xi_{j+\frac{1}{2},j}^{\frac{1}{2}}\right],$$

where

$$\Xi_{j\pm\frac{1}{2},j}^{\frac{1}{2}} = \mathfrak{F}_{j\pm\frac{1}{2},j}^{\frac{1}{2}}$$

with radial functions $F_{j\pm\frac{1}{2}}(kr)$ replaced by one, wherever they occur.

CHAPTER IV

NON-RELATIVISTIC TIME-DEPENDENT SCATTERING THEORY

1. Green's function for a non-free particle

The propagation of a free wave in non-relativistic wave mechanics may be described by the equation

$$\psi^{(0)}(\boldsymbol{x}, t) = \int \mathfrak{G}_{\mathrm{S}}^{(0)}(\boldsymbol{x} - \boldsymbol{x}', t - t')\, \psi^{(0)}(\boldsymbol{x}', t')\, \mathrm{d}^3\mathrm{x}' , \tag{1}$$

similar to that established in Part I, Chap. I.

We have changed the notation. Firstly, $^{(0)}$ denotes that the particle is not subject to forces. Secondly, $\mathfrak{G}_{\mathrm{S}}^{(0)}$ is not really identical with G_S of Part I, Chap. I, but is rather defined as

$$\mathfrak{G}_{\mathrm{S}}^{(0)}(\boldsymbol{x} - \boldsymbol{x}', t - t') = 0 , \quad \text{for} \quad t < t' , \tag{2}$$

$$\mathfrak{G}_{\mathrm{S}}^{(0)}(\boldsymbol{x} - \boldsymbol{x}', 0) = \delta(\boldsymbol{x} - \boldsymbol{x}') , \tag{3}$$

$$\left(\mathrm{i}\hbar \frac{\partial}{\partial t} + \frac{\hbar^2}{2m} \Delta\right) \mathfrak{G}_{\mathrm{S}}^{(0)}(\boldsymbol{x} - \boldsymbol{x}', t - t') = 0 , \quad \text{for} \quad t > t' . \tag{4}$$

Thus eq. (1) is valid only for $t > t'$.

The equation

$$\left(\mathrm{i}\hbar \frac{\partial}{\partial t} + \frac{\hbar^2}{2m} \Delta\right) \mathfrak{G}_{\mathrm{S}}^{(0)}(\boldsymbol{x} - \boldsymbol{x}', t - t') = \mathrm{i}\hbar\, \delta(\boldsymbol{x} - \boldsymbol{x}')\, \delta(t - t') \tag{5}$$

implies eq. (4) and the discontinuity * for $t = t'$.

---

* The Green's function is equal to $\delta(\boldsymbol{x} - \boldsymbol{x}')$ for $t = t'$ and vanishes for $t < t'$. Integrating the two sides of eq. (5) with respect to t over the interval $(t' - \varepsilon, t' + \varepsilon)$, and letting $\varepsilon \to 0^+$, the right-hand side gives

$$\mathrm{i}\hbar \lim_{\varepsilon \to 0} \delta(\boldsymbol{x} - \boldsymbol{x}') \int_{t'-\varepsilon}^{t'+\varepsilon} \delta(t - t')\, \mathrm{d}t = \mathrm{i}\hbar\, \delta(\boldsymbol{x} - \boldsymbol{x}') ,$$

while the left-hand side becomes

We now consider the case of a particle which is subject to forces. The wave function obeys the Schroedinger equation

$$\mathrm{i}\hbar \frac{\partial}{\partial t} \psi = (H^{(0)} + H^{(1)}) \psi , \tag{6}$$

with

$$H^{(0)} = -\frac{\hbar^2}{2m} \Delta , \quad H^{(1)} = \text{interaction Hamiltonian} .$$

For the present we do not make any assumption as to the form of the interaction, which may be static or a function of time.

A Green's function for particles subject to forces may be defined in analogy with $\mathfrak{G}_{\mathrm{S}}^{(0)}$ by the equations

$$\mathfrak{G}_{\mathrm{S}}(\boldsymbol{x}, \boldsymbol{x}'; t, t') = 0 , \quad (t < t'), \tag{7}$$

$$\mathfrak{G}_{\mathrm{S}}(\boldsymbol{x}, \boldsymbol{x}'; t, t) = \delta(\boldsymbol{x} - \boldsymbol{x}') , \tag{8}$$

$$\left(\mathrm{i}\hbar \frac{\partial}{\partial t} - H^{(0)} - H^{(1)}\right) \mathfrak{G}_{\mathrm{S}}(\boldsymbol{x}, \boldsymbol{x}'; t, t') = \mathrm{i}\hbar\, \delta(\boldsymbol{x} - \boldsymbol{x}')\, \delta(t - t') , \tag{9}$$

so that

$$\psi(\boldsymbol{x}, t) = \int \mathfrak{G}_{\mathrm{S}}(\boldsymbol{x}, \boldsymbol{x}'; t, t')\, \psi(\boldsymbol{x}', t')\, \mathrm{d}^3 x' , \quad (t > t') . \tag{10}$$

Eqs. (7), (8), (9) are equivalent to the integral equation *

$$\begin{aligned} \mathfrak{G}_{\mathrm{S}}(\boldsymbol{x}, \boldsymbol{x}'; t, t') &= \mathfrak{G}_{\mathrm{S}}^{(0)}(\boldsymbol{x} - \boldsymbol{x}', t - t') - \\ &- \frac{\mathrm{i}}{\hbar c} \int \mathfrak{G}_{\mathrm{S}}^{(0)}(\boldsymbol{x} - \boldsymbol{x}'', t - t'')\, H^{(1)}(\boldsymbol{x}'', t'')\, \mathfrak{G}_{\mathrm{S}}(\boldsymbol{x}'', \boldsymbol{x}'; t'', t')\, \mathrm{d}^4 x'' , \end{aligned} \tag{11}$$

$$\lim_{\varepsilon \to 0^+} \int_{t'-\varepsilon}^{t'+\varepsilon} \left(\hbar \mathrm{i} \frac{\partial}{\partial t} + \frac{\hbar^2}{2m} \Delta\right) \mathfrak{G}_{\mathrm{s}}^{(0)}(\boldsymbol{x} - \boldsymbol{x}', t - t')\, \mathrm{d}t =$$

$$= \mathrm{i}\hbar \lim_{\varepsilon \to 0^+} [\mathfrak{G}_{\mathrm{s}}^{(0)}(\boldsymbol{x} - \boldsymbol{x}', \varepsilon) - \mathfrak{G}_{\mathrm{s}}^{(0)}(\boldsymbol{x} - \boldsymbol{x}', -\varepsilon)] = \lim_{\varepsilon \to 0^+} \mathrm{i}\hbar\, \mathfrak{G}_{\mathrm{s}}^{(0)}(\boldsymbol{x} - \boldsymbol{x}', \varepsilon) .$$

* In fact, operating on the two sides of eq. (11) with

$$\left(\mathrm{i}\hbar \frac{\partial}{\partial t} - H^{(0)}\right),$$

one has

$$\left(\mathrm{i}\hbar \frac{\partial}{\partial t} - H^{(0)}\right) \mathfrak{G}_{\mathrm{s}}(\boldsymbol{x}, \boldsymbol{x}'; t, t') =$$

$$= \mathrm{i}\hbar\delta(\boldsymbol{x} - \boldsymbol{x}')\, \delta(t - t') - (\mathrm{i}/\hbar c) \int \hbar \mathrm{i} \delta(\boldsymbol{x} - \boldsymbol{x}'')\, \delta(t - t'')\, H^{(1)}(\boldsymbol{x}'', t'')\, \mathfrak{G}_{\mathrm{s}}(\boldsymbol{x}'', \boldsymbol{x}'; t'', t')\, \mathrm{d}^4 x''$$

$$= \mathrm{i}\hbar\delta(\boldsymbol{x} - \boldsymbol{x}')\, \delta(t - t') + H^{(1)}(\boldsymbol{x}, t)\, \mathfrak{G}_{\mathrm{s}}(\boldsymbol{x}, \boldsymbol{x}'; t, t').$$

Therefore eq. (9) is implied by (11). It is also easy to show that (7) follows from (11) and (2).

in which the integral is taken throughout space-time. In general, the dependence of $\mathfrak{G}_S$ on $\boldsymbol{x}, \boldsymbol{x}', t, t'$ is not through the intermediacy of $\boldsymbol{x} - \boldsymbol{x}'$ and $t - t'$, as for $\mathfrak{G}_S^{(0)}$. Only in the case of interactions which are invariant under space-time translations is the Green's function itself translation invariant as in the case of free particles.

Following Feynman *, we simplify the notation, indicating by 1 the space-time point $P_1 \equiv (\boldsymbol{x}_1, t_1)$, by 2 the point $P_2 \equiv (\boldsymbol{x}_2, t_2)$ etc. So we have

$$\mathfrak{G}_S(\boldsymbol{x}_2, \boldsymbol{x}_1; t_2, t_1) \equiv \mathfrak{G}_2(\mathrm{x}_2, \mathrm{x}_1) \equiv \mathfrak{G}_S(2, 1)\,,$$

and the integral equation for the Green's function reads

$$\mathfrak{G}_S(2, 1) = \mathfrak{G}_S^{(0)}(2, 1) - \frac{\mathrm{i}}{\hbar c} \int \mathfrak{G}_S^{(0)}(2, 3)\, H^{(1)}(3)\, \mathfrak{G}_S(3, 1)\, \mathrm{d}^4\mathrm{x}_3\,. \quad (11')$$

This equation can be solved by iteration,

$$\begin{aligned}\mathfrak{G}_S(2, 1) = \mathfrak{G}_S^{(0)}(2, 1) - (\mathrm{i}/\hbar c) \int \mathfrak{G}_S^{(0)}(2, 3)\, H^{(1)}(3)\, \mathfrak{G}_S^{(0)}(3, 1)\, \mathrm{d}^4\mathrm{x}_3 + \\ + (-\,\mathrm{i}/\hbar c)^2 \int \mathfrak{G}_S^{(0)}(2, 3)\, H^{(1)}(3)\, \mathfrak{G}_S^{(0)}(3, 4)\, H^{(1)}(4)\, \mathfrak{G}_S^{(0)}(4, 1)\, \mathrm{d}^4\mathrm{x}_3\, \mathrm{d}^4\mathrm{x}_4 + \dots\end{aligned} \quad (12)$$

In general $H^{(1)}$ is proportional to a parameter (*coupling constant*) characterizing the strength of the interaction (e.g., for an electromagnetic interaction

$$H^{(1)} = -\, e\, A_0 - \frac{ie\hbar}{mc} \boldsymbol{A} \cdot \boldsymbol{\nabla}_x + \frac{e^2}{2mc^2} A^2\,,$$

the coupling constant is $\sqrt{\alpha} = e(4\pi\hbar c)^{-\frac{1}{2}} = (137)^{-\frac{1}{2}}$), of which (12) is a power series. Convergence depends, though not entirely, on the value of the coupling constant.

Inserting (11′) for $\mathfrak{G}_S$ in eq. (10), we obtain the equation

$$\psi(2) = \psi^{(0)}(2) - (\mathrm{i}/\hbar c) \int \mathfrak{G}_S^{(0)}(2, 3)\, H^{(1)}(3)\, \psi(3)\, \eta(3, 1)\, \mathrm{d}^4\mathrm{x}_3\,, \quad (13)$$

where $\eta(3, 1)$ is the step-function defined by

$$\begin{aligned}\eta(3, 1) &= 1 \quad \text{for} \quad t_3 \geqslant t_1\,, \\ &= 0 \quad \text{for} \quad t_3 < t_1\,,\end{aligned}$$

and

$$\psi^{(0)}(2) = \int \mathfrak{G}_S^{(0)}(2, 1)\, \psi(1)\, \mathrm{d}^3\mathrm{x}_1\,.$$

---

* R. P. Feynman, *The Theory of Positrons*, Phys. Rev. **76** (1949), 749.

Iterating (13) one has

$$\psi(2) = \int \mathfrak{G}_S^{(0)}(2, 1)\, \psi(1)\, d^3x_1 -$$
$$- (i/\hbar c) \int \mathfrak{G}_S^{(0)}(2, 3)\, H^{(1)}(3)\, \mathfrak{G}_S^{(0)}(3, 1)\, \psi(1)\, d^4x_3\, d^3x_1 + \qquad (14)$$
$$+ (i/\hbar c)^2 \int \mathfrak{G}_S^{(0)}(2, 3)\, H^{(1)}(3)\, \mathfrak{G}_S^{(0)}(3, 4)\, H^{(1)}(4)\, \mathfrak{G}_S^{(0)}(4, 1)\, \psi(1)\, d^4x_3\, d^4x_4\, d^3x_1 + \ldots,$$

or

$$\psi(2) = \psi^{(0)}(2) - (i/\hbar c) \int \mathfrak{G}_S^{(0)}(2, 3)\, H^{(1)}(3)\, \psi^{(0)}(3)\, d^4x_3 + \qquad (14')$$
$$+ (i/\hbar c)^2 \int \mathfrak{G}_S^{(0)}(2, 3)\, H^{(1)}(3)\, \mathfrak{G}_S^{(0)}(3, 4)\, H^{(1)}(4)\, \psi^{(0)}(4)\, d^4x_3\, d^4x_4 + \ldots,$$

where

$$\psi^{(0)}(2) = \int \mathfrak{G}_S^{(0)}(2, 1)\, \psi(1)\, d^3x_1\,, \quad \psi^{(0)}(3) = \int \mathfrak{G}_S^{(0)}(3, 1)\, \psi(1)\, d^3x_1\,, \quad \ldots$$

This can be interpreted intuitively by noting that the first term on the right of (14′) describes free propagation of the wave function from 1 to 2 (i.e. propagation according to the free particle wave equation), the second term free propagation from 1 to 3, an interaction process of strength $H^{(1)}(3)$ at 3, and again free propagation from 3 to 2 etc.

The successive terms of the expansion (14′) may be represented by the Feynman diagrams

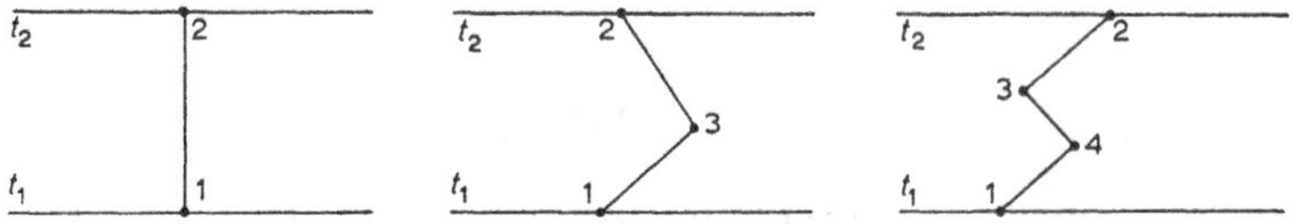

Fig. 10

etc.

By the property (2) of the Green's function, the "interaction points" are arranged in chronological order, and are comprised between the planes $t = t_1$ and $t = t_2$. For example, no diagram of the type

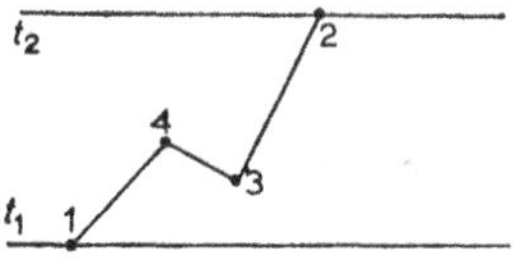

Fig. 11

is possible. It will be seen that this limitation does not exist in the relativistic theory.

We must now find the transition probability from a state i to a state f.

If the particle is represented, at the time t_1, by the wave function $\psi_i(1)$, the probability that, at the time t_2, it will be in the state described by the wave function $\psi_f(2)$ is given by the absolute square of the product of the two wave functions at the time t_2, i.e.

$$P_{i\to f} = |(\psi_f(2), \quad \psi_i(2))|^2 .$$

Here $\psi_i(2)$ is the wave function at the time t_2 arising from $\psi_i(1)$ by propagation according to the Green's function $\mathfrak{G}_S(2, 1)$.

Therefore

$$P_{i\to f} = | \int \psi_f^*(2)\ \psi_i(2)\ d^3x_2\ |^2 = | \int \psi_f^*(2)\ \mathfrak{G}_S(2, 1)\ \psi_i(1)\ d^3x_1\ d^3x_2\ |^2 .$$

The probability that the transition $i \to f$ may take place in the unit time interval is

$$\mathfrak{P}_{i\to f} = \frac{| \int \psi_f^*(2)\ \mathfrak{G}_S(2, 1)\ \psi_i(1)\ d^3x_1\ d^3x_2|^2}{t_2 - t_1} .$$

2. Static interaction. Time-independent theory

If the interaction is static, letting $t_1 \to -\infty$ and so replacing $\eta(3, 1)$ by one, eq. (10′) reduces to

$$\psi(2) = \psi^{(0)}(2) - (i/\hbar c) \int \mathfrak{G}_S^{(0)}(2, 3)\ H^{(1)}(\boldsymbol{x}_3)\ \psi(3)\ d^4x_3 . \qquad (15)$$

It is easily seen that, if $\psi^{(0)}(2)$ is a monochromatic solution of the free wave equation

$$\psi^{(0)}(2) = \psi^{(0)}(\boldsymbol{x}_2)\ e^{-i\omega t_2} ,$$

$\psi(2)$ must depend on the time in the same way,

$$\psi(2) = \psi(\boldsymbol{x}_2)\ e^{-i\omega t_2} ,$$

(energy conservation). In fact, for the second term of the iterated equation (14), one has

$$\int \mathfrak{G}_S^{(0)}(\boldsymbol{x}_2 - \boldsymbol{x}_3, t_2 - t_3)\ H^{(1)}(\boldsymbol{x}_3)\ \psi^{(0)}(3)\ d^4x_3 \propto \int_{-\infty}^{t_2} \mathfrak{G}_S^{(0)}(t_2 - t_3)\ e^{-i\omega t_3}\ dt_3 ,$$

and, putting $\tau = t_3 - t_2$, this becomes

$$e^{-i\omega t_2} \int_{-\infty}^{0} \mathfrak{G}_S^{(0)}(-\tau)\ e^{-i\omega\tau}\ d\tau .$$

One can deal in a similar way with the other terms.

From the integral equation (15) it follows that

$$\lim_{t_2 \to -\infty} \left(\psi(2) - \psi^{(0)}(2)\right) = 0\,, \tag{16}$$

and also

$$\lim_{t_2 \to -\infty} \left(H^{(0)}\,\psi(2) - H^{(0)}\,\psi^{(0)}(2)\right) = 0\,, \tag{16'}$$

while, from the assumption that the wave function is stationary, it follows that

$$\begin{aligned} \mathrm{i}\hbar\,\frac{\partial}{\partial t_2}\,\psi(2) &= \hbar\omega\,\psi(2)\,, \\ \mathrm{i}\hbar\,\frac{\partial}{\partial t_2}\,\psi^{(0)}(2) &= \hbar\omega\,\psi^{(0)}(2)\,. \end{aligned} \tag{16''}$$

Since

$$\begin{aligned} \left(\mathrm{i}\hbar\,\frac{\partial}{\partial t_2} - H^{(0)}\right)\psi(2) &= H^{(1)}(2)\,\psi(2)\,, \\ \left(\mathrm{i}\hbar\,\frac{\partial}{\partial t_2} - H^{(0)}\right)\psi^{(0)}(2) &= 0\,, \end{aligned} \tag{16'''}$$

using (16), (16′), (16″), (16‴) one finds

$$\lim_{t_2 \to -\infty} H^{(1)}(2)\,\psi(2) = 0\,.$$

As we have assumed that the interaction is static, this relation implies that, in the distant past, $\psi(2)$, and therefore also $\psi^{(0)}(2)$, is a wave packet vanishing in the region of space where the interaction $H^{(1)}(2)$ is active (see Fig. 12), which is in disaccord with the assumption that the incident wave is monochromatic.

To remedy this, we shall assume that $H^{(1)}$ is not strictly independent of

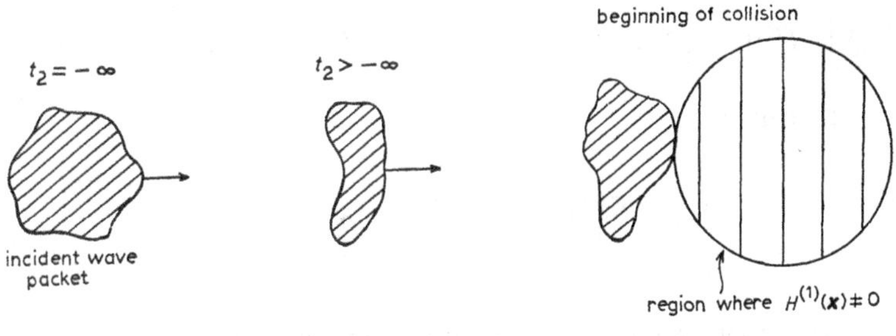

Fig. 12

time, but is switched on adiabatically at $t = -\infty$ and is switched off at $t = +\infty$.

This may be expressed analytically by writing

$$H^{(1)}(\boldsymbol{x}) \rightarrow H^{(1)}(\boldsymbol{x})\, \mathrm{e}^{-\varepsilon|t|}\,,$$

where ε is a positive quantity which will be made to tend to zero at the end of the calculation.

The *adiabatic switching-on* and *off* of the interaction makes it possible to have free particle wave functions in the distant past and future, since there $H^{(1)}(\boldsymbol{x}, \pm\infty) = 0$. This expresses the fact that the duration of an interaction of a particle with a centre of force is always limited. For instance, in a scattering experiment the interaction begins to act only when the target is exposed to the incident beam. However, as long as $\varepsilon \neq 0$, the scattered wave is not stationary.

We now multiply eq. (15) by $\mathrm{e}^{\mathrm{i}(E' t_2/\hbar) - \varepsilon|t_2|}$, and integrate over t_2 from $-\infty$ to $+\infty$, defining

$$\Psi_{E'}(\boldsymbol{x}_2) \equiv \lim_{\varepsilon\to 0} \int_{-\infty}^{+\infty} \mathrm{e}^{\mathrm{i}(E' t_2/\hbar) - \varepsilon|t_2|}\, \psi(2)\, \mathrm{d}t_2\,. \tag{17}$$

Thus

$$\begin{aligned}\Psi_{E'}(\boldsymbol{x}_2) &= 2\pi\hbar\delta(E' - E)\, \psi_E^{(0)}(\boldsymbol{x}_2) - \\ &- (\mathrm{i}/\hbar c) \lim_{\varepsilon\to 0} \int_{-\infty}^{+\infty} \mathrm{d}t_2\, \mathrm{e}^{\mathrm{i}E' t_2/\hbar}\, \mathrm{e}^{-\varepsilon|t_2|} \int \mathfrak{G}_{\mathrm{S}}^{(0)}(2, 3)\, \mathrm{e}^{-\varepsilon|t_3|}\, H^{(1)}(\boldsymbol{x}_3)\, \psi(3)\, \mathrm{d}^4 x_3\,.\end{aligned} \tag{18}$$

As the limits of the integration with respect to t_3 are $-\infty$ and t_2, one has, changing the order of integration for t_2 and t_3,

$$\int_{-\infty}^{+\infty} \mathrm{d}t_2 \int_{-\infty}^{t_2} \mathrm{d}t_3 = \int_{-\infty}^{+\infty} \mathrm{d}t_3 \int_{t_3}^{+\infty} \mathrm{d}t_2\,,$$

and thus the second term on the right-hand side of eq. (18) becomes

$$-(\mathrm{i}/\hbar c) \int H^{(1)}(\boldsymbol{x}_3)\, \psi(3)\, \mathrm{e}^{\mathrm{i}E' t_3/\hbar}\, \mathrm{e}^{-\varepsilon|t_3|}\, \mathrm{d}^4 x_3 \int_{t_3}^{\infty} \mathfrak{G}_{\mathrm{S}}^{(0)}(2, 3)\, \mathrm{e}^{\mathrm{i}E'(t_2 - t_3)/\hbar}\, \mathrm{e}^{-\varepsilon|t_2|}\, \mathrm{d}t_2\,.$$

On the other hand *

$$J(\boldsymbol{x}_2 - \boldsymbol{x}_3) \equiv \int_{t_3}^{\infty} \mathrm{d}t_2\, \mathfrak{G}_{\mathrm{S}}^{(0)}(2, 3)\, \mathrm{e}^{\mathrm{i}E'(t_2 - t_3)/\hbar}\, \mathrm{e}^{-\varepsilon|t_2|} = \frac{m}{2\pi\mathrm{i}\hbar}\, \frac{\mathrm{e}^{\mathrm{i}k'|\boldsymbol{x}_2 - \boldsymbol{x}_3|}}{|\boldsymbol{x}_2 - \boldsymbol{x}_3|}\,.$$

* For the evaluation of this integral use the integration variable $\tau = t_2 - t_3$. Noting that in the integrand $\mathrm{e}^{-\varepsilon|\tau + t_3|}$ can be replaced by $\mathrm{e}^{-\varepsilon\tau}$ (since ε is infinitely small), and

Thus the integral on the right hand side of (18) becomes

$$-\,(\mathrm{i}/\hbar)\int J(\boldsymbol{x}_2-\boldsymbol{x}_3)\,H^{(1)}(\boldsymbol{x}_3)\,\Psi_{E'}(\boldsymbol{x}_3)\,\mathrm{d}^3\mathrm{x}_3\,,$$

and that equation takes the form

$$\Psi_{E'}(\boldsymbol{x}_2)=2\pi\hbar\,\delta(E'-E)\,\psi_E^{(0)}(\boldsymbol{x}_2)-$$

$$-\frac{2m}{\hbar^2}\int\frac{\mathrm{e}^{\mathrm{i}k'|\boldsymbol{x}_2-\boldsymbol{x}_3|}}{4\pi\,|\boldsymbol{x}_2-\boldsymbol{x}_3|}\,H^{(1)}(\boldsymbol{x}_3)\,\Psi_{E'}(\boldsymbol{x}_3)\,\mathrm{d}^3\mathrm{x}_3\,.$$

Putting $\boldsymbol{x}_2\to\boldsymbol{x}$, $\boldsymbol{x}_3\to\boldsymbol{x}'$ and $\Psi_{E'}(\boldsymbol{x})=2\pi\hbar\,\delta(E'-E)\,\psi_E(\boldsymbol{x})$, we finally have

$$\psi_E(\boldsymbol{x})=\psi_{\mathrm{E}}^{(0)}(\boldsymbol{x})-(2m/\hbar^2)\int\mathfrak{G}_{\mathrm{S}}^{(0)}(\boldsymbol{x},\boldsymbol{x}')\,H^{(1)}(\boldsymbol{x}')\,\psi_E(\boldsymbol{x}')\,\mathrm{d}^3\mathrm{x}' \tag{19}$$

with

$$\mathfrak{G}_{\mathrm{S}}^{(0)}=\frac{\mathrm{e}^{\mathrm{i}k\,|\boldsymbol{x}-\boldsymbol{x}'|}}{4\pi\,|\boldsymbol{x}-\boldsymbol{x}'|}\,,\qquad k^2=2mE/\hbar^2\,.$$

Eq. (19) is identical with the basic eq. (3), Chap. I, of the time-independent scattering theory.

3. Cross section

Let $\psi_{\mathrm{i}}^{(0)}(1)$ and $\psi_{\mathrm{f}}^{(0)}(1)$ be two orthogonal free-particle wave functions at the time t_1. Their product (Schroedinger metric) is a constant of motion, and so

$$\big(\psi_{\mathrm{f}}^{(0)}(2),\,\psi_{\mathrm{i}}^{(0)}(2)\big)=\big(\psi_{\mathrm{f}}^{(0)}(1),\,\psi_{\mathrm{i}}^{(0)}(1)\big)=0\,.$$

We assume that at the time t_1 a particle is described by $\psi_{\mathrm{i}}^{(0)}(t_1)$. At time t_2, by virtue of the interaction, its wave function is not $\psi_{\mathrm{i}}^{(0)}(t_2)$, but $\psi_{\mathrm{i}}(t_2)$, which is not necessarily orthogonal to $\psi_{\mathrm{f}}^{(0)}(t_2)$.

---

inserting the explicit expression for $\mathfrak{G}_{\mathrm{s}}^{(0)}(2,3)$ (see Part I, Chap. I, § 2), one has

$$J=(2\pi)^{-3}\int\mathrm{e}^{\mathrm{i}\boldsymbol{k}\cdot(\boldsymbol{x}_2-\boldsymbol{x}_3)}\,\mathrm{d}^3\mathrm{k}\int_0^\infty\mathrm{e}^{\mathrm{i}(E'-E+\mathrm{i}\hbar\varepsilon)\tau/\hbar}\,\mathrm{d}\tau=(2\pi)^{-3}\,\mathrm{i}\hbar\int\frac{\mathrm{e}^{\mathrm{i}\boldsymbol{k}\cdot(\boldsymbol{x}_2-\boldsymbol{x}_3)}}{E'-E+\mathrm{i}\varepsilon}\,\mathrm{d}^3\mathrm{k}\,.$$

But E depends only on $|\boldsymbol{k}|$, so one can perform the integration over the solid angle obtaining

$$J=\frac{m}{2\pi^2\hbar}\frac{1}{|\boldsymbol{x}_2-\boldsymbol{x}_3|}\int_{-\infty}^{\infty}\frac{\mathrm{e}^{\mathrm{i}k|\boldsymbol{x}_2-\boldsymbol{x}_3|}}{k'^2-k^2+\mathrm{i}\varepsilon}\,k\,\mathrm{d}k=\frac{m}{2\pi\mathrm{i}\hbar}\frac{\mathrm{e}^{\mathrm{i}k'|\boldsymbol{x}_2-\boldsymbol{x}_3|}}{|\boldsymbol{x}_2-\boldsymbol{x}_3|}\,,$$

where k' is defined by

$$E'=\hbar^2k'^2/2m\,.$$

Therefore a transition probability from $\psi_i^{(0)}$ to $\psi_f^{(0)}$ exists during the time interval (t_1, t_2) and is given by

$$P_{i\to f} = \left|\left(\psi_f^{(0)}(2), \psi_i(2)\right)\right|^2 = \\ = \left|(-\,i/\hbar c)\int \psi_f^{(0)*}(2)\ \mathfrak{G}_S^{(0)}(2,3)\ H^{(1)}(3)\ \psi_i(3)\ d^3x_2\ d^4x_3\right|^2 \tag{20}$$

If $\psi_i^{(0)}(1)$ and $\psi_f^{(0)}(2)$ are stationary plane waves,

$$\psi_i^{(0)}(1) = \frac{1}{\sqrt{V}}\, e^{i(k_i\cdot x_1 - E_i t_1/\hbar)}$$

$$\psi_f^{(0)}(2) = \frac{1}{\sqrt{V}}\, e^{i(k_f\cdot x_2 - E_f t_2/\hbar)}$$

$$(V = \text{normalization volume; thus } \int_V \psi^{(0)*}\,\psi^{(0)}\, d^3x = 1),$$

one has

$$P_{i\to f} = \left|(1/\hbar c)\int \delta(\boldsymbol{k}-\boldsymbol{k}_f)\, e^{-i\boldsymbol{k}\cdot x_3}\, e^{-i\omega(k)(t_2-t_3)}\, H^{(1)}(3)\, e^{iE_f t_2/\hbar}\, \psi_i(3)\, d^3k\, d^4x_3\right|^2 V^{-1} \\ = \left|(1/\hbar c)\int e^{-i\boldsymbol{k}_f\cdot x_3}\, e^{iE_f t_3/\hbar}\, H^{(1)}(3)\, \psi_i(3)\, d^4x_3\right|^2 V^{-1}.$$

In the case of a static potential, taking the limit for $t_1 \to -\infty$ and $t_2 \to +\infty$ of the above expression divided by $t_2 - t_1$ and using eq. (17) one has the transition probability per unit time *

$$\mathfrak{P}_{i\to f} = \lim_{\substack{t_2\to+\infty \\ t_1\to-\infty}} \frac{(2\pi\hbar)^4\,[\delta(E_i - E_f)]^2\,|f(\boldsymbol{n})|^2}{m^2(t_2 - t_1)\,V^2}, \quad k\boldsymbol{n} = \boldsymbol{k}_f\,. \tag{21}$$

Here

$$f(\boldsymbol{n}) = -\,(\pi/2)^{\frac{1}{2}}\,(2m/\hbar^2)\int e^{-ik\boldsymbol{n}\cdot x_3}\, H^{(1)}(\boldsymbol{x}_3)\,\psi(\boldsymbol{x}_3)\, d^3x_3$$

is the scattering amplitude as given in Chap. I, § 2.

The "infinity" of $t_2 - t_1$ is cancelled by that of one of the δ functions. In fact

$$[\delta(E_i - E_f)]^2 = \delta(E_i - E_f)\Big[\lim_{\substack{t_2\to+\infty \\ t_1\to-\infty}} (1/2\pi\hbar)\int e^{i(E_i - E_f)\,t/\hbar}\, dt\Big]_{E_i = E_f} = \\ = \delta(E_i - E_f)\,(2\pi\hbar)^{-1} \lim_{\substack{t_2\to\infty \\ t_1\to-\infty}} \int_{t_1}^{t_2} dt = (2\pi\hbar)^{-1}\,\delta(E_i - E_f) \lim_{\substack{t_2\to\infty \\ t_1\to-\infty}} (t_2 - t_1)\,.$$

Thus, with the replacement

$$\delta(0)/(t_2 - t_1) \to (2\pi\hbar)^{-1},$$

* Here we follow B. A. Lippmann and J. Schwinger, Phys. Rev. **79** (1950), 469. See also J. Hamilton, *The Theory of Elementary Particles* (Oxford, 1959), p. 216.

eq. (21) becomes

$$\mathfrak{P}_{i\to f} = \frac{(2\pi\hbar)^3}{V^2 m^2} |f(\boldsymbol{n})|^2 \, \delta(E_f - E_i) \,. \tag{22}$$

Here $\delta(E_f - E_i)$ expresses energy conservation in the transition $i \to f$. Of course (22) is infinite for $E_i = E_f$ and zero for $E_i \neq E_f$, and is physically not of much use. However we can derive a formula for the transition probability from a state of energy E_i to states of energy comprised in a small interval containing E_i and momentum within the solid angle $d\Omega$ about $\boldsymbol{p}_f/|\boldsymbol{p}_f|$.

The number of such states equals that of the cells in phase space,

$$\rho(E_f)\, dE_f\, d\Omega = (2\pi\hbar)^{-3} V\, d^3 p_f \,.$$

Introducing polar coordinates, one has

$$\rho(E_f)\, d\Omega = \frac{V\, k_f^2\, dk_f}{(2\pi)^3\, dE_f}\, d\Omega \,.$$

The expression (22) for the transition probability per unit time, integrated with respect to E_f from $E_i - \eta$ to $E_i + \eta$ after multiplication by $\rho(E_f)\, d\Omega\, dE_f$, takes the form

$$\mathfrak{P}_{i\to f} = \frac{(2\pi\hbar)^3}{m^2} |f(\boldsymbol{n})|^2 \frac{\rho(E_i)\, d\Omega}{V^2} \,. \tag{23}$$

The cross section is obtained by dividing (23) by the current density of the incident wave, which, for the above normalization, is

$$j = \rho v = v/V \,,$$

where ρ and v denote respectively the density and the velocity of the incident particle.

Hence

$$\frac{d\sigma}{d\Omega} = \frac{(2\pi\hbar)^3}{m^2} |f(\boldsymbol{n})|^2 \frac{\rho(E_i)}{vV} \,.$$

Finally, noting that

$$\frac{(2\pi\hbar)^3}{m^2} \frac{\rho(E_i)}{vV} = 1 \,,$$

the differential cross section becomes

$$\frac{d\sigma}{d\Omega} = |f(\boldsymbol{n})|^2 \,.$$

as in Chap. I, § 5.

CHAPTER V

EMISSION AND ABSORPTION OF ELECTROMAGNETIC RADIATION

1. Green's function

Consider a particle of spin zero and charge $-e$ in an electrostatic field such that the potential energy $V = -e A_0$ is negative (attractive force).

We denote by $\psi_n^{(V)}(\boldsymbol{x})$ the wave functions of the bound states, the energies of which ($E_n < 0$) form a discrete spectrum, and by $\psi_E^{(V)}(\boldsymbol{x})$ the wave functions of the "scattering states", solutions of the equation *

$$\psi_E^{(V)}(\boldsymbol{x}) = \psi_E^{(0)}(\boldsymbol{x}) - (2m/\hbar^2) \int \mathfrak{G}_S^{(0)}(\boldsymbol{x}, \boldsymbol{x}') \, V(\boldsymbol{x}') \, \psi_E^{(V)}(\boldsymbol{x}') \, \mathrm{d}^3\mathrm{x}' \qquad (1)$$

with $\psi_E^{(0)}(\boldsymbol{x}) = (2\pi)^{-\frac{3}{2}} \, \mathrm{e}^{\mathrm{i}\boldsymbol{k}\cdot\boldsymbol{x}}$, whose (positive) energies form a continuous spectrum.

The functions $\psi_n^{(V)}$ of the discrete spectrum, together with the solutions of eq. (1), which we label by the wavevector of their "incident part", form a complete set of eigenfunctions of the Hamiltonian $H = H^{(0)} + V$. Completeness is expressed by the relation

$$\sum_n \psi_n^{(V)*}(\boldsymbol{x}) \, \psi_n^{(V)}(\boldsymbol{x}') + \int \psi_k^{(V)*}(\boldsymbol{x}) \, \psi_k^{(V)}(\boldsymbol{x}') \, \mathrm{d}^3\mathrm{k} = \delta(\boldsymbol{x} - \boldsymbol{x}') .$$

The Green's function $\mathfrak{G}_S^{(V)}(2, 1)$, by which the wave function at the time t_2 can be constructed from its values at the time t_1,

$$\psi^{(V)}(2) = \int \mathfrak{G}_S^{(V)}(2, 1) \, \psi^{(V)}(1) \, \mathrm{d}^3\mathrm{x}_1 ,$$

is

$$\mathfrak{G}_S^{(V)}(2, 1) = \eta(2,1) \sum_n \psi_n^{(V)}(\boldsymbol{x}_2) \, \psi_n^{(V)*}(\boldsymbol{x}_1) \, \mathrm{e}^{-\mathrm{i}E_n(t_2-t_1)/\hbar} +$$
$$+ \eta(2, 1) \int \psi_k^{(V)}(\boldsymbol{x}_2) \, \psi_k^{(V)*}(\boldsymbol{x}_1) \, \mathrm{e}^{-\mathrm{i}E(k)(t_2-t_1)/\hbar} \, \mathrm{d}^3\mathrm{k} \equiv \qquad (2)$$
$$\equiv \eta(2, 1) \, \mathop{\not{\Sigma}}_n \, \psi_n^{(V)}(\boldsymbol{x}_2) \, \psi_n^{(V)*}(\boldsymbol{x}_1) \, \mathrm{e}^{-\mathrm{i}E_n(t_2-t_1)/\hbar} .$$

It is easy to verify that this function satisfies the conditions (7), (8), (9) of the previous chapter with $H^{(1)}(\boldsymbol{x}) = -e A_0(\boldsymbol{x})$.

* This is eq. (3) of Chap. I, with $\mathfrak{V}(x) = 2mV(x)/\hbar^2$. The results of this section are, to a good approximation, also valid for electrons (neglecting spin effects and relativistic corrections).

We now propose to study transitions between eigenfunctions of the Hamiltonian $H = H^{(0)} + V$, caused by the interaction of the particle with an electromagnetic wave described by the vector potential

$$\boldsymbol{A} = \boldsymbol{A}_0 \, \mathrm{e}^{\mathrm{i}(\boldsymbol{k} \cdot \boldsymbol{x} - \omega t)} + \boldsymbol{A}_0^* \, \mathrm{e}^{-\mathrm{i}(\boldsymbol{k} \cdot \boldsymbol{x} - \omega t)} \,, \quad (\omega = c\,|\boldsymbol{k}|\,) \,, \tag{3}$$

obeying the transversality condition

$$\boldsymbol{k} \cdot \boldsymbol{A} = 0 \,. \tag{4}$$

NOTE 1. To help the reader we recall some properties of electromagnetic waves.

The electric and the magnetic field corresponding to the vector potential (3) are

$$\boldsymbol{E} = (\mathrm{i}\omega/c)\,[\boldsymbol{A}_0 \, \mathrm{e}^{\mathrm{i}(\boldsymbol{k} \cdot \boldsymbol{x} - \omega t)} - \boldsymbol{A}_0^* \, \mathrm{e}^{-\mathrm{i}(\boldsymbol{k} \cdot \boldsymbol{x} - \omega t)}] \,,$$

$$\boldsymbol{H} = \mathrm{i}\boldsymbol{k} \times [\boldsymbol{A}_0 \, \mathrm{e}^{\mathrm{i}(\boldsymbol{k} \cdot \boldsymbol{x} - \omega t)} - \boldsymbol{A}_0^* \, \mathrm{e}^{-\mathrm{i}(\boldsymbol{k} \cdot \boldsymbol{x} - \omega t)}] \,,$$

the time average of the Poynting vector is

$$\overline{\boldsymbol{S}} = 2\,\omega(\boldsymbol{A}_0 \cdot \boldsymbol{A}_0^*)\,\boldsymbol{k} \,,$$

and that of the energy density

$$\tfrac{1}{2}\,\overline{(\boldsymbol{E}^2 + \boldsymbol{H}^2)} = 2\,\omega^2\,\boldsymbol{A}_0 \cdot \boldsymbol{A}_0^*/c^2 \,,$$

(having used (4) and the formula $(\boldsymbol{a} \times \boldsymbol{b}) \cdot (\boldsymbol{c} \times \boldsymbol{d}) = (\boldsymbol{a} \cdot \boldsymbol{c})(\boldsymbol{b} \cdot \boldsymbol{d}) - (\boldsymbol{b} \cdot \boldsymbol{c})(\boldsymbol{a} \cdot \boldsymbol{d})$).

For a packet of waves all propagating in the direction of the unit vector $\boldsymbol{n}$, the vector potential is

$$\boldsymbol{A} = (2\pi)^{-\frac{1}{2}} \int_0^\infty \left[\boldsymbol{A}(\omega)\,\mathrm{e}^{\mathrm{i}(\boldsymbol{k} \cdot \boldsymbol{x} - \omega t)} + \boldsymbol{A}^*(\omega)\,\mathrm{e}^{-\mathrm{i}(\boldsymbol{k} \cdot \boldsymbol{x} - \omega t)}\right] \mathrm{d}\omega \,,$$

$$(\boldsymbol{k} = \omega\boldsymbol{n}/c) \,.$$

The energy transported by such a packet through a surface element of unit area, perpendicular to the direction of propagation, during the time interval ($t = -\infty$, $t = +\infty$) is

$$\mathscr{E} = c \int_{-\infty}^{+\infty} (\boldsymbol{E} \times \boldsymbol{H}) \cdot \boldsymbol{n} \, \mathrm{d}t = (c/2\pi) \int_0^\infty \mathrm{d}\omega \int_0^\infty \mathrm{d}\omega' (-\,\omega\omega'/c^2) \,\times$$

$$\times \int_{-\infty}^{+\infty} \boldsymbol{n} \cdot \{\boldsymbol{A}(\omega) \times [\boldsymbol{n} \times \boldsymbol{A}(\omega')]\, e^{i\boldsymbol{n}\cdot\boldsymbol{x}(\omega+\omega')/c}\, e^{-i(\omega+\omega')t} -$$
$$- \boldsymbol{A}(\omega) \times [\boldsymbol{n} \times \boldsymbol{A}^*(\omega')]\, e^{i\boldsymbol{n}\cdot\boldsymbol{x}(\omega-\omega')/c}\, e^{i(\omega'-\omega)t} - \qquad (\alpha)$$
$$- \boldsymbol{A}^*(\omega) \times [\boldsymbol{n} \times \boldsymbol{A}(\omega')]\, e^{-i\boldsymbol{n}\cdot\boldsymbol{x}(\omega-\omega')/c}\, e^{i(\omega-\omega')t} +$$
$$+ \boldsymbol{A}^*(\omega) \times [\boldsymbol{n} \times \boldsymbol{A}^*(\omega')]\, e^{-i\boldsymbol{n}\cdot\boldsymbol{x}(\omega+\omega')/c}\, e^{i(\omega+\omega')t}\}\, dt\,.$$

It is not difficult to recognize in (α) the δ functions $\delta(\omega + \omega')$ and $\delta(\omega - \omega')$. Then, as ω and ω' are both positive, $\delta(\omega + \omega') = 0$ for $\omega \neq 0 \neq \omega'$, whereas for $\omega = \omega' = 0$ the factor $\omega\omega'/c^2$ vanishes. Only the terms containing $\delta(\omega - \omega')$ are left. Integrating with respect to ω' and remembering that $\boldsymbol{A} \times (\boldsymbol{n} \times \boldsymbol{A}^*) = \boldsymbol{n}(\boldsymbol{A} \cdot \boldsymbol{A}^*)$, since each component is a transverse wave, one has

$$\mathscr{E} = 2c^{-1} \int \omega^2\, \boldsymbol{A}(\omega) \cdot \boldsymbol{A}^*(\omega)\, d\omega = \int I(\omega)\, d\omega\,.$$

Here $I(\omega)\, d\omega$ may be interpreted as the energy transported by the frequency band $(\omega, \omega + d\omega)$.

The Green's function for particles which are subject not only to the electrostatic field, but also to an electromagnetic wave, i.e. with the interaction Hamiltonian

$$H^{(1)} = -e A_0 + \left(-\frac{ie\hbar}{mc} \boldsymbol{A} \cdot \nabla_x + \frac{e^2}{2mc^2} A^2 \right) = V + H^{(1)}_{\text{em}}\,,$$

obeys the equation

$$\mathfrak{G}_S(2, 1) = \mathfrak{G}_S^{(V)}(2, 1) - (i/\hbar c) \int \mathfrak{G}_S^{(V)}(2, 3)\, H^{(1)}_{\text{em}}(3)\, \mathfrak{G}_S(3, 1)\, d^4x_3\,. \quad (5)$$

When an electromagnetic wave interacts with a particle which, at time $t_1 = -\infty$, is in a state described by $\psi_i^{(V)}$, the wave function develops according to the Schroedinger equation with Hamiltonian $H = H^{(0)} + V + H^{(1)}_{\text{em}}$ instead of $H = H^0 + V$, and, at time t_2, is given by

$$\psi_i(2) = \psi_i^{(V)}(2) - (i/\hbar c) \int \mathfrak{G}_S^{(V)}(2, 3)\, H^{(1)}_{\text{em}}(3)\, \psi_i(3)\, d^4x_3\,. \qquad (6)$$

2. Born approximation: first order in the electric charge

As has already been remarked, the integral on the right hand side of (6) may be evaluated by iteration. This gives a power series in the electric charge. Here we consider only the first term of this series. This amounts to evaluation of $\psi_i(2)$ by replacing, in eq. (6), $H^{(1)}_{\text{em}}$ by $(-ie\hbar/mc)\, \boldsymbol{A} \cdot \nabla_x$ and $\psi_i(3)$ by

$\psi_i^{(V)}(3)$, so that

$$\psi_i(2) = \psi_i^{(V)}(2) - (i/\hbar c) \int \mathfrak{G}_S^{(V)}(2,3) \left(-\frac{ie\hbar}{mc} A(3) \cdot \nabla_{x_3} \right) \psi_i^{(V)}(3)\, d^4x_3 .$$

To this approximation, the transition probability from the state with wave function $\psi_i^{(V)}$ to that with wave function $\psi_f^{(V)}$ (orthogonal to $\psi_i^{(V)}$), during the time interval ($t_1 = -\infty$, $t_2 = +\infty$), is given by the square of the modulus of

$$\begin{aligned}(\psi_f^{(V)}(2), \psi_i(2)) &= (-e/mc^2)(2\pi)^{-\frac{1}{2}} \int \psi_f^{(V)*}(x_3)\, e^{iE_f t_3/\hbar} \\ &\times \left[A(\omega)\, e^{i(k \cdot x_3 - \omega t_3)} + A^*(\omega)\, e^{-i(k \cdot x_3 - \omega t_3)} \right] \cdot \nabla_{x_3} \psi_i^{(V)}(x_3)\, e^{-iE_i t_3/\hbar}\, d\omega d^4x_3 \\ &= (2\pi)^{\frac{1}{2}}(-e\hbar/mc) \int \psi_f^{(V)*}(x_3) \left[A(\omega)\, e^{ik \cdot x_3}\, \delta(E_f - E_i - \hbar\omega) + \right. \\ &\left. + A^*(\omega)\, e^{-ik \cdot x_3}\, \delta(E_f - E_i + \hbar\omega) \right] \cdot \nabla_{x_3} \psi_i^{(V)}(x_3)\, d\omega\, d^3x_3 .\end{aligned}$$

The presence of $\delta(E_f - E_i \pm \hbar\omega)$ in this equation shows that energy is conserved in the transition. The energy for the jump from the initial to the final state of the particle is supplied by a photon of energy $\hbar\omega$, which may either be emitted or absorbed. Here, and in the following, the electromagnetic field is regarded as a classical field and an infinite "source" or "sink" of photons, of which it can supply or appropriate any number.

Thus two cases are possible:

(i) $E_f > E_i$, corresponding to an excitation from the energy level E_i to a higher level, by absorption of a photon.

In this case we put

$$E_f - E_i = \hbar\omega_{fi} ,$$

and the transition probability is then given by the square of the modulus of

$$(2\pi)^{\frac{1}{2}}(-e/mc) \int \psi_f^{(V)*}(x)\, e^{in \cdot x\omega_{fi}/c}\, A(\omega_{fi}) \cdot \nabla_x \psi_i^{(V)}(x)\, d^3x ; \tag{7}$$

(ii) $E_f < E_i$, corresponding to a de-excitation from an energy level to one lower by the emission of a photon.

Here we put

$$E_i - E_f = \hbar\omega_{fi} ,$$

and the transition probability is the absolute square of

$$(2\pi)^{\frac{1}{2}}(-e/mc) \int \psi_f^{(V)*}(x)\, e^{-in \cdot x\omega_{fi}/c}\, A^*(\omega_{fi}) \cdot \nabla_x \psi_i^{(V)}(x)\, d^3x . \tag{7'}$$

Summarizing, the total transition probability, relative to the time interval $(-\infty, +\infty)$, from the state i to the state f is given by

$$P_{i \to f} = (e/m\omega_{fi})^2 \frac{\pi}{c} I(\omega_{fi}) \left| \int \psi_f^{(V)*}(x)\, e^{\pm in \cdot x\omega_{fi}/c} (\varepsilon \cdot \nabla_x)\, \psi_i^{(V)}(x)\, d^3x \right|^2 , \tag{8}$$

where ε is the polarization vector of the electromagnetic field (unit vector in the direction of $A(\omega_{\mathrm{fi}})$) and $\omega_{\mathrm{fi}} = |E_{\mathrm{f}} - E_{\mathrm{i}}|/\hbar$.

As has already been shown (see *Note 1*), $I(\omega)\,\mathrm{d}\omega$ is the energy transported, through a surface element of unit area perpendicular to $\boldsymbol{n}$, by electromagnetic waves of frequency between ω and $\omega + \mathrm{d}\omega$ during the time interval ($t_1 = -\infty$, $t_2 = +\infty$). Therefore, $I(\omega_{\mathrm{fi}})$ is the value of the intensity for that particular value of ω ($= \omega_{\mathrm{fi}}$) which is required to excite the "atom" by absorption of a photon $\hbar\omega_{\mathrm{fi}}$, or to de-excite it by emission.

If $I(\omega)$ denotes the intensity per unit time, rather than that relative to the interval ($t_1 = -\infty$, $t_2 = +\infty$), eq. (8) is the formula for the transition probability per unit time.

3. Born approximation: second order in the electric charge

In the expansion

$$\begin{aligned}\psi(2) &= \psi_{\mathrm{i}}^{(V)}(2) - (\mathrm{i}/\hbar c) \int \mathfrak{G}_{\mathrm{S}}^{(V)}(2,3)\, H_{\mathrm{em}}^{(1)}(3)\, \psi_{\mathrm{i}}^{(V)}(3)\, \mathrm{d}^4x_3 + \\ &+ (-\mathrm{i}/\hbar c)^2 \int \mathfrak{G}_{\mathrm{S}}^{(V)}(2,3)\, H_{\mathrm{em}}^{(1)}(3)\, \mathfrak{G}_{\mathrm{S}}^{(V)}(3,4)\, H_{\mathrm{em}}^{(1)}(4)\, \psi_{\mathrm{i}}^{(V)}(4)\, \mathrm{d}^4x_3\, \mathrm{d}^4x_4\,,\end{aligned}$$

obtained by iterating eq. (6), the e^2 corrections to $\psi_{\mathrm{i}}^{(V)}(2)$ come both from the third term on the right side with $H_{\mathrm{em}}^{(1)}$ replaced by $(-\mathrm{i}e\hbar/mc)\,A \cdot \nabla_x$, and from the second term, since the interaction Hamiltonian also involves $(e^2/2mc^2)\,A^2$, which is of the second order.

This corresponds to the two Feynman diagrams

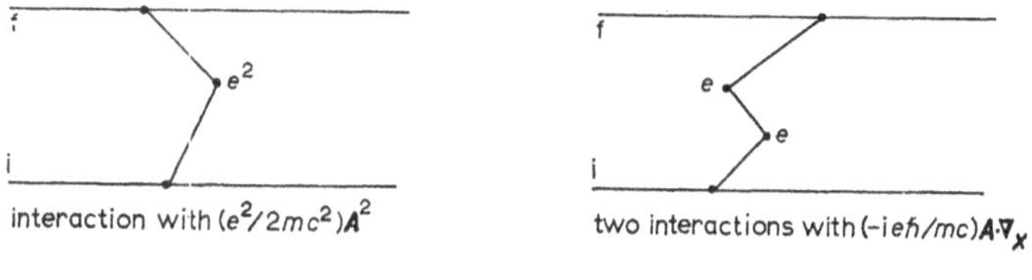

Fig. 13

Therefore the probability amplitude is

$$\left(\psi_{\mathrm{f}}^{(V)}(2), \psi_{\mathrm{i}}(2)\right) = (-\,\mathrm{i}/\hbar)\left(\mathfrak{M}_{\mathrm{fi}}(e) + \mathfrak{M}_{\mathrm{fi}}(e^2) + \ldots\right).$$

Here $\mathfrak{M}_{\mathrm{fi}}(e)$, which has already been evaluated, is

$$\mathfrak{M}_{\mathrm{fi}}(e) = \sqrt{2\pi}\,(e\hbar/\mathrm{i}mc) \int \psi_{\mathrm{f}}^{(V)*}(\boldsymbol{x})\, \mathrm{e}^{\mathrm{i}\boldsymbol{n}\cdot\boldsymbol{x}\omega_{\mathrm{fi}}/c}\, A(\omega_{\mathrm{fi}}) \cdot \nabla_x\, \psi_{\mathrm{i}}^{(V)}(\boldsymbol{x})\, \mathrm{d}^3\mathrm{x}\,, \quad (9)$$

if $E_{\mathrm{f}} > E_{\mathrm{i}}$. The expression for $E_{\mathrm{f}} < E_{\mathrm{i}}$ may be obtained from this by the replacement

$$e^{i\boldsymbol{n}\cdot\boldsymbol{x}\omega_{fi}/c} \to e^{-i\boldsymbol{n}\cdot\boldsymbol{x}\omega_{fi}/c}, \quad \boldsymbol{A} \to \boldsymbol{A}^* .$$

From now on, for simplicity, we shall study the case $E_f > E_i$ (excitation) only.

For $\mathfrak{M}_{fi}(e^2)$ we have

$$\mathfrak{M}_{fi}(e^2) = i\hbar\big(\psi_f^{(V)}(2), \{(-i/\hbar c)\int \mathfrak{G}_S^{(V)}(2,3)(e^2/2mc^2)\,\boldsymbol{A}^2(3)\,\psi_i^{(V)}(3)\,d^4x_3 + \\ + (e/mc^2)^2 \int \mathfrak{G}_S^{(V)}(2,3)\big(\boldsymbol{A}(3)\cdot\nabla_{x_3}\big)\,\mathfrak{G}_S^{(V)}(3,4)\big(\boldsymbol{A}(4)\cdot\nabla_{x_4}\big)\,\psi_i^{(V)}(4)\,d^4x_3\,d^4x_4\}\big), \tag{10}$$

where the first term comes from the interaction $(e^2/2mc^2)\,\boldsymbol{A}^2$ and the second from $(-ie\hbar/mc)\,\boldsymbol{A}\cdot\nabla_x$.

(i) Contribution from the $\dfrac{e^2}{mc^2}\boldsymbol{A}^2$ *interaction*

Calculation of the first term of (10) gives

$$i\hbar\left(\psi_f^{(V)}(2), -\frac{i}{\hbar c}\int \mathfrak{G}_S^{(V)}(2,3)\,\frac{e^2}{2mc^2}\,\boldsymbol{A}^2(3)\,\psi_i^{(V)}(3)\,d^4x_3\right) = \\ = \frac{e^2}{2mc^2}\int \psi_f^{(V)*}(\boldsymbol{x}_3)\,\psi_i^{(V)}(\boldsymbol{x}_3)\,\boldsymbol{A}(3)\cdot\boldsymbol{A}(3)\,e^{i\omega_{fi}t_3}\,dt_3\,d^3x_3 . \tag{11}$$

On the other hand,

$$\int_{-\infty}^{\infty} \boldsymbol{A}(3)\cdot\boldsymbol{A}(3)\,e^{i\omega_{fi}t_3}\,dt_3 = \int_0^\infty\int_0^\infty \big[\boldsymbol{A}(\omega)\cdot\boldsymbol{A}(\omega')\,e^{i\boldsymbol{n}\cdot\boldsymbol{x}_3(\omega+\omega')/c}\,\delta(\omega+\omega'-\omega_{fi}) + \\ + \boldsymbol{A}(\omega)\cdot\boldsymbol{A}^*(\omega')\,e^{i\boldsymbol{n}\cdot\boldsymbol{x}_3(\omega-\omega')/c}\,\delta(\omega-\omega'-\omega_{fi}) + \\ + \boldsymbol{A}^*(\omega)\cdot\boldsymbol{A}(\omega')\,e^{i\boldsymbol{n}\cdot\boldsymbol{x}_3(\omega'-\omega)/c}\,\delta(\omega-\omega'+\omega_{fi}) + \\ + \boldsymbol{A}^*(\omega)\cdot\boldsymbol{A}^*(\omega')\,e^{-i\boldsymbol{n}\cdot\boldsymbol{x}_3(\omega+\omega')/c}\,\delta(\omega+\omega'+\omega_{fi})\big]\,d\omega\,d\omega' .$$

This reduces to

$$e^{i\boldsymbol{n}\cdot\boldsymbol{x}_3\omega_{fi}/c}\left[\int_0^{\omega_{fi}} \boldsymbol{A}(\omega)\cdot\boldsymbol{A}(\omega_{fi}-\omega)\,d\omega + \int_{\omega_{fi}}^{\infty} \boldsymbol{A}(\omega)\cdot\boldsymbol{A}^*(\omega-\omega_{fi})\,d\omega + \\ + \int_0^\infty \boldsymbol{A}^*(\omega)\cdot\boldsymbol{A}(\omega+\omega_{fi})\,d\omega\right],$$

since $\delta(\omega+\omega'+\omega_{fi}) = 0$. In fact $\omega \geqslant 0$, $\omega' \geqslant 0$ and $\omega_{fi} = (E_f - E_i)/\hbar > 0$. Introduction of the new integration variable $\omega' = \omega + \omega_{fi}$, into the third integral shows that this is equal to the second. Substituting in (11) and putting

$x_3 \to x$ we finally obtain

$$\frac{e^2}{2mc^2}\left(\int \psi_{\mathrm{f}}^{(V)*}(x)\, \mathrm{e}^{\mathrm{i}n\cdot x\omega_{\mathrm{fi}}/c}\, \psi_{\mathrm{i}}^{(V)}(x)\, \mathrm{d}^3\mathrm{x}\right) \times$$
$$\times\left(\int_0^{\omega_{\mathrm{fi}}} A(\omega)\cdot A(\omega_{\mathrm{fi}}-\omega)\,\mathrm{d}\omega + 2\int_{\omega_{\mathrm{fi}}}^{\infty} A(\omega)\cdot A^*(\omega-\omega_{\mathrm{fi}})\,\mathrm{d}\omega\right). \tag{12}$$

This expression may be interpreted physically as follows. The integral

$$\int_0^{\omega_{\mathrm{fi}}} A(\omega)\, A(\omega_{\mathrm{fi}}-\omega)\,\mathrm{d}\omega$$

represents an interaction process with the electromagnetic wave with the absorption of two photons, in such a manner that the sum of their energies is equal to $E_{\mathrm{f}} - E_{\mathrm{i}}$ (Fig. 14)

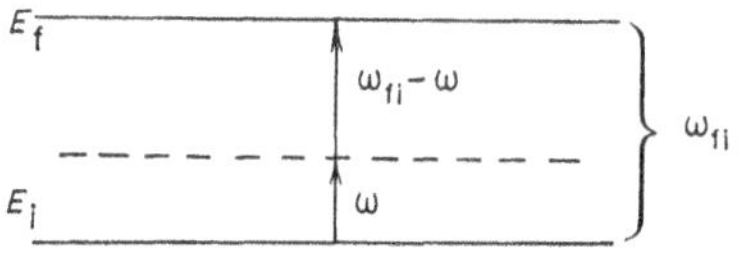

Fig. 14

The integration with respect to ω, between the limits 0 and ω_{fi}, takes into account all possible processes of this type, which are compatible with energy conservation.

Similarly the integral

$$\int_{\omega_{\mathrm{fi}}}^{\infty} A(\omega)\cdot A^*(\omega-\omega_{\mathrm{fi}})\,\mathrm{d}\omega$$

corresponds to the absorption of a photon of energy $\hbar\omega$ and emission of one of energy $\hbar(\omega - \omega_{\mathrm{fi}})$ (Fig. 15), the integration over ω being between limits compatible with energy conservation.

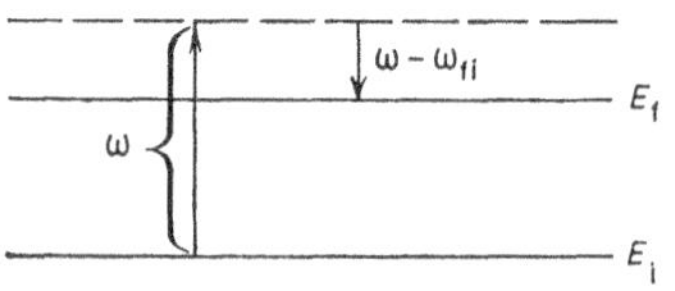

Fig. 15

It is easy to understand that, if $A(\omega - \omega_{\mathrm{fi}})$ corresponds to absorption of a photon, $A^*(\omega - \omega_{\mathrm{fi}})$ must correspond to the emission of one. In fact, comparing (7) and (7′), one sees that the emission amplitude may be obtained from the absorption amplitude, changing the sign of the exponent and replacing A by A^*. Writing the exponential $e^{i\boldsymbol{n}\cdot\boldsymbol{x}\omega_{fi}/c}$ in (12) in the form $e^{i\boldsymbol{n}\cdot\boldsymbol{x}\omega}\,e^{-i\boldsymbol{n}\cdot\boldsymbol{x}(\omega-\omega_{fi})/c}$ the comparison with (7) and (7′) becomes clearer.

The factor 2 in the second term of (12) accounts for the fact that the absorption of the photon of energy $\hbar\omega$ may either precede or follow the emission of the photon of energy $\hbar(\omega-\omega_{fi})$.

Therefore, together with the diagram in Fig. 15, one has the alternative diagram shown in Fig. 16 *

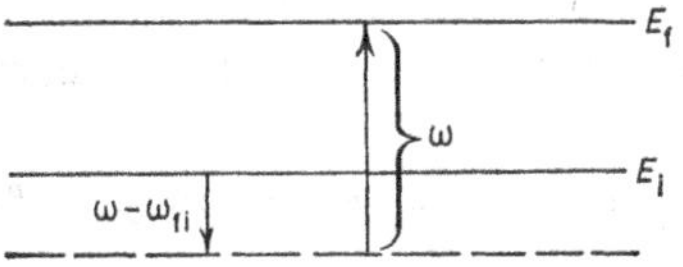

Fig. 16

(ii) Contribution from the double $-(ie\hbar/mc)\,\boldsymbol{A}\cdot\nabla_x$ *interaction*

By a simple manipulation the second term of (10) takes the form

$$i\hbar(e/mc)^2\int\psi_f^{(V)*}(\boldsymbol{x}_3)\left[e^{iE_ft_3/\hbar}\left(\boldsymbol{A}(3)\cdot\nabla_{x_3}\right)\mathfrak{G}_S^{(V)}(3,4)\times\right.$$
$$\left.\times\left(\boldsymbol{A}(4)\cdot\nabla_{x_4}\right)e^{-iE_it_4/\hbar}\right]\psi_i^{(V)}(\boldsymbol{x}_4)\,d^3x_3\,d^3x_4\,dt_3\,dt_4\,. \tag{13}$$

Now we use the expression (2) for $\mathfrak{G}_S^{(V)}$ and denote the l component of $\boldsymbol{A}(3)$ by $A_l(3)$ and the m component of $\boldsymbol{A}(4)$ by $A_m(4)$, introduce the adiabatic switching, and omit functions which do not depend on t_3 and t_4. The integral over t_3 and t_4 becomes

$$\sum_n\int_{-\infty}^{\infty} e^{iE_ft_3/\hbar}A_l(3)\,e^{-iE_n(t_3-t_4)/\hbar}A_m(4)\,e^{-iE_it_4/\hbar}\,\eta(t_3-t_4)\,e^{-\varepsilon'|t_3|}\,e^{-\varepsilon|t_4|}\,dt_3\,dt_4\,. \tag{14}$$

Here ε' and ε, not necessarily equal, are small positive quantities which will be made to tend to zero at the end.

The integral over t_4 gives

---

* Remember that the factor 2 arose in fact from the sum of two equal terms, corresponding to the diagrams shown in Figs. (15) and (16) respectively.

$$\int_{-\infty}^{t_3} e^{iE_n t_4/\hbar} A_m(4)\, e^{-iE_i t_4/\hbar}\, e^{-\varepsilon|t_4|}\, dt_4 =$$

$$\frac{1}{\sqrt{2\pi}} \int_{-\infty}^{+\infty} d\omega'\, A_m(\omega')\, e^{i n\cdot x_4 \omega'/c} \int_{-\infty}^{t_3} e^{i(E_n - E_i - \hbar\omega')t_4/\hbar}\, e^{-\varepsilon|t_4|}\, dt_4\,,$$

where we defined $A_m(-\omega) = A^*_m(\omega)$ (which is in agreement with the interpretation of A^* as corresponding to an emission process), and may be split thus

$$\int_{-\infty}^{t_3} = \int_{-\infty}^{0} + \int_{0}^{t_3} = \int_{-\infty}^{0} + \eta(t_3)\int_{0}^{t_3} + \eta(-t_3)\int_{0}^{t_3}\,.$$

Then, remembering eq. (12) of Chap. I, the integral (14) takes the form

$$\sum_n\!\!\!\!\!\!\int \frac{i\hbar}{2\pi} \int A_l(\omega)\, A_m(\omega')\, e^{i n\cdot(\omega x_3 + \omega' x_4)/c}\, e^{i(E_f - E_n - \hbar\omega)t_3/\hbar}\, e^{-\varepsilon'|t_3|} \times$$

$$\times \Bigg[P\left(\frac{1}{E_i + \hbar\omega' - E_n}\right) - i\,\pi\delta\,(E_i + \hbar\omega' - E_n) +$$

$$+ P\left(\frac{1}{E_i + \hbar\omega' - E_n}\right)\left(e^{i(E_n - E_i - \hbar\omega')t_3/\hbar}\, e^{-\varepsilon|t_3|} - 1\right) +$$

$$+ i\,\pi\delta\,(E_i + \hbar\omega' - E_n)\,\theta(t_3)\left(e^{-\varepsilon|t_3|} - 1\right)\Bigg]\, dt_3\, d\omega\, d\omega'\,,$$

where

$$\theta(t_3) = \begin{cases} 1 & \text{for} \quad t_3 > 0\,, \\ -1 & \text{for} \quad t_3 < 0\,. \end{cases}$$

Finally, letting ε and ε' tend to zero, and performing integrations with respect to ω and ω' where they are possible, we obtain

$$\sum_n\!\!\!\!\!\!\int \Bigg[\pi A_l\left(\frac{E_f - E_n}{\hbar}\right) A_m\left(\frac{E_n - E_i}{\hbar}\right) e^{i n\cdot[x_3(E_f - E_n) + x_4(E_n - E_i)]/\hbar c} +$$

$$+ i\hbar^2 \int A_l(\omega)\, A_m(\omega')\, e^{i n\cdot(x_3\omega + x_4\omega')/c}\, P\left(\frac{1}{E_i + \hbar\omega' - E_n}\right) \times$$

$$\times\ \delta\,(E_i + \hbar\omega + \hbar\omega' - E_f)\, d\omega\, d\omega' \Bigg]\,.$$

Inserting this in (13), the e^2 correction originating from $(-\,ie\hbar/mc)\,\boldsymbol{A}\cdot\boldsymbol{\nabla}_x$ takes the form

$$
\begin{aligned}
& i\hbar\left(\frac{e}{mc}\right)^2 \mathop{\sum\!\!\!\!\!\!\int}_n \Bigg\{ \pi \int \psi_{\mathrm{f}}^{(V)*}(\boldsymbol{x}_3)\left[\boldsymbol{A}\left(\frac{E_{\mathrm{f}}-E_n}{\hbar}\right)\cdot\boldsymbol{\nabla}_{x_3}\psi_n^{(V)}(\boldsymbol{x}_3)\right] e^{i\boldsymbol{n}\cdot\boldsymbol{x}_3(E_{\mathrm{f}}-E_n)/c\hbar}\times \\
& \times \psi_n^{(V)*}(\boldsymbol{x}_4)\left[\boldsymbol{A}\left(\frac{E_n-E_{\mathrm{i}}}{\hbar}\right)\cdot\boldsymbol{\nabla}_{x_4}\psi_{\mathrm{i}}^{(V)}(\boldsymbol{x}_4)\right] e^{i\boldsymbol{n}\cdot\boldsymbol{x}_4(E_n-E_{\mathrm{i}})/\hbar c}\,\mathrm{d}^3\mathrm{x}_3\,\mathrm{d}^3\mathrm{x}_4 + \\
& -i\hbar^2\int P\left(\frac{1}{E_{\mathrm{i}}+\hbar\omega'-E_n}\right)\delta\left(E_{\mathrm{i}}+\hbar\omega+\hbar\omega'-E_{\mathrm{f}}\right)\times \\
& \times \left(\int \psi_{\mathrm{f}}^{(V)*}(\boldsymbol{x}_3)\,\boldsymbol{A}(\omega)\cdot\boldsymbol{\nabla}_{x_3}\psi_n^{(V)}(\boldsymbol{x}_3)\,e^{i\boldsymbol{n}\cdot\boldsymbol{x}_3\omega/c}\times\right. \\
& \left.\times\, \psi_n^{(V)*}(\boldsymbol{x}_4)\,\boldsymbol{A}(\omega')\cdot\boldsymbol{\nabla}_{x_4}\psi_{\mathrm{i}}^{(V)}(\boldsymbol{x}_4)\,e^{i\boldsymbol{n}\cdot\boldsymbol{x}_4\omega'/c}\,\mathrm{d}^3\mathrm{x}_3\,\mathrm{d}^3\mathrm{x}_4\right)\mathrm{d}\omega\,\mathrm{d}\omega'\Bigg\}.
\end{aligned}
\tag{15}
$$

The first term in braces is the product of two matrix elements similar to those of the first order approximation. It may be interpreted as representing two transitions; one from the state i to an intermediate state n with absorption of a photon, followed by another transition from n to the final state f, also with absorption of a photon. Such transitions * may be represented by the graph shown in Fig. 17.

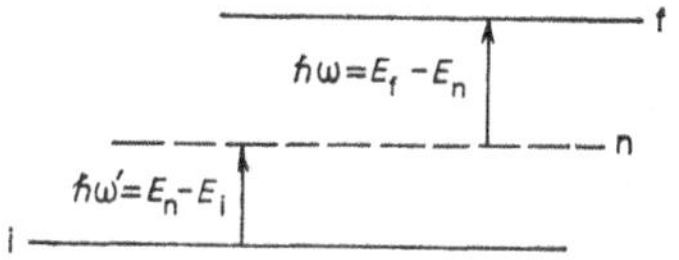

Fig. 17

It should be emphasized that the two transitions are subject to energy conservation. In the transition from i to n the photon absorbed has the energy $\hbar\omega' = E_n - E_{\mathrm{i}}$, and similarly for the second transition.

The first term of (15) arises from the introduction of the adiabatic switching, as is easily recognized by inspecting the calculation leading to (15) **. Its occurrence is peculiar to collision processes, and it has no counterpart in the perturbation theory of stationary states by time independent interactions.

---

* By virtue of the convention $\boldsymbol{A}(-\,\omega) = \boldsymbol{A}^*(\omega)$, it is easy to recognize that the same considerations also apply to the case $E_{\mathrm{f}} < E_{\mathrm{i}}$, in which two photons are emitted. In this case also eq. (15) is valid.

** In fact this term comes from the second term on the right-hand side of

In the next section its relation with *radiation damping* will be discussed.

The second term of (15) represents also two successive transitions $\mathrm{i} \to n \to \mathrm{f}$. Now, however, energy is not conserved in the transitions to and from intermediate states.

In fact

$$\omega' \neq \frac{E_n - E_\mathrm{i}}{\hbar}, \quad \omega \neq \frac{E_\mathrm{f} - E_n}{\hbar},$$

as is guaranteed by the principal value

$$P\left(\frac{1}{E_\mathrm{i} + \hbar\omega' - E_n}\right),$$

which excludes the singularity $\hbar\omega' = E_n - E_\mathrm{i}$ (and therefore also $\hbar\omega = E_\mathrm{f} - E_n$; notice the δ function $\delta\,(E_\mathrm{i} + \hbar\omega + \hbar\omega' - E_\mathrm{f})$).

Energy is conserved only in the transition $\mathrm{i} \to \mathrm{f}$, since

$$\hbar(\omega + \omega') = E_\mathrm{f} - E_\mathrm{i}$$

in virtue of $\delta\,(\hbar\omega + \hbar\omega' + E_\mathrm{i} - E_\mathrm{f})$.

The second term of (15) can be obtained by using the expression for second order corrections as given by the familiar perturbation theory of stationary states.

To summarize, the transition probability from the state i to the state f is given by

$$\begin{aligned} P_{\mathrm{i}\to\mathrm{f}} &= |(\psi_\mathrm{f}^{(V)}(2), \psi_\mathrm{i}(2))|^2 = \hbar^{-2}\, |\mathfrak{M}_{\mathrm{fi}}(e) + \mathfrak{M}_{\mathrm{fi}}(e^2) + \ldots|^2 \\ &= \hbar^{-2}\, \{|\mathfrak{M}_{\mathrm{fi}}(e)|^2 + 2\,\mathrm{Re}\,[\mathfrak{M}_{\mathrm{fi}}(e)^*\, \mathfrak{M}_{\mathrm{fi}}(e^2)] + \ldots\}, \end{aligned}$$

where $\mathfrak{M}_{\mathrm{fi}}(e)$ and $\mathfrak{M}_{\mathrm{fi}}(e^2)$ are given by (9) and (10) respectively.

4. Radiation damping

For simplicity the following notation will be introduced. We denote by

$$< b, \omega\, |\mathfrak{M}(e)|\, a >,$$

$$\lim_{\varepsilon \to 0^+} \int_{-\infty}^{\infty} \frac{\mathrm{d}x}{x \pm \mathrm{i}\varepsilon} = P\int_{-\infty}^{\infty} \frac{\mathrm{d}x}{x} \mp \mathrm{i}\pi \int_{-\infty}^{\infty} \delta(x)\, \mathrm{d}x\,,$$

and, would not be present if $\int_{-\infty}^{\infty} \frac{\mathrm{d}x}{x}$ were defined as a principal value.

the transition amplitude, to first order in e, from the state a to the state b with the emission of a photon of energy $\hbar\omega$, and by

$$< b \,|\mathfrak{M}(e)|\, a, \omega > ,$$

that from the state a to the state b with absorption of a photon of energy $\hbar\omega$.

According to (7) and (7′)

$$< b \,|\mathfrak{M}(e)|\, a, \omega > =$$
$$= \sqrt{2\pi}(e\hbar/mci) \int \psi_b^{(V)*}(\boldsymbol{x})\, \mathrm{e}^{\mathrm{i}\boldsymbol{n}\cdot\boldsymbol{x}\omega/c}(\boldsymbol{A}(\omega)\cdot\boldsymbol{\nabla}_x)\, \psi_a^{(V)}(\boldsymbol{x})\, \mathrm{d}^3\mathrm{x}\,,$$
$$< b, \omega \,|\mathfrak{M}(e)|\, a > =$$
$$= \sqrt{2\pi}(e\hbar/mci) \int \psi_b^{(V)*}(\boldsymbol{x})\, \mathrm{e}^{-\mathrm{i}\boldsymbol{n}\cdot\boldsymbol{x}\omega/c}(\boldsymbol{A}^*(\omega)\cdot\boldsymbol{\nabla}_x)\, \psi_a^{(V)}(\boldsymbol{x})\, \mathrm{d}^3\mathrm{x}\,.$$

Note, however, that here we do not require that $\pm\, \hbar\omega = E_b - E_a$.

The transition amplitudes for the above processes, taking place while other photons $\hbar\omega_1$, $\hbar\omega_2$ ..., act as spectators (i.e. without interacting with the atomic system), and are therefore present both in the initial and in the final state, are denoted by

$$< b, \omega_1, \omega_2 \ldots \,|\mathfrak{M}(e)|\, a, \omega, \omega_1, \omega_2 \ldots > ,$$
$$< b, \omega, \omega_1, \omega_2 \ldots \,|\mathfrak{M}(e)|\, a, \omega_1, \omega_2 \ldots > .$$

Adopting this notation, it is not difficult to see that the second term of (15) takes the form

$$-\frac{1}{2\pi}\int \mathrm{d}\omega\, P \sum\!\!\!\!\!\!\int \frac{< \mathrm{F} \,|\mathfrak{M}(e)|\, N > < N \,|\mathfrak{M}(e)|\, \mathrm{I} >}{E_N - E_\mathrm{I}}\,. \tag{16}$$

In eq. (16), I denotes a 'state' of the system consisting of the electromagnetic radiation and of the atomic system in the state i, so that

$$E_\mathrm{I} = E_\mathrm{i} + E_\mathrm{rad}\,.$$

Thus the total energy of the state I is the sum of the energy of the atomic system in the state i and the energy of the radiation (regarded as a source or sink of photons).

N is an intermediate state consisting of the atomic system in the state n and of the electromagnetic radiation decreased (or increased) by one photon, and thus

$$E_N = E_n + E_\mathrm{rad} \mp \hbar\omega\,.$$

Note that $E_N \neq E_\mathrm{I}$ (no energy conservation in the intermediate states!).

Finally, if the transition i $\rightarrow$ f is accompanied by the emission (or absorption) of 0, 1, 2, photons, F consists of the atomic system in the state f and of the electromagnetic radiation plus (or minus) 0, 1, 2 photons.

Therefore

$$E_{\mathrm{F}} = E_{\mathrm{f}} + E'_{\mathrm{rad}} \,.$$

As is easy to see, the transition I $\rightarrow$ F occurs with energy conservation.

The contribution to $\mathfrak{M}_{\mathrm{fi}}(e^2)$ from the first term of (15) may be written in the form

$$(-\,\mathrm{i}/2\hbar) \sum_n\!\!\!\!\!\!\int \; < \mathrm{F}\, |\mathfrak{M}(e)|\, N > < N\, |\mathfrak{M}(e)|\, \mathrm{I} > .$$

As follows from the discussion in the preceding section, here one has

$$E_{\mathrm{I}} = E_N = E_{\mathrm{F}} \,.$$

It may be shown that this term is connected with the width of spectral lines and with radiation damping.

In fact, on evaluating the amplitude of the probability that the atomic system, which is in the state i at the time $t_1 = -\,\infty$, will still be in the same state at the time $t_2 = +\,\infty$, we find, omitting for simplicity the interaction in A^2 and the term of the first order in e,

$$\lim_{\substack{t_2 \to +\infty \\ t_1 \to -\infty}} (\psi_{\mathrm{i}}^{(V)}(2), \psi_{\mathrm{i}}(2)) =$$

$$= 1 + \frac{\mathrm{i}}{\hbar} \left[\frac{1}{2\pi} \int \mathrm{d}\omega\, P \sum_n\!\!\!\!\!\!\int \frac{< \mathrm{I}\, |\mathfrak{M}(e)|\, N > < N\, |\mathfrak{M}(e)|\, \mathrm{I} >}{E_N - E_{\mathrm{I}}} + \right. \tag{17}$$

$$\left. + (\mathrm{i}/2\hbar) \sum_n\!\!\!\!\!\!\int \; < \mathrm{I}\, |\mathfrak{M}(e)|\, N > < N\, |\mathfrak{M}(e)|\, \mathrm{I} > \right] + \ldots$$

From this result it appears that $\psi_{\mathrm{i}}(\boldsymbol{x}_2)$ consists of a component orthogonal, and one proportional to $\psi_{\mathrm{i}}^{(V)}(\boldsymbol{x}_2)$. The coefficient of proportionality for the latter is given by (17). Thus the component of $\psi_{\mathrm{i}}(2)$ which is proportional to $\psi_{\mathrm{i}}^{(V)}(2) = \psi_{\mathrm{i}}^{(V)}(\boldsymbol{x}_2)\, \mathrm{e}^{-\mathrm{i}E_{\mathrm{i}}t_2/\hbar}$ has not the time dependence

$$\mathrm{e}^{-\mathrm{i}E_{\mathrm{i}}(t_2 - t_1)/\hbar} \,,$$

but rather

$$\left[1 - \frac{\mathrm{i}}{\hbar} \Delta E_{\mathrm{i}}(t_2 - t_1) + \ldots \right] \mathrm{e}^{-\mathrm{i}E_{\mathrm{i}}(t_2 - t_1)/\hbar} \,,$$

where

$$\Delta E_{\mathrm{i}} = -\frac{1}{2\pi}\int \left\{ P \sum_n \frac{< \mathrm{I}\,|\mathfrak{M}(e)|\,N > < N\,|\mathfrak{M}(e)|\,\mathrm{I} >}{E_N - E_{\mathrm{I}}} \right\} \mathrm{d}\omega -$$

$$- \frac{\mathrm{i}}{2\hbar} \sum_n < \mathrm{I}\,|\mathfrak{M}(e)|\,N > < N\,|\mathfrak{M}(e)|\,\mathrm{I} > . \tag{18}$$

The matrix elements $< |\mathfrak{M}(e)| >$ are normalized in such a way that, in this expression, they are amplitudes of the transition probability per unit time, whereas before they referred to the time interval $(-\infty, +\infty)$.

If $t_2 - t_1$ is not too large, we may write

$$[1 - (\mathrm{i}/\hbar)\,\Delta E_{\mathrm{i}}(t_2 - t_1) + \ldots]\, \mathrm{e}^{-\mathrm{i}E_{\mathrm{i}}(t_2 - t_1)/\hbar} = \mathrm{e}^{-\mathrm{i}(E_{\mathrm{i}} + \Delta E_{\mathrm{i}})(t_2 - t_1)/\hbar} ,$$

so that ΔE_{i} may be interpreted as a correction to the energy of the state i due to the interaction of the atomic system with the electromagnetic field.

As one sees from (18), ΔE_{i} consists of a real part

$$-\frac{1}{2\pi}\int \mathrm{d}\omega\, P \sum_n \frac{< \mathrm{I}\,|\mathfrak{M}(e)|\,N > < N\,|\mathfrak{M}(e)|\,\mathrm{I} >}{E_N - E_{\mathrm{I}}} =$$

$$= -\frac{1}{2\pi}\int \mathrm{d}\omega\, P \sum_n \frac{|< \mathrm{I}\,|\mathfrak{M}(e)|\,N >|^2}{E_N - E_{\mathrm{I}}} ,$$

and an imaginary part

$$(-\,\mathrm{i}/2\hbar) \sum_n |< \mathrm{I}\,|\mathfrak{M}(e)|\,N >|^2 .$$

Owing to the latter, the amplitude of the wave function of the state i decreases with time due to the factor

$$\mathrm{e}^{-\gamma(t_2 - t_1)/2} ,$$

with

$$\gamma = \hbar^{-2} \sum_n |< \mathrm{I}\,|\mathfrak{M}(e)|\,N >|^2 = \sum_n p_{\mathrm{i}n} .$$

Here $p_{\mathrm{i}n}$ denotes the transition probability from the state i to the state n. Time-independent perturbation theory does not account for this effect.

Because of the decreasing exponential, the energy of the state i is not sharp and the level i has a width proportional to $\hbar\gamma$.

In fact

$$\Delta E\, \Delta t \gtrsim \hbar .$$

Since here

$$\Delta t \simeq 1/\gamma ,$$

one has

$$\Delta E \simeq \hbar\gamma .$$

The level is wider the greater the probability that interaction with the radiation field may cause transitions from it to any other level.

5. Conservation of physical quantities in a transition

Green's functions possess the same invariance properties as the corresponding wave equations. Therefore $\mathfrak{G}_{\mathrm{S}}^{(0)}$ is invariant under space and time translations, space rotations and inversion. On the other hand, $\mathfrak{G}_{\mathrm{S}}^{(V)}$ with $V = V(r)$ (central static potential) is invariant under time translations (but not under space translations), and under space rotations and inversion about the centre of force.

Finally $\mathfrak{G}_{\mathrm{S}}$, relative to the Hamiltonian $H = H^{(0)} + V(r) + H_{\mathrm{em}}^{(1)}$, where $H_{\mathrm{em}}^{(1)}$ is the interaction with a given electromagnetic wave has, in general, no invariance property at all.

However, one may show that the probability amplitude

$$\mathfrak{M}_{\mathrm{fi}} \propto \lim_{\substack{t_2 \to \infty \\ t_1 \to -\infty}} (\psi_{\mathrm{f}}^{(V)}(2), \psi_{\mathrm{i}}(2)) = \lim_{\substack{t_2 \to \infty \\ t_1 \to -\infty}} \int_{t_1}^{t_2} \dots ,$$

discussed in the previous sections, is invariant under time translations, in the sense that

$$\lim_{\substack{t_2 \to \infty \\ t_1 \to -\infty}} \left(\int_{t_1+\tau}^{t_2+\tau} \dots - \int_{t_1}^{t_2} \dots \right) = 0 . \tag{19}$$

This property will be connected with the fact that each term of the expansion of $\mathfrak{M}_{\mathrm{fi}}$ may be interpreted as describing a transition from the state i of the atomic system together with a certain number of photons, to the final state f of the atomic system plus photons, in such a way that

$$\begin{aligned} &E_{\mathrm{i}} + \sum(\textit{energies of photons initially present}) = \\ &= E_{\mathrm{f}} + \sum(\textit{energies of photons finally present}) . \end{aligned} \tag{20}$$

Notice that, omitting space integrations and the interaction in A^2, and denoting by τ_n a generic time variable, the structure of a term of the expansion of $\mathfrak{M}_{\mathrm{fi}}$ is

$$\lim_{\substack{t_2 \to \infty \\ t_1 \to -\infty}} \int \dots \int_{t_1}^{t_2} \mathrm{e}^{\mathrm{i}E_{\mathrm{f}}\tau_n/\hbar}\, \mathrm{e}^{\pm \mathrm{i}\omega_n\tau_n}\, \mathrm{e}^{-\mathrm{i}E_n(\tau_n-\tau_{n-1})/\hbar}\, \eta(\tau_n - \tau_{n-1}) \times$$

$$\times\ e^{\pm i\omega_{n-1}\tau_{n-1}}\, e^{-iE_{n-1}(\tau_{n-1}-\tau_{n-2})/\hbar} \ldots e^{-iE_1(\tau_2-t_1)/\hbar}\, e^{-iE_i t_1/\hbar}\, d\tau_2 \ldots d\tau_n\,. \quad (21)$$

In this expression the terms $e^{\pm i\omega_n\tau_n}$ originate from the factors $A(\tau_n)$. Moreover $e^{-i\omega_n\tau_n}$ is multiplied by $A(\omega_n)$ and, therefore, corresponds to the absorption of a photon, while $e^{i\omega_n\tau_n}$ is multiplied by $A^*(\omega_n)$ and corresponds to the emission of a photon.

Space-time diagrams describing (21) are of the type shown in Fig. 18.

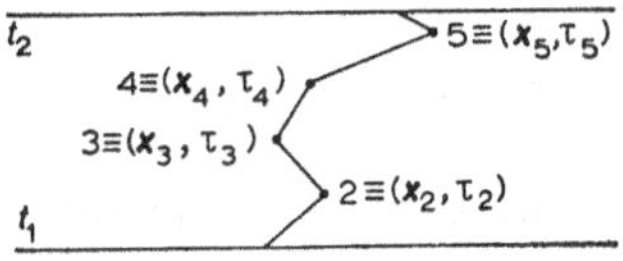

Fig. 18

At the points 2, 3, 4, 5 the matter wave interacts with the electromagnetic field which loses (or acquires) a photon.

To prove (19) we notice that, by a change of integration variables,

$$\lim_{\substack{t_2\to\infty\\ t_1\to-\infty}} \int_{t_1+\tau}^{t_2+\tau} e^{iE_f\tau_n/\hbar}\, e^{\pm i\omega_n\tau_n}\, e^{-iE_n(\tau_n-\tau_{n-1})/\hbar}\, \eta(\tau_n-\tau_{n-1}) \times$$

$$\times\ e^{\pm i\omega_{n-1}\tau_{n-1}}\, e^{-iE_{n-1}(\tau_{n-1}-\tau_{n-2})/\hbar} \ldots e^{-iE_1(\tau_2-t_1)/\hbar}\, e^{-iE_i t_1/\hbar}\, d\tau_2 \ldots d\tau_n =$$

$$= \lim_{\substack{t_2\to\infty\\ t_1\to-\infty}} \int_{t_1}^{t_2} (\text{integrand above})\, e^{i(E_f/\hbar\pm\omega_n\pm\omega_{n-1}+\cdots-E_i/\hbar)\tau}\, d\tau_2 \ldots d\tau_n\,. \quad (22)$$

In the integrand on the left of (22), we introduce the new integration variables

$$\tau_2' = \tau_2\,, \quad \tau_3' = \tau_3-\tau_2\,, \quad \tau_4' = \tau_4-\tau_3 \ldots \quad \tau_n' = \tau_n-\tau_{n-1}\,.$$

One finds that the integral with respect to τ_2' is essentially

$$\int_{-\infty}^{\infty} e^{i(E_f-E_i\pm\hbar\omega_n\pm\hbar\omega_{n-1}\pm\cdots)\tau_2'/\hbar}\, d\tau_2' = \quad (23)$$

$$= 2\pi\hbar\delta(E_f - E_i \pm \hbar\omega_n \pm \hbar\omega_{n-1} \pm \ldots)\,,$$

so that the exponential on the right-hand side of (22) is equal to one.

This completes the proof of eq. (19). It is clear that, since $-\ \hbar\omega_n$

corresponds to the absorption, and $\hbar\omega_n$ to the emission of a photon, (23) expresses the energy conservation as in (20). The connection between energy conservation and invariance under time translations of the transition probability is therefore proved *.

Now we assume that the electromagnetic waves (all propagating in the direction $\boldsymbol{n}$) are circularly polarized**, and that the initial and final wave functions $\psi_{\mathrm{i}}^{(V)}$, $\psi_{\mathrm{f}}^{(V)}$ are eigenfunctions of the component of the angular momentum along $\boldsymbol{n}$ (taken as the z axis), i.e.

$$L_3\, \psi_{\mathrm{f}}^{(V)} = \hbar m_{\mathrm{f}}\, \psi_{\mathrm{f}}^{(V)}\,, \quad L_3\, \psi_{\mathrm{i}}^{(V)} = \hbar m_{\mathrm{i}}\, \psi_{\mathrm{i}}^{(V)}\,.$$

Then, from the invariance of the transition amplitude under rotations about the z axis, we can derive the conservation of the z component of the angular momentum, provided that a "spin" $s_3 = +\hbar (s_3 = -\hbar)$ is attributed to each photon with right (left) circular polarization.

Considering for simplicity only the case of right circular polarization, one has

$$\boldsymbol{A}(\omega) \propto \boldsymbol{e}_1 + \mathrm{i}\, \boldsymbol{e}_2\,, \quad \boldsymbol{A}^*(\omega) \propto \boldsymbol{e}_1 - \mathrm{i}\, \boldsymbol{e}_2\,,$$

where

$$\boldsymbol{e}_1 \equiv (1, 0, 0,)\,, \quad \boldsymbol{e}_2 \equiv (0, 1, 0)\,, \quad \boldsymbol{n} \equiv (0, 0, 1)\,.$$

A generic term of the expansion of the transition amplitude is of the type

$$\begin{aligned} \mathfrak{M}_{\mathrm{fi}} = \int \psi_{\mathrm{f}}^{(V)*}(\boldsymbol{x}_n) \ldots (\boldsymbol{e}_1 \pm \mathrm{i}\boldsymbol{e}_2) \cdot \nabla_{\boldsymbol{x}_n} \psi_n(\boldsymbol{x}_n) \times \\ \times (\boldsymbol{e}_1 \pm \mathrm{i}\boldsymbol{e}_2) \cdot \nabla_{\boldsymbol{x}_{n-1}} \psi_{n-1}(\boldsymbol{x}_{n-1}) \ldots \psi_{\mathrm{i}}^{(V)}(\boldsymbol{x}_1)\, \mathrm{d}^3\mathrm{x}_1 \ldots \mathrm{d}^3\mathrm{x}_n \end{aligned} \tag{24}$$

(the interaction in A^2 will be considered below).

Expressing $\boldsymbol{x}_\alpha (\alpha = 1, \ldots n)$ in terms of $\boldsymbol{x}'_\alpha = R^{-1}\boldsymbol{x}$ (R being the transformation corresponding to a rotation ϕ around the z axis) we have

$$\psi_{\mathrm{f}}^{(V)*}(\boldsymbol{x}_n) = \mathrm{e}^{\mathrm{i}m_{\mathrm{f}}\phi}\, \psi_{\mathrm{f}}^{(V)*}(\boldsymbol{x}'_n)\,, \quad \psi_{\mathrm{i}}^{(V)}(\boldsymbol{x}_1) = \mathrm{e}^{-\mathrm{i}m_{\mathrm{i}}\phi}\, \psi_{\mathrm{i}}^{(V)}(\boldsymbol{x}'_1)\,,$$

(see Part I, Chap. IV, § 3), while

$$(\boldsymbol{e}_1 \pm \mathrm{i}\boldsymbol{e}_2) \cdot \nabla_{\boldsymbol{x}_\alpha} = \mathrm{e}^{\mp \mathrm{i}\phi} \left(\frac{\partial}{\partial x'_\alpha} \pm \mathrm{i} \frac{\partial}{\partial y'_\alpha} \right),$$

and

$$\mathrm{d}^3\mathrm{x}_1 \ldots \mathrm{d}^3\mathrm{x}_n = \mathrm{d}^3\mathrm{x}'_1 \ldots \mathrm{d}^3\mathrm{x}'_n\,.$$

---

* One can say briefly that the transition probability relative to the time interval (t_1, t_2) is the same as that for the interval $(t_1 + \tau, t_2 + \tau)$, provided that the time interval is sufficiently long. This is strictly true only for an infinitely long interval.

** The polarization is not necessarily the same (right or left) for all Fourier components.

On the other hand, the dots in (24) denote plane waves propagating along the z axis. They depend only on the third components of $\boldsymbol{x}_\alpha (\alpha = 1, 2, \dots n)$, which are left unchanged by the above transformation. Therefore, by the introduction of rotated integration variables the transition amplitude becomes

$$e^{i\phi[m_f - m_i \pm 1 \pm 1 \cdots]} \mathfrak{M}_{fi} .$$

Here a minus sign indicates the absorption of a photon, a plus the emission of a photon.

On requiring that the new from of the transition amplitude be identical to the old one. one has

$$m_f + 1 + 1 \dots = m_i + 1 + 1 \dots$$

There are as many 1's on the left and right as there are photons emitted and absorbed, respectively. This equation may therefore be written as

$$\hbar m_f + (\sum s_3)_f = \hbar m_i + (\sum s_3)_i$$

with $s_3 = \hbar$ (right polarization).

Taking into account both left and right polarized photons, we would have

$$m_f + \begin{pmatrix}\text{photons emitted with}\\ \text{right polarization}\end{pmatrix} - \begin{pmatrix}\text{photons emitted with}\\ \text{left polarization}\end{pmatrix} = \tag{25}$$

$$m_i + \begin{pmatrix}\text{photons absorbed with}\\ \text{right polarization}\end{pmatrix} - \begin{pmatrix}\text{photons absorbed with}\\ \text{left polarization}\end{pmatrix} .$$

The results (20) and (25) remain valid if the interaction $(e^2/2mc^2) A^2$ is also taken into account. In fact, this interaction may cause the atomic system to

(i) emit two photons with opposite spin,
(ii) absorb two photons with opposite spin,
(iii) absorb a photon of a certain spin and emit one with the same spin, whereas it may not
(iv) emit (absorb) two photons with the same spin,
(v) absorb a photon with a certain spin and emit one with opposite spin.

In fact, in the cases (i), (ii), (iii), the contribution of the interaction $(e^2/2mc^2) A^2$ to the probability amplitude is proportional to

$$(e_1 \pm ie_2)(e_1 \mp ie_2) = 2 ,$$

while in the cases (iv) and (v) it is proportional to

$$(e_1 \pm ie_2)(e_1 \pm ie_2) = 0 .$$

6. Approximations of electric dipole, magnetic dipole and electric quadrupole

According to (7) and (7′) of § 2, the first order transition amplitude is

$$-\ (2\pi)^{\frac{1}{2}}\,\frac{e}{mc}\int \psi_{\mathrm{f}}^{(V)*}(\boldsymbol{x})\,\mathrm{e}^{\pm\mathrm{i}\boldsymbol{k}\cdot\boldsymbol{x}}\begin{bmatrix} A(\omega)\cdot\nabla_x \\ \\ A^*(\omega)\cdot\nabla_x\end{bmatrix}\psi_{\mathrm{i}}^{(V)}(\boldsymbol{x})\,\mathrm{d}^3\mathrm{x}\,. \qquad (26)$$

If $\psi_{\mathrm{f}}^{(V)*}(\boldsymbol{x})\,\nabla_x\,\psi_{\mathrm{i}}^{(V)}(\boldsymbol{x})$ is negligible for $|\boldsymbol{x}| > a$ (a being the radius of the Bohr orbit for the initial or final state of the atomic system) and if the wavelength of the electromagnetic wave is much larger than a,

$$|\boldsymbol{k}|\,a \ll 1\,,$$

the exponential in (26) can be expanded in a power series

$$\mathrm{e}^{\pm\mathrm{i}\boldsymbol{k}\cdot\boldsymbol{x}} = 1 \pm \mathrm{i}\,\boldsymbol{k}\cdot\boldsymbol{x} + \dots \qquad (27)$$

of which one may consider, to a good approximation, only a limited number of terms.

6.1. *Electric dipole approximation* (E1)

Taking only the first term of the expansion (27), (26) becomes

$$-\,(2\pi)^{\frac{1}{2}}\frac{e}{mc}\int\psi_{\mathrm{f}}^{(V)*}\begin{bmatrix} A\cdot\nabla_x \\ \\ A^*\cdot\nabla_x\end{bmatrix}\psi_{\mathrm{i}}^{(V)}\,\mathrm{d}^3\mathrm{x} = (2\pi)^{\frac{1}{2}}(e/\mathrm{i}m\hbar c)\begin{bmatrix} A \\ \\ A^*\end{bmatrix}\cdot\int\psi_{\mathrm{f}}^{(V)*}\,\boldsymbol{p}\,\psi_{\mathrm{i}}^{(V)}\,\mathrm{d}^3\mathrm{x}\,,$$

where $\boldsymbol{p} = -\,\mathrm{i}\hbar\nabla_x$ is the momentum operator.

It is not difficult to see that, if

$$(H^{(0)} + V)\,\psi_{\mathrm{f}}^{(V)} = E_{\mathrm{f}}\,\psi_{\mathrm{f}}^{(V)}\,, \quad (H^{(0)} + V)\,\psi_{\mathrm{i}}^{(V)} = E_{\mathrm{i}}\,\psi_{\mathrm{i}}^{(V)}\,,$$

also *

$$\int\psi_{\mathrm{f}}^{(V)*}\,\boldsymbol{p}\,\psi_{\mathrm{i}}^{(V)}\,\mathrm{d}^3\mathrm{x} = (\mathrm{i}m/\hbar)(E_{\mathrm{f}} - E_{\mathrm{i}})\int\psi_{\mathrm{f}}^{(V)*}\,\boldsymbol{x}\,\psi_{\mathrm{i}}^{(V)}\,\mathrm{d}^3\mathrm{x}\,.$$

The transition amplitude (order e and approximation E1)

$$\mathfrak{M}_{\mathrm{fi}}(e) \simeq \frac{\sqrt{2\pi}}{\hbar^2 c}\,(E_{\mathrm{f}} - E_{\mathrm{i}})\begin{bmatrix} A \\ \\ A^*\end{bmatrix}\cdot\int\psi_{\mathrm{f}}^{(V)*}\,e\,\boldsymbol{x}\,\psi_{\mathrm{i}}^{(V)}\,\mathrm{d}^3\mathrm{x}\,, \qquad (28)$$

---

* In fact

$$\int\psi_{\mathrm{f}}^{(V)*}(H^{(0)} + V)\,\boldsymbol{x}\,\psi_{\mathrm{i}}^{(V)}\,\mathrm{d}^3\mathrm{x} = E_{\mathrm{f}}\int\psi_{\mathrm{f}}^{(V)*}\,\boldsymbol{x}\,\psi_{\mathrm{i}}^{(V)}\,\mathrm{d}^3\mathrm{x}\,,$$
$$\int\psi_{\mathrm{f}}^{(V)*}\,\boldsymbol{x}(H^{(0)} + V)\,\psi_{\mathrm{i}}^{(V)}\,\mathrm{d}^3\mathrm{x} = E_{\mathrm{i}}\int\psi_{\mathrm{f}}^{(V)*}\,\boldsymbol{x}\,\psi_{\mathrm{i}}^{(V)}\,\mathrm{d}^3\mathrm{x}\,,$$

and, subtracting,

$$(E_{\mathrm{f}} - E_{\mathrm{i}})\int\psi_{\mathrm{f}}^{(V)*}\,\boldsymbol{x}\,\psi_{\mathrm{i}}^{(V)}\,\mathrm{d}^3\mathrm{x} = \int\psi_{\mathrm{f}}^{(V)*}\,[H^{(0)} + V,\,\boldsymbol{x}]\,\psi_{\mathrm{i}}^{(V)}\,\mathrm{d}^3\mathrm{x} = $$
$$= (\hbar/m\mathrm{i})\int\psi_{\mathrm{f}}^{(V)*}\,\boldsymbol{p}\,\psi_{\mathrm{i}}^{(V)}\,\mathrm{d}^3\mathrm{x}\,.$$

is therefore proportional to the matrix element of the *electric dipole* (hence the name).

There exist selection rules for E1 transitions. In fact, changing the integration variable in (28) from $\boldsymbol{x}$ to $\boldsymbol{x}' = -\boldsymbol{x}$, if

$$\psi_{\rm f}^{(V)}(-\boldsymbol{x}) = P_{\rm f}\,\psi_{\rm f}^{(V)}(\boldsymbol{x})\,, \quad \psi_{\rm i}^{(V)}(-\boldsymbol{x}) = P_{\rm i}\,\psi_{\rm i}^{(V)}(\boldsymbol{x})\,,$$

one has

$$\int \psi_{\rm f}^{(V)*}(\boldsymbol{x}')\,\boldsymbol{x}'\,\psi_{\rm i}^{(V)}(\boldsymbol{x}')\,{\rm d}^3{\rm x}' = -P_{\rm i}P_{\rm f}\int \psi_{\rm f}^{(V)*}(\boldsymbol{x})\,\boldsymbol{x}\,\psi_{\rm i}^{(V)}(\boldsymbol{x})\,{\rm d}^3{\rm x}\,.$$

Therefore the matrix element for an E1 transition is different from zero only if

$$P_{\rm f} = -P_{\rm i}\,. \tag{29}$$

If $\psi_{\rm i}^{(V)}$ and $\psi_{\rm f}^{(V)}$ are eigenfunctions of the angular momentum, i.e. their angular parts are spherical harmonics of order $l_{\rm i}$ and $l_{\rm f}$ respectively, since $(\boldsymbol{x}/r)\,Y_{l_{\rm i}}$ is a linear combination of $Y_{l_{\rm i}+1}$ and $Y_{l_{\rm i}-1}$, and so

$$\int Y_{l_{\rm f}}^*\,\boldsymbol{x}\,Y_{l_{\rm i}}\,{\rm d}\Omega = 0$$

unless

$$l_{\rm f} = l_{\rm i} \pm 1\,,$$

one finds that, for E1 transitions, $\Delta l = \pm 1$. This agrees with (29), as is seen at once.

Finally, we assume that $\psi_{\rm i}^{(V)}$ and $\psi_{\rm f}^{(V)}$ are eigenfunctions of L_3 belonging to the eigenvalues $\hbar m_{\rm i}$ and $\hbar m_{\rm f}$ respectively.

If the vector potential $\boldsymbol{A}$ has (right/left) circular polarization, and propagation takes place along the z axis, the E1 matrix element is proportional to

$$\begin{pmatrix} \boldsymbol{e}_1 \pm {\rm i}\,\boldsymbol{e}_2 \\ \boldsymbol{e}_1 \mp {\rm i}\,\boldsymbol{e}_2 \end{pmatrix} \cdot \int \psi_{\rm f}^{(V)*}\,\boldsymbol{x}\,\psi_{\rm i}^{(V)}\,{\rm d}^3{\rm x}\,. \tag{30}$$

(The first line refers to absorption of a photon, the second to emission. Upper and lower signs refer to right and left circular polarization respectively.)

(30) may also be written in the form

$$\int \psi_{\rm f}^{(V)*}\begin{pmatrix} x \pm {\rm i}\,y \\ x \mp {\rm i}\,y \end{pmatrix}\psi_{\rm i}^{(V)}\,{\rm d}^3{\rm x} \propto \int_0^{2\pi} {\rm e}^{-{\rm i}m_{\rm f}\phi}\begin{pmatrix} {\rm e}^{\pm{\rm i}\phi} \\ {\rm e}^{\mp{\rm i}\phi} \end{pmatrix}{\rm e}^{{\rm i}m_{\rm i}\phi}\,{\rm d}\phi\,,$$

showing that it is different from zero only if

$$\begin{pmatrix} m_{\rm f} = m_{\rm i} \pm 1 \\ m_{\rm f} = m_{\rm i} \mp 1 \end{pmatrix}.$$

This means that the absorption (emission) of a photon with (right/left) circular polarization causes the magnetic quantum number to (increase (decrease)/decrease (increase)) by one unit, in agreement with the general result obtained in § 5, eq. (25).

If, on the other hand, $\boldsymbol{n}$ is perpendicular to, and $\boldsymbol{A}$ is linearly polarized along the z axis, the E1 matrix element is proportional to

$$\int_0^{2\pi} e^{-i[m_f - m_i]\phi} \, d\phi ,$$

and the selection rule is

$$m_f = m_i .$$

6.2. Approximation of magnetic dipole (M1) *and of electric quadrupole* (E2)

The contribution to the transition amplitude from the second term of the expansion of $e^{\pm i\boldsymbol{k}\cdot\boldsymbol{x}}$ is

$$\sqrt{2\pi}\,\frac{ie}{mc}\begin{bmatrix} -\boldsymbol{A} \\ \boldsymbol{A}^* \end{bmatrix} \cdot \int \psi_f^{(V)*}(\boldsymbol{k}\cdot\boldsymbol{x})\, \nabla_x \psi_i^{(V)} \, d^3x .$$

We have

$$(\boldsymbol{k}\cdot\boldsymbol{x})\nabla_x = \tfrac{1}{2}\big[(\boldsymbol{k}\cdot\boldsymbol{x})\nabla_x - \boldsymbol{x}(\boldsymbol{k}\cdot\nabla_x)\big] + \tfrac{1}{2}\big[(\boldsymbol{k}\cdot\boldsymbol{x})\nabla_x + \boldsymbol{x}(\boldsymbol{k}\cdot\nabla_x)\big] . \quad (31)$$

The first term on the right may be written as

$$\tfrac{1}{2}[(\boldsymbol{k}\cdot\boldsymbol{x})\nabla_x - \boldsymbol{x}(\boldsymbol{k}\cdot\nabla_x)] = -\tfrac{1}{2}\boldsymbol{k}\times(\boldsymbol{x}\times\nabla_x) = -i\,\boldsymbol{k}\times\boldsymbol{L}/2\hbar ,$$

where $\boldsymbol{L}$ is the orbital angular momentum. This, since $\boldsymbol{\mu}_L = -e\boldsymbol{L}/2mc$ is the orbital magnetic moment, gives the *magnetic dipole transition* amplitude

$$\frac{\sqrt{2\pi}}{\hbar}\begin{bmatrix} \boldsymbol{A} \\ -\boldsymbol{A}^* \end{bmatrix} \cdot \left(\boldsymbol{k}\times\int \psi_f^{(V)*}\, \boldsymbol{\mu}_L\, \psi_i^{(V)} \, d^3x\right) .$$

Similarly the second term of (31)

$$\tfrac{1}{2}\big[(\boldsymbol{k}\cdot\boldsymbol{x})\nabla_x + \boldsymbol{x}(\boldsymbol{k}\cdot\nabla_x)\big]_n = \tfrac{1}{2}k_m(x_m\nabla_n + x_n\nabla_m) ,$$

since *

---

* In fact

$$(E_f - E_i)\int \psi_f^{(V)*}\, x_m x_n \psi_i^{(V)}\, d^3x = \int \psi_f^{(V)*}[H^{(0)} + V, x_m x_n]\,\psi_i^{(V)}\, d^3x .$$

On the other hand

$$[H^{(0)} + V, x_m x_n] = [\boldsymbol{p}^2/2m, x_m]\,x_n + x_m\,[\boldsymbol{p}^2/2m, x_n] = (\hbar/mi)(p_m x_n + x_m p_n) =$$
$$= (\hbar/mi)(x_m p_n + x_n p_m - i\hbar\delta_{mn}) .$$

Thus, if $\psi_f^{(V)}$ and $\psi_i^{(V)}$ are orthogonal, one has

$$(E_f - E_i)\int \psi_f^{(V)*} x_m x_n \psi_i^{(V)}\, d^3x = (-\hbar^2/m)\int \psi_f^{(V)*}(x_m\nabla_n + x_n\nabla_m)\,\psi_i^{(V)}\, d^3x .$$

$$\int \psi_{\rm f}^{(V)*}(x_m \nabla_n + x_n \nabla_m)\psi_{\rm i}^{(V)}\, {\rm d}^3{\rm x} = -(m/\hbar^2)(E_{\rm f} - E_{\rm i}) \int \psi_{\rm f}^{(V)*}\, x_m\, x_n\, \psi_{\rm i}^{(V)}\, {\rm d}^3{\rm x}\,,$$

gives a contribution to the transition amplitude which is proportional to the matrix element of the *quadrupole moment* between the initial and final wave functions.

Note that, in fact, the electric quadrupole moment tensor is

$$Q_{mn} = -\tfrac{1}{2}\, e(x_m\, x_n - \tfrac{1}{3}\, \delta_{mn}\, \boldsymbol{x}^2)\,.$$

However, the contribution from the term $\frac{1}{3}\,\delta_{mn}\,\boldsymbol{x}^2$, since the indices of δ_{mn} are saturated by those of A_m and k_n, gives the scalar product $\boldsymbol{k}\cdot\boldsymbol{A}$, which is zero because of the electromagnetic wave being transverse.

The contribution to the E2 transition amplitude may therefore be written as follows

$$\sqrt{2\pi}\, \frac{\rm i}{\hbar^2 c}\,(E_{\rm f} - E_{\rm i}) \begin{bmatrix} -A_n\, k_m \\ \\ A_n^*\, k_m \end{bmatrix} \int \psi_{\rm f}^{(V)*}\, Q_{mn}\, \psi_{\rm i}^{(V)}\, {\rm d}^3{\rm x}\,,$$

where the indices m and n run from 1 to 3.

As is easy to show, the parity selection rule

$$P_{\rm i} = P_{\rm f}$$

holds for both magnetic dipole and electric quadrupole transitions *.

* Other selection rules for M1 and E2 transitions may be found, for instance, in the article by Bethe and Salpeter, *loc. cit.*, p. 364.

CHAPTER VI

TIME-DEPENDENT RELATIVISTIC SCATTERING THEORY

1. Dirac particles

We will treat first the case of Dirac particles. Since the Dirac equation is of the first order in the time, like the Schroedinger equation, we can use the non-relativistic theory developed in Chap. IV as a pattern.

Thus it would seem natural to introduce the retarded Green's function $\mathfrak{G}_{\mathrm{D\,ret}}(x, x')$ defined by

$$\mathfrak{G}_{\mathrm{D\,ret}}(x, x') = 0\,, \quad \text{for} \quad t < t'\,,$$
$$\left[\mathfrak{G}_{\mathrm{D\,ret}}(x, x')\right]_{t=t'} = \delta(\boldsymbol{x} - \boldsymbol{x}')\,,$$
$$\left(\frac{1}{c}\frac{\partial}{\partial t} + \boldsymbol{\alpha}\cdot\nabla_x + \mathrm{i}\,\beta\,\kappa + \mathrm{i}\,\beta\frac{\mathfrak{V}}{\hbar c}\right)\mathfrak{G}_{\mathrm{D\,ret}}(x, x') = \delta(x - x')\,,$$

which, for free particles and for $t > t'$, is equal to the Green's function $\mathrm{G_D}(x, x')$ defined in Part II, Chap. V.

Then the wave function, obeying the equation ($\beta\,\mathfrak{V}^{\dagger}\beta = \mathfrak{V}$)

$$\left(\frac{1}{c}\frac{\partial}{\partial t} + \boldsymbol{\alpha}\cdot\nabla_x + \mathrm{i}\,\beta\,\kappa + \mathrm{i}\,\beta\frac{\mathfrak{V}}{\hbar c}\right)\psi(x) = 0\,, \tag{1}$$

could be determined at a point 2 ($\boldsymbol{x}_2$, t_2) from its values in entire space at a previous time t_1 by the formula

$$\psi(2) = \int \mathfrak{G}_{\mathrm{D\,ret}}(2, 1)\,\psi(1)\,\mathrm{d}^3\mathrm{x}_1\,, \quad \text{for} \quad t_2 > t_1\,,$$

analogous to (10) of Chap. IV. An integral equation for $\mathfrak{G}_{\mathrm{D\,ret}}$, similar to eq. (11) of Chap. IV, would be easily established.

This approach, however, must be modified on account of the possibility that a wave function which, at the time t_1, possesses only positive frequencies, may acquire negative frequencies on propagating to t_2 according to eq. (1).

Although the occurrence of negative frequencies is an unavoidable feature of all relativistic theories, they can only be interpreted satisfactorily in a theory which accounts for antiparticles as well as for particles, for the production and annihilation of pairs of particles and antiparticles, in short, in a theory in which the number of particles is not a constant of motion.

To construct such a theory, following Feynman,* instead of the Green's function $\mathfrak{G}_{\mathrm{D\,ret}}$ another Green's function $\mathfrak{G}_{\mathrm{D}}$ must be used, which does not vanish for $t < t'$, satisfying the equation

$$\left(\frac{1}{c}\frac{\partial}{\partial t} + \boldsymbol{\alpha}\cdot\nabla_x + \mathrm{i}\,\beta\,\kappa + \mathrm{i}\,\beta\frac{\mathfrak{V}}{\hbar c}\right)\mathfrak{G}_{\mathrm{D}}(x, x') = \delta(x - x')\,, \tag{2}$$

and the condition that, for $t > t'$, it must only involve positive frequencies and, for $t < t'$, only negative frequencies.

For free particles such a Green's function $\mathfrak{G}^{(0)}_{\mathrm{D}}$ is easily constructed by subtracting from $\mathfrak{G}^{(0)}_{\mathrm{D\,ret}}$ its negative frequency part both for $t > t'$ and for $t < t'$.

This gives

$$\mathfrak{G}^{(0)}_{\mathrm{D}}(x, x') =$$

$$= \begin{cases} \left(-\dfrac{1}{c}\dfrac{\partial}{\partial t} + \boldsymbol{\alpha}\cdot\nabla_x + \mathrm{i}\,\beta\,\kappa\right)\dfrac{1}{(2\pi)^3}\dfrac{c}{2\mathrm{i}}\displaystyle\int \mathrm{e}^{\mathrm{i}\boldsymbol{k}\cdot(x-x')-\mathrm{i}\omega(t-t')}\dfrac{1}{\omega(\boldsymbol{k}^2)}\,\mathrm{d}^3\mathrm{k}\,, & (t > t'), \\[2ex] \left(-\dfrac{1}{c}\dfrac{\partial}{\partial t} + \boldsymbol{\alpha}\cdot\nabla_x + \mathrm{i}\,\beta\,\kappa\right)\dfrac{1}{(2\pi)^3}\dfrac{c}{2\mathrm{i}}\displaystyle\int \mathrm{e}^{\mathrm{i}\boldsymbol{k}\cdot(x-x')-\mathrm{i}\omega(t'-t)}\dfrac{1}{\omega(\boldsymbol{k}^2)}\,\mathrm{d}^3\mathrm{k}\,, & (t < t')\,. \end{cases}$$

It is easy to show that

$$\left(\frac{1}{c}\frac{\partial}{\partial t} + \boldsymbol{\alpha}\cdot\nabla_x + \mathrm{i}\,\beta\,\kappa\right)\mathfrak{G}^{(0)}_{\mathrm{D}}(x, x') = \delta(x - x')\,,$$

and that

$$\lim_{\varepsilon\to 0^+}\left[\mathfrak{G}^{(0)}_{\mathrm{D}}(x, x')\right]_{t=t'\pm\varepsilon} = \pm\,(2\pi)^{-3}\int \mathrm{e}^{\mathrm{i}\boldsymbol{k}\cdot(x-x')}\,\Lambda^{\pm}(\boldsymbol{k})\,\mathrm{d}^3\mathrm{k}\,,$$

where

$$\Lambda^{\pm}(\boldsymbol{k}) = \tfrac{1}{2}\left(I \pm \Lambda(\boldsymbol{k})\right),$$

with

$$\Lambda(\boldsymbol{k}) = \frac{\boldsymbol{\alpha}\cdot\boldsymbol{k} + \beta\,\kappa}{(\boldsymbol{k}^2 + \kappa^2)^{\frac{1}{2}}}\,,$$

are essentially the projection operators for positive and negative energy given in Part II, Chap. V, *Note 1*.

Using this Green's function, a free particle wave function at a space-time point 2 is determined by its positive frequency components at a time $t_1 < t_2$,

* R. P. Feynman, *The Theory of Positrons*, Phys. Rev. **76** (1949), 749.

and by its negative frequency components at a time $t_1' > t_2$ by the relation *

$$\begin{aligned}\psi^{(0)}(2) = &- \int \mathfrak{G}_{\mathrm{D}}^{(0)}(2, 1')\, \Lambda^- \psi^{(0)}(1')\, \mathrm{d}^3\mathrm{x}_1' \\ &+ \int \mathfrak{G}_{\mathrm{D}}^{(0)}(2, 1)\, \Lambda^+ \psi^{(0)}(1)\, \mathrm{d}^3\mathrm{x}_1 . \quad (t_1' > t_2 > t_1) .\end{aligned} \tag{3}$$

In fact, if t_1, t_2, t_1' are all different, $\psi^{(0)}(2)$ clearly obeys the free particle Dirac equation, since $\mathfrak{G}_{\mathrm{D}}^{(0)}(2, 1)$ and $\mathfrak{G}_{\mathrm{D}}^{(0)}(2, 1')$ do so.

Denoting by $\psi^{(0)}(\boldsymbol{k}, \pm)$ the Fourier transforms of the positive and negative frequency parts of $\psi^{(0)}(\boldsymbol{x}_2, 0)$, the right side of eq. (3), for $t_1' \to t_2$ and $t_1 \to t_2$, gives

$$\begin{aligned}&(2\pi)^{-3} \int e^{i\boldsymbol{k}\cdot(\boldsymbol{x}_2 - \boldsymbol{x}_1')}\, \Lambda^-(\boldsymbol{k}) \int \psi^{(0)}(\boldsymbol{k}', +)\, e^{i\boldsymbol{k}'\cdot\boldsymbol{x}_1'}\, \mathrm{d}^3\mathrm{x}_1'\, \mathrm{d}^3\mathrm{k}\, \mathrm{d}^3\mathrm{k}' + \\ &+ (2\pi)^{-3} \int e^{i\boldsymbol{k}\cdot(\boldsymbol{x}_2 - \boldsymbol{x}_1)}\, \Lambda^+(\boldsymbol{k}) \int \psi^{(0)}(\boldsymbol{k}', -)\, e^{i\boldsymbol{k}'\cdot\boldsymbol{x}_1}\, \mathrm{d}^3\mathrm{x}_1\, \mathrm{d}^3\mathrm{k}\, \mathrm{d}^3\mathrm{k}' = \psi^{(0)}(2) .\end{aligned}$$

The physical meaning of eq. (3) will be explained below. For the present, with the help of $\mathfrak{G}_{\mathrm{D}}^{(0)}$ we may write, for the Green's function $\mathfrak{G}_{\mathrm{D}}$ of interacting particles, the integral equation

$$\mathfrak{G}_{\mathrm{D}}(x, x') = \mathfrak{G}_{\mathrm{D}}^{(0)}(x, x') - (\mathrm{i}/\hbar c) \int \mathfrak{G}_{\mathrm{D}}^{(0)}(x, x'')\, \beta\, \mathfrak{V}(x'')\, \mathfrak{G}_{\mathrm{D}}(x'', x')\, \mathrm{d}^4x'' . \tag{4}$$

This equation, as is easy to see, is equivalent to the inhomogeneous differential equation (2), and to the condition that $\mathfrak{G}_{\mathrm{D}}(x, x')$ has only positive frequencies for $t > t'$ and negative frequencies for $t < t'$.

On iterating eq. (4) one has

$$\begin{aligned}\mathfrak{G}_{\mathrm{D}}(2, 1) = &\mathfrak{G}_{\mathrm{D}}^{(0)}(2, 1) - (\mathrm{i}/\hbar c) \int \mathfrak{G}_{\mathrm{D}}^{(0)}(2, 3)\, \beta\, \mathfrak{V}(3)\, \mathfrak{G}_{\mathrm{D}}^{(0)}(3, 1)\, \mathrm{d}^4x_3 \\ &+ (\mathrm{i}/\hbar c)^2 \int \mathfrak{G}_{\mathrm{D}}^{(0)}(2, 3)\, \beta\, \mathfrak{V}(3)\, \mathfrak{G}_{\mathrm{D}}^{(0)}(3, 4)\, \beta\, \mathfrak{V}(4)\, \mathfrak{G}_{\mathrm{D}}^{(0)}(4, 1)\, \mathrm{d}^4x_3\, \mathrm{d}^4x_4 ,\end{aligned} \tag{5}$$

whose terms one may picture by Feynman diagrams.

Thus the first term corresponds to the diagrams

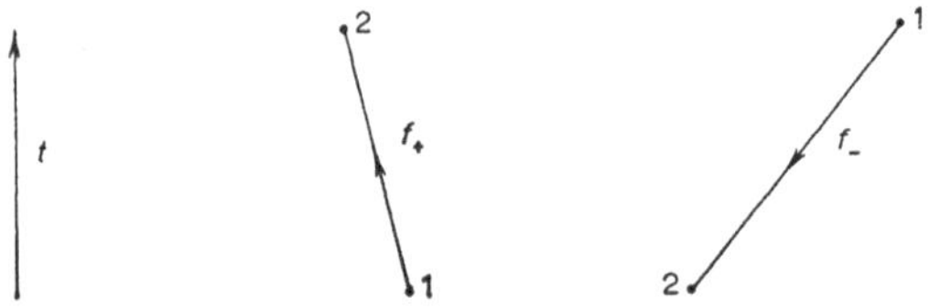

Fig. 19

* Later we will prove that

$\psi^{(0)}(2) = - \int \mathfrak{G}_{\mathrm{D}}^{(0)}(2, 1')\, \psi^{(0)}(1')\, \mathrm{d}^3\mathrm{x}'_1 + \int \mathfrak{G}_{\mathrm{D}}^{(0)}(2, 1)\, \psi^{(0)}(1)\, \mathrm{d}^3\mathrm{x}_1, \qquad (t_1' > t_2 > t_1) .$

This is equivalent to eq. (3) since, by the definition of the Green's function, only positive frequencies propagate in the future, and only negative frequencies in the past.

where f_+ denotes propagation by positive frequencies and f_- by negative frequencies, and the vertical arrow indicates the time axis. Similarly the second term corresponds to such diagrams as

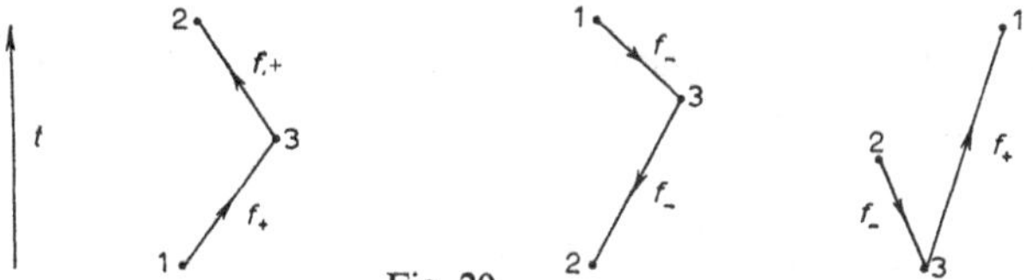

Fig. 20

and the third to

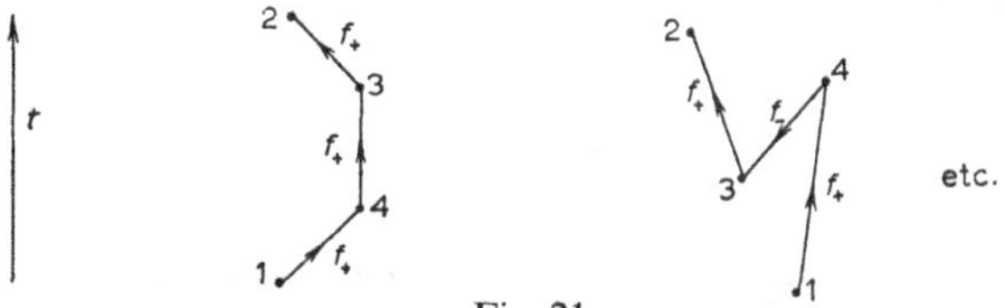

Fig. 21

For the interpretation of lines running backwards in time, a reformulation of the theory becomes necessary. Whereas a line running forward in time is taken to represent the propagation of a positive energy particle, one running backwards is made to correspond to the propagation along it, in the forward direction, of an antiparticle.

Thus, denoting a particle (e.g. electron) by p and an antiparticle (positon) by $\bar{p}$, the diagram

(a)

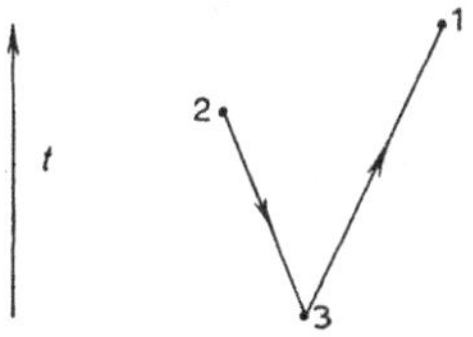

Fig. 22a

is equivalent to

(b)

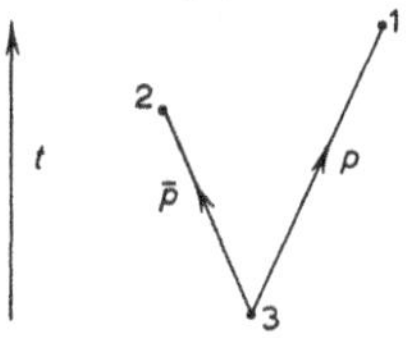

Fig. 22b

in which the interaction produces a particle-antiparticle pair at 3.

Hence the diagram

(a)

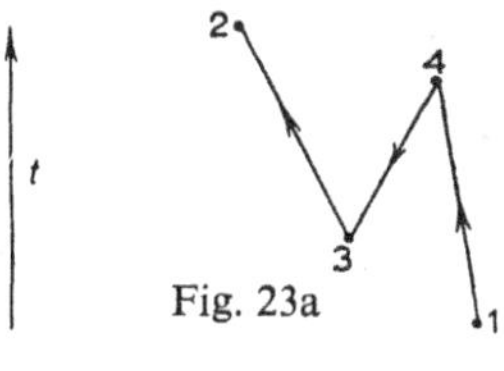

Fig. 23a

means

(b)

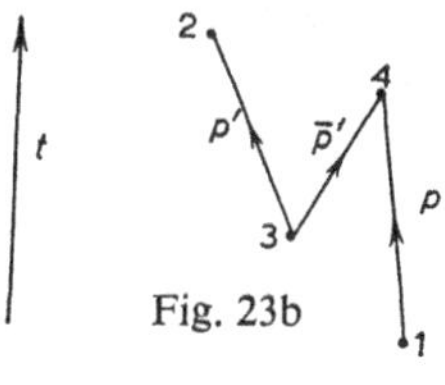

Fig. 23b

that is, production at 3 of a pair p', $\bar{p}'$, and annihilation of $\bar{p}'$ with p at 4.

2. Scattering processes

Let us now study the propagation of wave functions. Multiplying eq. (2), written in the covariant form

$$\left(\gamma_\mu \frac{\partial}{\partial x_\mu^{(2)}} + \kappa + \frac{\mathfrak{V}(2)}{\hbar c}\right) \mathfrak{G}_\mathrm{D}(2, 3) = -\,\mathrm{i}\,\beta\, \delta(\mathsf{x}_2 - \mathsf{x}_3)\,,$$

from the right by

$$\beta\left(-\,\gamma_\mu \frac{\overleftarrow{\partial}}{\partial x_\mu^{(3)}} + \kappa + \frac{\mathfrak{V}(3)}{\hbar c}\right),$$

(the arrow means that $\partial/\partial x_\mu{}^{(3)}$ acts on the left), one has

$$\left(\gamma_\mu \frac{\partial}{\partial x_\mu^{(2)}} + \kappa + \frac{\mathfrak{V}(2)}{\hbar c}\right) \mathfrak{G}_\mathrm{D}(2, 3)\,\beta\left(-\,\gamma_\mu \frac{\overleftarrow{\partial}}{\partial x_\mu^{(3)}} + \kappa + \frac{\mathfrak{V}(3)}{\hbar c}\right) =$$

$$= -\,\mathrm{i}\left(\gamma_\mu \frac{\partial}{\partial x_\mu^{(2)}} + \kappa + \frac{\mathfrak{V}(2)}{\hbar c}\right) \delta(\mathsf{x}_2 - \mathsf{x}_3)\,.$$

From this it follows that

$$\mathfrak{G}_\mathrm{D}(2, 3)\,\beta\left(-\,\gamma_\mu \frac{\overleftarrow{\partial}}{\partial x_\mu^{(3)}} + \kappa + \frac{\mathfrak{V}(3)}{\hbar c}\right) = -\,\mathrm{i}\,\delta(\mathsf{x}_2 - \mathsf{x}_3)\,.$$

Multiplying this equation from the right by $\psi(3)$, and the Dirac equation

$$\left(\gamma_\mu \frac{\partial}{\partial x_\mu^{(3)}} + \kappa + \frac{\mathfrak{V}(3)}{\hbar c}\right) \psi(3) = 0\,,$$

from the left by $\mathfrak{G}_\mathrm{D}(2, 3)\beta$, subtracting and integrating over the region of space-time comprised between the planes $t = t_1$ and $t = t_1'$, we obtain

$$\int_{t_1}^{t_1'} \left[\mathfrak{G}_\mathrm{D}(2, 3)\,\beta\left(-\,\gamma_\mu \frac{\overleftarrow{\partial}}{\partial x_\mu^{(3)}} + \kappa + \frac{\mathfrak{V}(3)}{\hbar c}\right) \psi(3) -\right.$$

$$\left. -\,\mathfrak{G}_\mathrm{D}(2, 3)\,\beta\left(\gamma_\mu \frac{\overrightarrow{\partial}}{\partial x_\mu^{(3)}} + \kappa + \frac{\mathfrak{V}(3)}{\hbar c}\right) \psi(3)\right] \mathrm{d}^4\mathsf{x}_3 = -\,\mathrm{i}\,\psi(2)\,.$$

The integrand on the left may be written as

$$\frac{\partial}{\partial x_\mu^{(3)}}[-\ \mathfrak{G}_D(2,3)\,\beta\,\gamma_\mu\,\psi(3)]\,.$$

Hence, by Gauss' theorem in four-dimensional space, we have *

$$\psi(2) = -\int \mathfrak{G}_D(2,1')\,\psi(1')\,d^3x_1' + \int \mathfrak{G}_D(2,1)\,\psi(1)\,d^3x_1\,, \tag{6}$$

where $t_1' > t_2 > t_1$.

Thus, by the properties of the Green's function, the wave function at 2 is determined by its positive frequency components at an earlier time and by its negative frequency components at a later time.

We assume that the interaction is active only in the time interval $(T, T + \Delta T)$ with $T > t_1$ and $T + \Delta T < t_1'$. Then $\psi(1)$ and $\psi(1')$ are free particle wave functions, which we denote by $\psi^{(0)}(1)$ and $\psi^{(0)}(1')$. Clearly, in eq. (6) they may be replaced by their positive and negative frequency parts $\psi^{(0,+)}(1)$ and $\psi^{(0,-)}(1')$ respectively.

We are now in a position to calculate the amplitudes for physical processes.

The probability amplitude of finding a free particle with wave function $\psi_f^{(0,+)}$ at time t_2, if a particle with wave function $\psi_i^{(0,+)}$ existed at time t_1, is given by the product

$$\begin{aligned}(\psi_f^{(0,+)}(t_2), \psi(t_2)) &= \int (\psi_f^{(0,+)}(2))^\dagger\,\psi(2)\,d^3x_2 = \\ &= \int (\psi_f^{(0,+)}(2))^\dagger\,\mathfrak{G}_D(2,1)\,\psi_i^{(0,+)}(1)\,d^3x_1\,d^3x_2\,, \quad (t_2 = t_1')\,,\end{aligned}$$

where we have obtained $\psi(2)$ by putting in eq. (6)

$$\psi(1) = \psi_i^{(0,+)}(1)\,, \quad \psi(1') \to \Lambda^-\,\psi(1') = 0\,.$$

If the iterated form (5) of the Green's function is inserted, this transition amplitude is put in the form of a power series of the strength of $\mathfrak{V}$, whose terms correspond to the Feynman diagrams

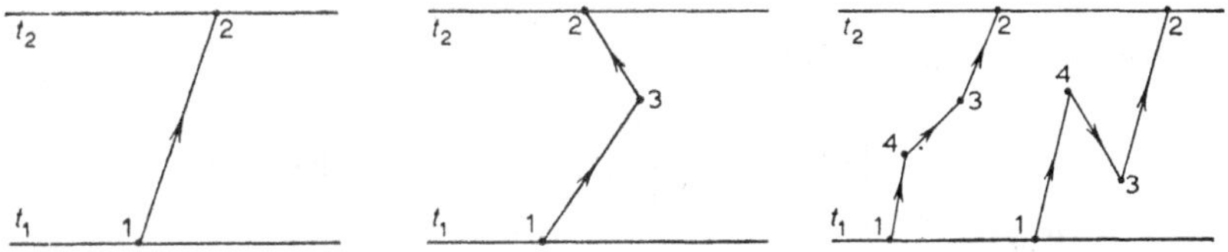

Fig. 24

* The equation given in the footnote on p. 292, is a special case of (6) for $\mathfrak{V} = 0$.

Similarly, the probability amplitude for the transition of an antiparticle from the state i at time t_1 to the state f at time t_1' is given by

$$(\psi_{-i}^{(0,-)}(2)\,,\quad \psi(2))\,,\quad (t_2 = t_1)\,,$$

where $\psi(2)$ is given by eq. (6) with

$$\psi(1') = \psi_{-f}^{(0,-)}\,,\quad \psi(1) \to \Lambda^+\,\psi(1) = 0\,,$$

and $\psi_{-i}^{(0,-)}$ and $\psi_{-f}^{(0,-)}$ are negative energy solutions of the free particle Dirac equation, corresponding to momenta and angular momenta opposite to those of the antiparticles in the states i and f.

For example, if the antiparticle in i has momentum $\hbar \boldsymbol{k}_i$ and spin parallel to the momentum,

$$\psi_{-i}^{(0,-)} = v(\boldsymbol{k}_i, \uparrow)\, e^{-i\boldsymbol{k}_i \cdot x + i\omega_i t}\,.$$

Finally the probability amplitude for a particle-antiparticle pair in the states f and f′ respectively, to be produced from a vacuum, is given by

$$(\psi_f^{(0,+)}(t_1'), \psi(t_1')) = -\int (\psi_f^{(0,+)}(2))^\dagger\, \mathfrak{G}_D(2, 1')\, \psi_{-f'}^{(0,-)}(1')\, d^3x_1'\, d^3x_2\,,$$

$(t_2 = t_1')$.

Here $\psi_f^{(0,+)}$ is the wave function of the particle produced, and $\psi(2)$ is given by eq. (6) with

$$\psi(1') = \psi_{-f'}^{(0,-)}\,,\quad \psi(1) = 0\,,$$

(f′ = state of the antiparticle, while − f′ labels the corresponding negative energy wave function).

3. Scattering of spinless particles

A similar theory can be constructed for wave functions obeying the Klein-Gordon equation

$$(\Box - \kappa^2)\,\psi = \mathfrak{V}\,\psi\,. \tag{7}$$

We define the Green's function $\mathfrak{G}_{KG}(x, x')$ obeying the inhomogeneous equation

$$\begin{aligned}[\Box - \kappa^2 - \mathfrak{V}(x)]\, \mathfrak{G}_{KG}(x, x') &= \delta(x - x') = \\ &= [\Box' - \kappa^2 - \mathfrak{V}(x')]\, \mathfrak{G}_{KG}(x, x')\,.\end{aligned} \tag{8}$$

For free particles this is given by

$$\mathfrak{G}_{KG}^{(0)}(x, x') = \begin{cases} (2\pi)^{-3}\dfrac{c}{2i}\displaystyle\int \dfrac{1}{\omega}\, e^{ik\cdot(x-x') - i\omega(t-t')}\, d^3k\,, & t > t'\,, \\[2ex] (2\pi)^{-3}\dfrac{c}{2i}\displaystyle\int \dfrac{1}{\omega}\, e^{ik\cdot(x'-x) - i\omega(t'-t)}\, d^3k\,, & t < t'\,. \end{cases}$$

Then, from (7) and (8) one has,

$$\psi(2) = \int_{t_1}^{t_1'} [\psi(3)\, \Box_3\, \mathfrak{G}_{\mathrm{KG}}(2,3) - \mathfrak{G}_{\mathrm{KG}}(2,3)\, \Box_3\, \psi(3)]\, \mathrm{d}^4 x_3 =$$

$$= \int_{t_1}^{t_1'} \frac{\partial}{\partial x_\mu^{(3)}} \left[\mathfrak{G}_{\mathrm{KG}}(2,3) \frac{\overleftrightarrow{\partial}}{\partial x_\mu^{(3)}} \psi(3) \right] \mathrm{d}^4 x_3\,,$$

where

$$\frac{\overleftrightarrow{\partial}}{\partial x_\mu^{(3)}} = \frac{\overleftarrow{\partial}}{\partial x_\mu^{(3)}} - \frac{\overrightarrow{\partial}}{\partial x_\mu^{(3)}}\,.$$

Hence, by Gauss' theorem,

$$\psi(2) = -\frac{1}{c} \int \mathfrak{G}_{\mathrm{KG}}(2,1') \frac{\overleftrightarrow{\partial}}{\partial t_1'} \psi(1')\, \mathrm{d}^3 x_1' + \frac{1}{c} \int \mathfrak{G}_{\mathrm{KG}}(2,1) \frac{\overleftrightarrow{\partial}}{\partial t_1} \psi(1)\, \mathrm{d}^3 x_1\,, \quad (9)$$

with $t_1' > t_2 > t_1$.

The integral equation for $\mathfrak{G}_{\mathrm{KG}}$

$$\mathfrak{G}_{\mathrm{KG}}(2,1) = \mathfrak{G}_{\mathrm{KG}}^{(0)}(2,1) + \int \mathfrak{G}_{\mathrm{KG}}^{(0)}(2,3)\, \mathfrak{V}(3)\, \mathfrak{G}_{\mathrm{KG}}(3,1)\, \mathrm{d}^4 x_3\,,$$

may be solved by iteration, and the corresponding Feynman diagrams can be interpreted in terms of particles and antiparticles as for the Dirac case *.

The calculation of probability amplitudes is the same as for Dirac particles, but for a different definition of the product of two wave functions, which, for spin zero particles, is given by the Feshbach-Villars metric.

4. Formal developments towards field theory

This book would hardly be complete if the connection between Feynman's scattering theory and the methods of field theory were not, at least briefly, outlined.

A translation of Feynman's prescriptions for the evaluation of transition amplitudes into the language of field theory, yet unknown to the reader, may be made as follows.

Let $\psi^{(0)}$ be a free particle wave function represented in the form

$$\psi^{(0)} = \sum_\alpha [A(\alpha)\, \psi^{(0)}(x;\alpha) + B(\alpha)\, \psi^{(0)}(x; -\alpha)]\,. \quad (10)$$

* For example a negative pion (π^-) is the antiparticle of a positive pion (π^+).

Here α and $-\alpha$ label the functions of a complete orthonormal set of free particle solutions of positive and negative frequencies respectively. For instance α may be the four-momentum of a particle of spin zero, in which case

$$\psi^{(0)}(x; \pm k) = e^{\pm i(k \cdot x - \omega t)}$$

and the sum reduces to an integral over $\boldsymbol{k}$.

Similarly for Dirac particles

$$\psi^{(0)}(x; k, \lambda) = u(k, \lambda)\, e^{i(k \cdot x - \omega t)}\,, \quad \psi^{(0)}(x; -k, \lambda) = v(k, \lambda)\, e^{-i(k \cdot x - \omega t)}\,.$$

We now introduce the states $|\varepsilon, \alpha>$ of a particle $(\varepsilon = +1)$ and an antiparticle $(\varepsilon = -1)$, having the 'quantum numbers' α, and assume that such states are orthonormal

$$< \varepsilon', \alpha' | \varepsilon, \alpha > = \delta_{\alpha'\alpha}\, \delta_{\varepsilon'\varepsilon}\,.$$

We may also introduce the vacuum state $|\mathrm{vac}>$, $(<\mathrm{vac}|\mathrm{vac}> = 1)$, as well as the operators $\mathrm{a}(\alpha)$, $\mathrm{a}^\dagger(\alpha)$ and $\mathrm{b}(\alpha)$, $\mathrm{b}^\dagger(\alpha)$, having the properties

$$\begin{aligned}
&\mathrm{a}(\alpha)\,|+, \alpha'> = \mathrm{b}(\alpha)\,|-, \alpha'> = \delta_{\alpha\alpha'}\,|\mathrm{vac}>\,,\\
&<+, \alpha'|\,\mathrm{a}^\dagger(\alpha) = <-, \alpha'|\,\mathrm{b}^\dagger(\alpha) = \delta_{\alpha'\alpha} < \mathrm{vac}|\,,\\
&\mathrm{b}(\alpha)\,|+, \alpha'> = 0\,, \qquad \mathrm{a}(\alpha)\,|-, \alpha'> = 0\,,\\
&<+, \alpha'|\,\mathrm{b}^\dagger(\alpha) = 0\,, \qquad <-, \alpha'|\,\mathrm{a}^\dagger(\alpha) = 0\,,\\
&\mathrm{a}(\alpha)\,|\mathrm{vac}> = \mathrm{b}(\alpha)\,|\mathrm{vac}> = 0\,,\\
&<\mathrm{vac}|\,\mathrm{a}^\dagger(\alpha) = <\mathrm{vac}|\,\mathrm{b}^\dagger(\alpha) = 0\,.
\end{aligned} \tag{11}$$

The wave function $\psi^{(0)}$ will now be replaced by the operator

$$\psi^{(0)} = \sum_{\alpha} \left[\mathrm{a}(\alpha)\, \psi^{(0)}(x;\alpha) + \mathrm{b}^\dagger(\alpha)\, \psi^{(0)}(x; -\alpha)\right].$$

Let us define the generalized transition amplitude

$$(\psi^{(0)}, \psi) = \int \psi^{(0)\dagger}(x)\, \psi(x)\, \mathrm{d}^3 x\,, \quad \text{(Dirac particles)}\,,$$

$$(\psi^{(0)}, \psi) = \frac{\hbar i}{2mc^2} \int \left[\psi^{(0)\dagger}(x) \frac{\partial \psi(x)}{\partial t} - \frac{\partial \psi^{(0)}(x)^\dagger}{\partial t} \psi(x)\right] \mathrm{d}^3 x\,, \quad \text{(spin zero particles)},$$

where $\psi(x)$ is given by eqs. (6) or (9).

Thus

$$\begin{aligned}
(\psi^{(0)}(2), \psi(2)) = \int \psi^{(0)\dagger}(2) \big[&- \int \mathfrak{G}_{\mathrm{D}}(2, 1')\, \psi^{(0)}(1')\, \mathrm{d}^3 x_1' + \\
&+ \int \mathfrak{G}_{\mathrm{D}}(2, 1)\, \psi^{(0)}(1)\, \mathrm{d}^3 x_1\big]\, \mathrm{d}^3 x_2\,, \quad \text{(Dirac particles)},
\end{aligned}$$

and similarly for spin zero particles.

The order in which the operators a, $a^\dagger$, b, $b^\dagger$ appear in these expressions must, at first, be preserved. Now, on assuming that any two of the operators introduced above anticommute or commute according as one is dealing with Dirac or spin zero particles, respectively, transition amplitudes for the processes treated in the previous sections may be defined as follows:

(i) particle scattering $< +, \mathrm{f}\,|(\psi^{(0)}, \psi)|\, +, \mathrm{i} >$,

(ii) antiparticle scattering $< -, \mathrm{f}\,|(\psi^{(0)}, \psi)|\, -, \mathrm{i} >$,

(iii) pair creation (particle in state f′, antiparticle in state f) $< +, \mathrm{f}'; -, \mathrm{f}\,|(\psi^{(0)}, \psi)|\, \mathrm{vac} >$,

(iv) pair annihilation (particle in state i′, antiparticle in state i) $< \mathrm{vac}\,|(\psi^{(0)}, \psi)|\, +, \mathrm{i}'; -, \mathrm{i} >$.

Using the properties (11) it can readily be shown that the absolute values of the transition amplitudes so defined and those of the corresponding amplitudes given by Feynman's method coincide.

We now recall that, for the spin zero particle wave function

$$\psi(\boldsymbol{x}, t) = (2\pi)^{-\frac{3}{2}} \int \left[A(\boldsymbol{k})\, e^{i(\boldsymbol{k}\cdot\boldsymbol{x}-\omega t)} + B(\boldsymbol{k})\, e^{-i(\boldsymbol{k}\cdot\boldsymbol{x}-\omega t)}\right] d^3k\,, \tag{12}$$

the energy and the charge take the form

$$E = (\hbar/mc^2) \left[\int \hbar\omega\, |A(\boldsymbol{k})|^2\, \omega\, d^3k + \int \hbar\omega\, |B(\boldsymbol{k})|^2\, \omega\, d^3k\right], \tag{13}$$

$$Q = (\hbar/mc^2) \left[\int |A(\boldsymbol{k})|^2\, \omega\, d^3k - \int |B(\boldsymbol{k})|^2\, \omega\, d^3k\right]. \tag{14}$$

In field theory, the energy and the charge of a field of spin zero particles are given by expressions similar to (13) and (14),

$$E = \int [a^\dagger(\boldsymbol{k})\, a(\boldsymbol{k}) + b^\dagger(\boldsymbol{k})\, b(\boldsymbol{k})]\, \hbar\omega\, d^3k\,,$$
$$Q = \int [a^\dagger(\boldsymbol{k})\, a(\boldsymbol{k}) - b^\dagger(\boldsymbol{k})\, b(\boldsymbol{k})]\, d^3k\,.$$

These are operators.

From the relations

$$a(\boldsymbol{k})|\boldsymbol{k}'> = \delta(\boldsymbol{k} - \boldsymbol{k}')|\,\mathrm{vac}>, \text{etc.}$$

which are the natural extension of eqs. (11) to the case of continuous quantum numbers $\boldsymbol{k}$, one has the matrix elements of E and Q between one-particle and one-antiparticle states

$$< +, \boldsymbol{k}'\,|E|\, +, \boldsymbol{k} > = < -, \boldsymbol{k}'\,|E|\, -, \boldsymbol{k} > = \hbar\omega\, \delta\,(\boldsymbol{k}' - \boldsymbol{k})\,,$$
$$< +, \boldsymbol{k}'\,|Q|\, +, \boldsymbol{k} > = - < -, \boldsymbol{k}'\,|Q|\, -, \boldsymbol{k} > = \delta\,(\boldsymbol{k}' - \boldsymbol{k})\,.$$

From these one sees that a formalism using operator, instead of c-number, wave functions, and the prescription for obtaining physical quantities by taking matrix elements between one-particle and one-antiparticle states, leads to positive energies for both particles and antiparticles, to positive charge for the former and negative charge for the latter.

For particles of spin one-half, the field theoretical expressions for the energy and the charge are obtained by replacing in eqs. (16), Part II, Chap. V

$$|A(\boldsymbol{k}, \lambda)|^2 \text{ by } \mathrm{a}^\dagger(\boldsymbol{k}, \lambda)\,\mathrm{a}(\boldsymbol{k}, \lambda)$$

and

$$|B(\boldsymbol{k}, \lambda)|^2 \text{ by } \mathrm{b}(\boldsymbol{k}, \lambda)\,\mathrm{b}^\dagger(\boldsymbol{k}, \lambda)$$

$(= -\,\mathrm{b}^\dagger\,(\boldsymbol{k}, \lambda)\,\mathrm{b}(\boldsymbol{k}, \lambda)$, since the operators a, $\mathrm{b}^\dagger$ etc. anticommute).

Thus one has

$$<+\,,\boldsymbol{k}',\lambda'\,|E|\,+\,,\boldsymbol{k},\lambda> \;=\; <-\,,\boldsymbol{k}',\lambda'\,|E|\,-\,,\boldsymbol{k},\lambda> \;=\; \hbar\omega\,\delta(\boldsymbol{k}'-\boldsymbol{k})\,\delta_{\lambda'\lambda}\,,$$
$$<+\,,\boldsymbol{k}',\lambda'\,|Q|\,+\,,\boldsymbol{k},\lambda> \;=\; -<-\,,\boldsymbol{k}',\lambda'\,|Q|\,-\,,\boldsymbol{k},\lambda> \;=\; \delta(\boldsymbol{k}'-\boldsymbol{k})\,\delta_{\lambda'\lambda}\,.$$

The energy of a Dirac anti-particle is positive, while the charge is negative.

This simple-minded tentative modification of the one-particle theory is only meant to show the rôle played by negative frequencies in field theory.

When the reader eventually pursues this subject, he will recognize a relationship between the operators a, $\mathrm{a}^\dagger$, b, $\mathrm{b}^\dagger$ and the so-called creation and destruction operators, and thus, perhaps, better appreciate how and why wave mechanics has naturally developed into field theory.

REFERENCES

GENERAL REFERENCES:

DIRAC, P. A. M., *The Principles of Quantum Mechanics*, 4th ed. (Oxford University Press, Oxford, 1958).
HEISENBERG, W., *The Physical Principles of the Quantum Theory* (University of Chicago Press, Chicago, 1930).
PAULI, W., Encyclopedia of Physics, Vol. 5, Part 1, S. Flügge editor (Springer-Verlag, Berlin, 1958).
PERSICO, E., *Fundamentals of Quantum Mechanics* (Prentice-Hall, New York, 1954).

OTHER REFERENCES:

BARGMANN, V. and WIGNER, E. P., Proc. Natl. Acad. Sci. U.S. **34** (1948) 211.
BELINFANTE, F. J., Physica **6** (1939) 887.
BETHE, H. A. and SALPETER, E. E., Encyclopedia of Physics, Vol. 35, Part 1, S. Flügge editor (Springer, Berlin, 1957).
BLATT, J. M. and WEISSKOPF, V. F., *Theoretical Nuclear Physics* (Wiley, New York, 1952).
BOGOLIUBOV, N. N. and SHIRKOV, D. V., *Introduction to the Theory of Quantized Fields* (Interscience, New York, 1959).
BROGLIE, L. DE, Nature **112** (1923) 540; Ann.d. Phys. (10), **3** (1925) 22; Thesis (Paris, 1924).
CASE, K. M., Phys. Rev. **95** (1554) 1323.
CINI, M., and TOUSCHEK, B., Nuovo Cimento **7** (1958) 422.
CORINALDESI, E., *Particles and Symmetries*, Nuclear Physics **7** (1958) 305.
DIRAC, P. A. M., Proc. Roy. Soc. (London) **117** (1928) 610; and **118** (1928) 341.
FESHBACH, H. and VILLARS, F., Revs. Mod. Phys. **30** (1958) 24.
FEYNMAN, R. P., *The Theory of Positrons*, Phys. Rev. **76** (1949) 749.
FEYNMAN, R. P. and GELL-MANN, M., Phys. Rev. **109** (1958) 193.
FOLDY, L. L. and WOUTHUYSEN, S. A., Phys. Rev. **78** (1950) 29.
FRADKIN, D. M. and GOOD, JR., R. H., Revs. Mod. Phys. **33** (1961) 343.
GOOD, JR., R. H., *Theory of Particles with Zero Rest Mass*, in Lectures on Theoretical Physics, Vol. I, Boulder 1958 (Interscience, New York, 1959).
GUTH, E. and MULLIN, C. J., Phys. Rev. (L) **83** (1951) 667.
HAMILTON, J., *The Theory of Elementary Particles* (Oxford, 1959).
HEITLER, W., *The Quantum Theory of Radiation* (Oxford, 1954).
HILGEVOORD, J. and WOUTHUYSEN, S. A., Nucl. Phys. **40** (1963) 1.
JOST, R. and PAIS, A., Phys. Rev. **82** (1951) 840.
KRAMERS, H. A., Proc. of Amsterdam Academy **40** (1937) 814; and Hand. u. Jahrb. d. Chem. Physik I (1934) 63, 64.
LANDAU, L. D. and LIFSHITZ, E. M., *Quantum Mechanics* (Pergamon Press, London, 1958).
LEE, T. D. and YANG, C. N., Phys. Rev. **105** (1957) 1671.
LIPPMANN, B. A. and SCHWINGER, J., Phys. Rev. **79** (1950) 469.
MADELUNG, E., *Die Mathematischen Hilfsmittel des Physikers* (Springer, Berlin, 1950).
MAJORANA, E., Nuovo Cimento **14** (1937) 171.
MARGENAU, H. and MURPHY, G. M., *The Mathematics of Physics and Chemistry* (Van Nostrand, New York, 1956).
MATHEWS, P. M. and SANKARANARAYANAN, A., Progr. of Theor. Phys. **26** (1961) 1.
MOTT, N. F. and MASSEY, H. S. W., *The Theory of Atomic Collisions* (Oxford, 1949).
NEWTON, T. D. and WIGNER, E. P., Rev. Mod. Phys. **21** (1949) 400.
NOETHER, E., Nachr. d. Kgl. d. Wiss. (Göttingen) (1918) p. 235.

PEIERLS, R. E., *Quantum Theory of Solids* (Oxford, 1955).
ROSENFELD, L., Mém. de l'Acad. Roy. de Belgique **6** (1940) 30.
SAKATA, S. and TAKETANI, M., Proc. Phys. Math. Soc. (Japan) **22** (1940) 757.
SCHIFF, L. I., SNYDER, H. and WEINBERG, J., Phys. Rev. **57** (1940) 315.
SCHWARTZ, L., *Théorie des Distributions* (Hermann, Paris, 1950).
SCHWEBER, S., *Relativistic Quantum Field Theory* (Row, Peterson and Co., Evanston, 1961).
SOMMERFELD, A., *Atombau und Spectrallinien* (Braunschweig, 1951).
SOMMERFELD, A., *Partial Differential Equations in Physics* (Academic Press, New York, 1949).
SNYDER, H. and WEINBERG, J., Phys. Rev. **57** (1940) 307.
STECH, B., Z. f. Phys. **144** (1956) 214.
VAN DER WAERDEN, B. L., *Die Gruppentheoretische Methode in der Quantenmechanik* (Springer, Berlin, 1932).
WATSON, G. N., *Theory of Bessel Functions* (Cambridge University Press, Cambridge, 1952).
WEST, D., *Mesonic Atoms,* Reports on Progress in Physics **21** (1958) 271.
WEYL, H., Zs. für Physik **56** (1929) 330.

SUBJECT INDEX